T0182100

New Ways and Needs for Exploiting Nuclear Energy

Didier Sornette • Wolfgang Kröger •
Spencer Wheatley

New Ways and Needs for Exploiting Nuclear Energy

 Springer

Didier Sornette
Professor of Entrepreneurial Risks
Department of Management,
Technology and Economics
ETH Zürich
Zürich, Switzerland

Wolfgang Kröger
Professor Emeritus for Safety Technology
and Analysis
ETH Zürich
Zürich, Switzerland

Spencer Wheatley
Department of Management,
Technology and Economics
ETH Zürich
Zürich, Switzerland

ISBN 978-3-030-07384-8   ISBN 978-3-319-97652-5   (eBook)
https://doi.org/10.1007/978-3-319-97652-5

This Springer imprint is published by the registered company Springer Nature Switzerland AG
The registered company address is: Gewerbestrasse 11, 6330 Cham, Switzerland

Preface

With strongly rising demographics, development, and improving living standards globally, our societies are confronted with a common understanding that energy security together with decarbonization is key, while electrification is of growing importance. Accordingly, our energy systems are undergoing a transition, driven by socio-technological changes, globalization, and urbanization, largely aiming at renewable sources. In this context, nuclear fission has proven to be a reliable low-carbon energy source; however, the early enthusiasm for nuclear energy in the 1960s and 1970s has waned, and views on the current state and future role of nuclear energy differ and diverge by country. It is a deeply politicized issue, where many fundamentalist views are held, and is subject to misperceptions and arguably singular aversions. Nuclear fission has tremendous potential, but currently deployed technologies and systems dating from the 1970s and 1980s leave room for improvement, especially once acknowledging the need for raising public acceptability and trust.

In view of this, in 2015, one of the authors wrote an article noting strategic and technological reasons for a significant increase in the funding of evolutionary and revolutionary nuclear technologies and systems. In particular, it was argued that the need for dense energy sources mirroring the concentration of population in urban centers and the requirement of nuclear stewardship over long timescales motivates the quest for new designs to make nuclear energy clean and intrinsically safe. To scrutinize this proposal, as well as to take it to the next level by formulating more stringent requirements for nuclear energy, it was recognized that further work, solidly anchored in the most current technology, should be undertaken. This was the motivation of this book, which revisits the science and technological evidence and presents a balanced non-partisan view on the current state and future potentials of nuclear energy.

The approach was to be comprehensive and fair, which benefited from the different disciplines mastered by the authors, drawing on work by major scientific bodies from different fields, as well as generous feedback from a number of deeply qualified reviewers. The result is a scientific book with a strong technical basis and a global system's view. The book is well informed by major global trends, the latest developments in energy systems, and comprehensive sustainability assessments. Further, assessments and prescriptions about nuclear reactor technologies and fuel cycle strategies for the future address the delicate issues of perception, fear, and uncertainty. For instance, a major emphasis is placed on fundamentally increased safety and reduced risks in nuclear energy and how to achieve it—an area of deep common interest and utmost concern of the authors.

We hope that this book can contribute to serious dialogues about energy strategies based on scientific evidence and informed assessments. In particular, the book offers comprehensive information on the different and, in some cases, radically new ways that civilian use of nuclear energy can serve the needs of our rapidly growing and evolving world and boost related research and development to foster sustainable economic growth.

Zürich, Switzerland

Didier Sornette
Wolfgang Kröger
Spencer Wheatley

Executive Summary

The history of mankind is that of its ascent to unprecedented levels of comfort, productivity, and consumption—enabled by the increased mastery of the basic stocks and flows of energy. For instance, the more than one billion citizens residing in the OECD countries, on average, each consume the energy equivalent of nearly 200 full-time human slaves performing raw manual labor. Intensive use of electricity in particular has enabled the third and fourth industrial revolutions; the latter is ongoing with the progressive fusion of the physical and digital worlds. Such innovations are part and parcel of a global urbanization trend. And while about 20% of the global population are at the forefront of this development, the remaining 80% wish to attain the same standard of efficiency and prosperity. This requires a massive growth in the overall supply of electricity, in particular in concentrated form to power future mega-cities and e-mobility.

This miraculous trajectory is confronted by the consensus that anthropogenic emissions are harmful and must decrease, requiring decarbonization of the energy system, currently relying heavily (about 70%) on fossil fuels. More generally deemed necessary is a decoupling of economic growth and development from the environment, stipulating limited increases in primary energy demand. The OECD International Energy Agency (IEA) indicates that the overall quantity of low-carbon electricity generation, and its share of total electricity generation, will need to more than double by 2050 to meet the widely accepted "2 °C target" for global warming. The OECD Nuclear Energy Agency (NEA) suggests that an increase of nuclear capacity by 2.3 times will be necessary to accomplish this.

Nuclear power is an incomparably dense form of energy, and is mature, competitive, and proven to be continuously available, but has been left out of many energy scenarios and strategies. However, there are serious concerns about whether intermittent and seasonal sources like wind and solar alone will be sufficient to decarbonize electricity supplies and stabilize the grid. The mature field of indicator-based sustainability assessment provides a rigorous systematic framework to balance the pros and cons of the various existing energy technologies, via lifecycle assessments and weighting criteria covering the environment, economy, and society, as the three pillars of sustainability. In such a framework, nuclear power is ranked favorably, but, as strong emphasis is often put on radioactive wastes and risk aversion, renewables are usually top-ranked.

However, besides large benefits, nuclear fission has the downside that the physical process generates a surplus of neutrons and radioactive fission products, leading to major design challenges. Therefore, objectives and provisions exist to ensure the control of reactivity, confinement of radioactive substances, and decay heat removal as well as long-term waste management, which are supported by mature methods and increasingly stringent safety requirements. Operators, regulators, and risk analysts often claim that the risk of well-designed and operated power plants with light-water reactors, which dominate the current worldwide fleet of roughly 450 units in operation in 31 countries, is justifiably small. The operating experience has accumulated to more than fifteen thousand reactor-years, with typical capacity factor now at about 80%. Another almost 60 facilities are under construction in 15 countries; at the same time, some countries are phasing out nuclear while promoting renewables, and often relying on carbon-based energy in the intermediate term. New and future generation designs of nuclear plants, mostly evolutionary, aim at further—in some cases radical—reduction of risk of core melt accidents as well as advanced fuel cycles, including the use of thorium rather than uranium, which use resources more efficiently and reduce reliance on husbandry of long-lived wastes (notably actinides) to a historical timescale, but may introduce other risks. In contrast, most current nuclear fuel cycles end with long-term storage, relying on the deep geological repository concept, subject to a range of geo-scientific and social uncertainties. Potential human doses due to long-term potential releases from a repository are calculated to be extremely small, much below natural radioactivity. Although no operating civil disposal facility for highly radioactive wastes exists yet, plans and works are advanced in some countries, especially Finland, where construction has begun in 2016.

Energy and its provision effectively constitute a global public good, and a local accident in a commercial nuclear power facility can lead to global consequences for the fleet and society, such that nuclear power might be perceived as a "public bad". The potential extreme (low probability and high consequence) risks and the importance of strong safety culture, regulation, and governance have been recognized early and to an increasing extent over time. As a result, exceptional and leading efforts into comprehensive risk and reliability analysis have and continue to take place within the nuclear community. Indeed, authoritative organizations provide standards, best practices, and platforms for exchange at the international level. And strong regulatory bodies provide assurances and verification of safety at the national level. However, the adequacy of existing control mechanisms can be questioned, because minimum standards are not trivial to impose internationally since they can only be applied in states with their consent. Further, case studies of major accidents show the pervasiveness of distorted information and sometimes risk information concealment—highly problematic in this safety-sensitive domain. Consulting well-established governance concepts, we put forth more specific improvements and organizational principles, relating to stronger and obligatory monitoring and standards, with adequate controls at the international level, to help promote safe civilian operation of nuclear power.

For nuclear power to be acceptable to the public, plants must not only be exceptionally safe, but be perceived as such. Nuclear accidents are rare but can and have occurred and may be highly costly. A common but wrong opinion is that, if the external cost of nuclear accidents were included in the price, then nuclear power would become economically unfeasible. Quantifying the severity of the consequences of nuclear accidents on a rough integral cost basis, and balancing severity with low core damage accident probabilities, indicates that the average external cost of nuclear accidents is similar to that of modern renewables and far less than carbon-based energy.

Probabilistic safety analysis (PSA) has evolved to become the standard method used to complement the deterministic approach to safety and to estimate the probability of accidents and their consequences, with a focus on severe plant states, as well as health and environmental impact and certain direct costs. This methodology has developed substantially, and level 1 (plant core damage) and sometimes level 2 (containment response) are now used for regulatory purposes. But it is not without defects, notably when quantifying overall risk (level 3: off-site consequences of accidents). More broadly, impacts of potential accidental releases are highly uncertain due to the difficulty of estimating the effect of low doses of radiation, and the use of linear no threshold (LNT) may overestimate latent health effects. Statistical assessment of accident precursors identifies a meaningful reduction in risk over time and convergence of "theoretical" level 1 PSA figures to match experience—especially in the USA and North and Western Europe. However, the potential for major accidents remains, even if with very small probability, leaving the pragmatic door open to seriously explore technologies with radically less extreme potential consequences.

Nuclear power stations have become more efficient and safer due to learning from valuable operational experience, largely in reaction to major and near accidents. The consequences of nuclear accidents are shown to be comparable to other severe industrial accidents, although perceived as singular due to fear of radiation. Further, the dependence upon reliable human performance at various levels and—most importantly—the need for long-term social stability to manage such critical infrastructures and long-term secure disposal of waste emphasize the importance of further safety improvements, for which unprecedentedly demanding criteria are proposed. The overall goal is to tame nuclear risks by excluding mechanisms that lead to severe accidents and avoiding extremely long stewardship times as far as possible, by design.

To satisfy the criteria, key design features ("building blocks") are viewed and revisited and need to be combined in a radically new way, to come up with "revolutionary" or even "exotic" system designs. To check whether such designs are feasible, we track many of the most recent developments of reactor concepts, differing by coolant including liquid metals such as sodium and lead; molten salt and gas; neutron spectra (fast to thermal); power level (from large to small modular) and varying power density; and fundamental design features and purposes, including breeding of fuel and burning of waste. Furthermore, we explore ways to extend fuel reserves, including the use of thorium. As a deliverable, we provide scores for the selected designs against the set of stringent criteria and indicate a high potential for far-reaching improvements compared to most advanced light-water reactors (LWR), with small modular lines of the best versions as being most attractive. However, none of the candidate concepts fulfill all requirements completely; notably the goals of avoiding reactivity-induced accidents and of enhanced proliferation resistance appear most challenging. There is also a potential for new concept-specific accidents, which deserve special attention but appear manageable. Further, the more complex and unfamiliar a design, the more one enters into a "starting from scratch" situation, whose challenges and unknowns should not be underestimated. A purely deterministic safety approach is tempting, in that we would like to absolutely exclude the possibility of severe accidents; however, caution is warranted.

Global trends clearly indicate a growing demand for a larger share and overall amount of concentrated decarbonized electricity. With such high stakes, to rely on dilute wind and solar (or other renewable) energy sources as the only feasible solutions is a critical error, violating the energy system strategic design principle of diversity. We argue for the need of keeping the nuclear option open—supported by future revolutionary safe and clean nuclear technologies, and efficient and proliferation-resistant fuel cycles including the use of thorium—which would become acceptable to an otherwise often nuclear-averse society. We also acknowledge the real problem of stewardship of already existing high-grade nuclear waste over timescales eclipsing that of a stable society. To address this, substantial ongoing national and international RD&D programs exist, although funding is historically low, and the sufficiency of current activities to meet expected demand at an acceptable level of emissions has been questioned. Moreover, in some areas, there is the risk of stagnation or loss of essential human-capital capabilities and know-how.

The promising concepts and designs identified in this book provide the impulse to get us over the existing hurdles, but the scope is ambitious, and time delay from R&D to commercial deployment can be long. Therefore, we call for an urgent increase in funding of at least two orders of magnitude—i.e., hundreds of billions of USD per year, for a broad international civilian "super-Apollo" program on nuclear energy systems. We emphasize that such a large-scale public program is not unprecedented in size although, unlike our proposal, many historical examples have a military use as their origin. We hope that our shared "enemy"—climate change, food and water scarcity, pollution, and so on—all requiring abundant cheap and clean energy to combat, can be recognized in its historical essence and allow for a proactive and ambitious civilian program. Experience indicates that such investments in fundamental technology enable otherwise unattainable revolutionary innovations with massive beneficial spillovers for generations to the private sector and the public.

Acknowledgements We are grateful to numerous colleagues for their help and constructive feedback. In particular, we would like to acknowledge comprehensive comments from Euan Mearns and Bruno Pellaud, as well as comments from Horst-Michael Prasser and Christian Streffer. Additional contributions from Denis Bonnelle, Peter Cauwels, Dominique Escande, and Bertrand Roehner are also acknowledged. We thank our master students, Daniel Fischmann, Lan Chen, Ecem Yildiz, and Magali Chever, for their help in the bibliography and data search needed for the book.

Contents

Chapter 1
Strategic Aspects of Energy

Abstract The history of mankind is that of its ascent to unprecedented levels of comfort, productivity, and consumption—enabled by the increased mastery of the basic stocks and flows of energy. Intensive use of electricity in particular has enabled the 3rd and 4th industrial revolutions, the latter is on-going with the progressive fusion of the natural and digital worlds. Such innovations drive and are supported by a global urbanization trend, leading to futuristic mega-cities. And while about 20% of the global population are at the forefront of this development, the remaining 80% wish to attain the same standard. This requires a massive growth in the overall supply of electricity, in particular in concentrated form to power mega-cities and e-mobility. This miraculous trajectory is confronted by the consensus that anthropogenic CO_2 emissions must decrease, requiring de-carbonization of the energy system, and more generally a decoupling of economic growth and development from CO_2 emission and departure from unsustainable practices. Modern renewables are often proposed as the way forward. Nuclear power, on the other hand, is by far the densest available form of energy and, unlike intermittent renewables, it has proven to be continuously available. But, it has been left out of many energy scenarios and strategies. There are serious concerns about whether sources like wind and solar alone will be sufficient to power expansion of human populations and prosperity. The mature field of sustainability analysis provides a rigorous systematic way to balance the pros and cons of the existing energy technologies, via lifecycle assessments and weighting criteria covering the environment, society, and economy. In such a framework, nuclear power is ranked favorably but, as a strong emphasis is often put on husbandry of radioactive wastes and dread of radiations, renewables are usually top-ranked by politicians and the general public.

1.1 Role of Energy in the Development of Humanity

1.1.1 History of Energy and Human Development

Energy is the entity that can be transformed into work or heat and vice-versa (as well as mass, according to Einstein's special relativity theory). Work allows us to

© Springer Nature Switzerland AG 2019
D. Sornette et al., *New Ways and Needs for Exploiting Nuclear Energy*,
https://doi.org/10.1007/978-3-319-97652-5_1

1 man-day = 35 Watt for 8 hrs = 1 Mega Joule
OECD citizens use 0.2 Giga Joule per capita per year = 4.7 tons of oil equivalent
———➤ 178 "energy slave" man-days per OECD citizen per day

Fig. 1.1 Comparison of the power (energy per unit time) and energy of a typical man, of a horse and of a 200 horse-power tractor. The amount of energy given by the work of a man over 8 hours is equal to the energy contained in 23.5 g of oil. The power of 35 W, that a typical man can deploy over an 8 hour period, is comparable to that consumed by a light bulb. A 'ton of oil equivalent' corresponds to approximately 42×10^9 J or 11,630 kWh. With only 33% efficiency, the 1.2 billion members of the OECD therefore have about 178 man-days of energy supporting each of them. Adapted from Euan Mearns, Energy Matters (http://euanmearns.com/energy-and-mankind-part-3/)

transform and shape all things around us for our convenience. Heat provides comfortable dwellings and can also be transformed into mechanical work (and other forms of energy) via machines (from the steam engines to modern turbines).

One cannot overstate the importance of harnessing and mastering the use of energy to human evolution and the development of modern society. Initially, a human being had to rely on its own biological energy to produce work, namely gather food, fish and hunt as well as construct shelter and tools. Then, large scale use of the energy of many human beings channelled towards one's own goal powered the construction of the pyramids and the propulsion of galleys among many other activities, sometimes in the form of slavery and sometimes in the form of loyalty or social constructs. The drafting of animals to work fields, haul timber and carriage led to a jump by about one order of magnitude in energy use (and thus work done), allowing the growth of trade specialisations such as farmers who produced surplus food. The occurrence of surplus food was the catalyst of societies that can support large numbers of people doing all kinds of different crafts, and is thus the cradle of civilisations, see Fig. 1.1.

During most of mankind's history, from the tens of thousands of years in the past until the first industrial revolution, biomass, wood in particular, was the primary source of energy transformed into heating. The continuous growth of the use of coal since the second part of the eighteenth century led coal to overtake wood as the main source of heating and mechanical work (with the steam engine) by the end of the nineteenth century. Serious drilling of oil (so-called rock oil) began in the second half of the nineteenth century, overtaking coal in total energy output around 1965. But rather than replacing wood use, coal energy consumption was added to that of wood, the later remaining stable and even increasing in the recent decades. Similarly,

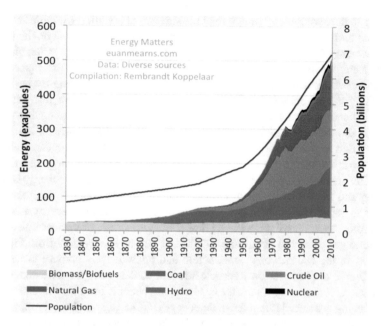

Fig. 1.2 Global energy and population: history of energy transitions for almost 200 years. The brown line shows in comparison the human population (right scale). Courtesy of Euan Mearns, Energy Matters (http://euanmearns.com/energy-and-mankind-part-3/)

the emergence of oil and then of gas did not displace wood or coal but added to the energy pool available for the development of human civilisation, with coal production and consumption continuing its ascent, as can be observed in Fig. 1.2. Truth be said, some products were actually displaced with the emergence of oil, such as whale oil,[1] which had for hundreds of years set the standard for high-quality illumination, with a peak production of 18 millions of gallons in the U.S. extracted from the whales of the oceans in the 1850s, when favourable tax laws and the progressive development of technology allowed kerosene to take over a depleted resource.[2]

In fact, in the developed world, fossil fuels literally saved the whales as well as our forests from the annihilation that occurred in pre-industrial civilisations,[3] and helped humanity to avoid energy starvation; all the while leading directly to the double-edged sword of 7.6 billion living humans on Earth in 2018. All these past "energy transitions" have been voluntary and gone towards higher energy quality,

---

[1] Daniel Yergin, *The Prize: The Epic Quest for Oil, Money & Power*, Free Press; Reissue edition (December 23, 2008).

[2] The whale schools of the Atlantic ocean had been decimated, so that whaling ships were forced to sail farther and farther away, around Cape Horn and into the distant outstretches of the Pacific ocean.

[3] Jared Diamond, Collapse: how societies choose to fail or succeed, Penguin Books; Revised edition (January 4, 2011).

obeying to the evolutionary principle of the survival of the fittest (or more accurately "survival of the sufficiently fit"). In contrast, it is interesting to reflect on the fact that the "new energy transition" presented by Green supporters and governments is the first to attempt substituting dense with dilute energy. It is also the first that does not follow an evolutionary path but seems imposed upon a population by politicians, with the goal to replace "all that has gone before", rather than just adding an additional source to the stack. However, in the path towards sustainable growth, the logic should be that both nuclear power, as well as dilute sources like wind and solar, should replace fossil fuels, as their use declines through either natural or imposed causes. While the early enthusiasm for civil nuclear energy in the 1960s and 1970s seemed to support this view, the present status of nuclear energy is widely dividing and divergent across different countries. In this context, the goal of this book is to revisit the science and technological evidence, and present a balanced non-partisan view on what could be (should be?) the future of nuclear power.

Natural gas started to contribute significantly to the energy pool from the 1960s and later. In the decades of reconstruction and booming productivity growth after World War II until the 1970s, hydroelectric and nuclear energies were added to the mix. Hydroelectric energy is relatively cheap and benefits from the immense advantage of being a controllable renewable energy (unlike solar and wind energies that depend instantaneously on the weather or time of day). Nuclear (fission) energy emerged in some cases for military reasons (US, UK, Soviet Union, France, China) while, for other countries that are short of indigenous supplies of fossil fuels (Sweden, Finland, France, Japan and South Korea), it was added for strategic energy independence. France is the country that has embraced the most nuclear power in relative terms, with about 75% of its electricity (which is 39% of its total energy consumption) generated by 58 nuclear power reactors. Note that France's nuclear program had roots in military applications but underwent massive expansion since 1980 with the aim of energy independence.

During the course of the twentieth century, each person living now in an OECD[4] country enjoyed a massive increase in their energy consumption and living standards, with the equivalent of 4.7 tonnes of oil per year used on average by citizens in 2010. Using the energy comparison provided in Fig. 1.1, this corresponds to the equivalent of 178 energy slaves working for each one of us around the clock and every day of the week (24/7) for our benefits and comfort. In the European Union in 2015,[5] the final end use of energy was dominated by three main categories: transport (33.1%), households (25.4%) and industry (25.3%). Services accounted for 13.6%, and agriculture and forestry for 2.2%. Notwithstanding its relatively small share, the importance of fossil fuels and energy in our food cannot be overstated. Indeed, modern agriculture is heavily dependent on fossil energy, the global oil and gas[6]

---

[4]OECD: Organisation for Economic Co-operation and Development, which has 35 member countries as of 2017.

[5]Eurostat: consumption of energy (http://ec.europa.eu/eurostat/statistics-explained/index.php/Consumption_of_energy).

[6]Gas may be just as important as oil for its use in manufacture of ammonium nitrate fertilizers.

production, refining and delivery systems. Since the 1960s, major increases in food production have occurred as a result of fertilizers, pesticides and machinery production as well as crop management, all based on petroleum energy uses. Virtually all of the processes in the modern food system are now dependent upon fossil energy. One can go so far as stating that we literally "eat oil". [7,8,9] Fossil energy enters at all levels, from agricultural production, to industrial refining of foodstuff, packaging, transportation, conservation, consumption and waste management. [10,11,12] There have been regular discussions about the coming "peak oil" and its implications for the human food chain. [13]

These main sources of energy (biomass, coal, oil, gas and to a lesser extent nuclear and hydro) have created and continue to support the developed world as we know it. Everything we now take for granted as the hallmarks of civilization—namely health, longevity, security, well-being and comfort, education for all, wealth, savings and pensions, as well as modern warfare—result from and rely on the availability of cheap concentrated energy. With the still quasi-exponential growth of the human population, [14] the intensive use of energy is increasingly coming with environmental degradation and an unsustainable ecological footprint, [15,16] which needs to be addressed.

In standard economic theory, the role of energy in production, gross domestic product (GDP) growth and welfare is often relegated to a secondary role if even mentioned, while the emphasis is more on capital and credit, productivity gains, new technologies such as information technology and the transition to artificial intelligence. This widely shared view derives from the recurrent observations in many developed countries[17] that (1) the ratio of energy use to GDP has been continuously

---

[7]Green, B. M., 1978. Eating Oil – Energy Use in Food Production. Westview Press, Boulder, CO (1978).

[8]Walter Youngquist, Geodestinies: the inevitable control of Earth resources over nations and individuals, Natl Book Co (June 1, 1997).

[9]Dale Allen Pfeiffer, Eating fossil fuels (oil, food and the coming crisis in agriculture), New Society Publishers, Gabriola Island, Canada (2006).

[10]Andersson, K. Ohlsson, P and Olsson, P., Screening life cycle assessment (LCA) of tomato ketchup: a case study, Journal of Cleaner Production 6, 277-288 (1998).

[11]Jeremy Woods, Adrian Williams, John K. Hughes, Mairi Black and Richard Murphy, Energy and the food system, Phil. Trans. R. Soc. B 365, 2991-3006 (2010).

[12]Norman J. Church, Why our food is so dependent on oil, Resilience, April 1, 2005 (http://www.resilience.org/stories/2005-04-01/why-our-food-so-dependent-oil/).

[13]K. Aleklett, Peeking at peak oil, Springer New York Heidelberg Dordrecht London (2012).

[14]Andreas D. Hüsler and Didier Sornette, Human population and atmospheric carbon dioxide growth dynamics: diagnostics for the future, Eur. Phys. J. Special Topics 223, 2065–2085 (2014).

[15]Wackernagel et al., Tracking the ecological overshoot of the human economy, Proc. Natl. Acad. Sci. USA 99 (14), 9266-9271 (2002).

[16]Arjen Y. Hoekstra and Thomas O. Wiedmann, Humanity's unsustainable environmental footprint, Science 344, 1114-1117 (2014).

[17]Zsuzsanna Csereklyei, M. d. Mar Rubio Varas and David I. Stern, Energy and Economic Growth: The Stylized Facts, CCEP Working Paper 1417, November 2014.

declining over the last decades (for the U.S. over the period 1950–1998, at an annual rate of 1.4% on average), (2) the ratio of energy cost relative to GDP has continuously declined (for the U.S at about 1% per annum), (3) the relative price of energy to labor has been steadily declining and (4) the energy/capital ratio has been steadily falling. With energy cost and real price of energy falling, and in the presence of ample supply, energy has become an "old economy" component, not really worth considering for the future development and elevation of mankind. This is well-captured by the weight of 5% given by mainstream economics, as the share of energy in the total cost of the production factors (capital, labor and energy). Moreover, standard economics views energy consumption as a consequence of growth, rather than the other way around. But this reducing view fails to account for the positive feedback loop between energy and economic development, such as increasing availability of energy and its declining costs leading to increasing demand and the development of novel uses, further driving the production of energy up and its declining prices thanks to economies of scale and learning effects.[18]

Against this backdrop, even in developed countries, per capita energy use has been increasing (in the U.S. at an average annual rate of about 1% over the last five decades). Moreover, econometric analyses of growth in the USA, Japan, and Germany between 1960 and 1996 show that energy drives about 50% of economic growth.[19] In particular, Ayres and Warr[20] show that GDP growth in the U.S. until the 1970s has been achieved mainly by historical improvements in the efficient use of energy conversion to physical work, with the progressive appearance of information and other virtual technology as drivers in the most recent decades.

1.1.2 Origins of the Various Sources of Energy

The sources of energy mentioned above can be classified as coming from two main sources: the supernova precursor(s) of our solar system and the sun, both being decomposed into stored components and flows. The most commonly accepted scenario is that the Solar System formed from the gravitational collapse of the pre-solar nebula, a fragment of a giant molecular cloud. The atomic composition of this cloud reflected in part the nucleosynthesis of relatively light elements (lighter than iron) from hydrogen and helium performed in the interior of stars pre-existing the sun, which were delivered to the interstellar medium at the end of the life of low mass stars, when they ejected their outer envelope before they collapsed to form

---

[18]Robert U. Ayres and Benjamin Warr, The economic growth engine (how energy and work drive material prosperity), Edward Elgar, Cheltenham, UK and Northampton, MA, USA (2009).

[19]C. Hall, D. Lindenberger, R. Kummel, T. Kroeger, and W. Eichhorn, The need to reintegrate the natural sciences with economics, BioScience 51, 663-673 (2001).

[20]Robert U. Ayres and Benjamin Warr, Accounting for growth: the role of physical work, Structural Change and Economic Dynamics 16, 181-209 (2005).

white dwarfs. Larger stars explode at the end of their life as supernovas, leading to the nucleosynthesis of heavier elements. For type II supernova events, elements heavier than iron and nickel are synthesized, including the heaviest elements known, such as the naturally radioactive elements uranium ^{238}U and thorium ^{232}Th, as well as potassium ^{40}K ranking third, after ^{232}Th and ^{238}U, as a source of radiogenic heat. Thus, the atomic composition of the solar system and of Earth reflects the abundance created by these events that preceded the formation of the Solar System, about 4.6 billion years ago. Moreover, the gravitational collapse of the pre-solar nebula led, by conservation of angular momentum, to the rotations of the Earth and to its trajectory around the sun.

We can thus classify the following energy sources as stores of the primordial energy from the remnant of previous stars and supernova explosions in the composition of the present Earth.

Supernova and Primordial Pre-solar Nebula Energy Sources

1. Fossil energy stocks

    (a) uranium and thorium (the two prominent nuclear fission sources) and other heavy elements heavier than iron created by supernovas before the formation of the solar system;
    (b) geothermal energy: a large part of the heat in the interior of the Earth, which powers mantle convection and plate tectonics, i.e. the formation of mountains and ocean ridges, is generated by nuclear fission,[21] itself burning the fissile elements created by supernovas prior to the solar system formation. The other contributions are a left-over from the Earth's formation and other not well-understood sources such as tidal friction within the Earth mantle and core.

2. Energy flows

    (a) Tidal energy[22]: results from the rotation of the Earth on its axis and from the orbits of the Earth-Moon and Earth-Moon-Sun systems. These rotations of massive objects stored the angular momentum pre-existing in the pre-solar nebula. The tidal energy can be progressively collected, either with stream devices making use of the kinetic energy of moving water to power turbines or with dams and hydraulic turbines that make use of the gravitational potential energy difference between high and low tides.[23] Compared with

---

[21]The KamLAND collaboration. Partial radiogenic heat model for Earth revealed by geoneutrino measurements, Nature Geoscience 4, 647-651 (2011).

[22]Tidal energy is one of the oldest forms of energy used by humans. So-called tide mills were found on the Spanish, French and British coasts, dating back to 787 A.D. They consisted of a storage pond, filled by the incoming tide through a sliding gate and emptied during the outgoing tide through a water wheel.

[23]The Atlantropa project pushed to the extreme the concept of garnering tidal energy to power the whole of Europe. Devised by the German architect Herman Sörgel in the 1920s, the central feature of Atlantropa was a hydroelectric dam to be built across the Strait of Gibraltar, which would have provided enormous amounts of hydroelectricity. By preventing evaporation in the Mediterranean

other renewable energies such as wind and solar, one advantage of tidal energy is its high predictability based on its 12-hour cycle. But, its environmental impacts, costs and durability remain to be assessed given its early development stage.

(b) Hydroelectric power: while the water comes from rain, through the sun powered cycle of evaporation and condensation, the possibility to store it at altitude in lakes and barrages is made possible by the existence of mountains and more generally of relief, which are direct consequences of plate tectonics, itself a child of the mantle convection generated by the heat, itself coming from the formation of the Earth and its nuclear fissile elements.

The other sources of energy, while obviously linked with the pre-solar nebula with respect to the origin of the elements, are the direct transformation of the specific energy radiated by the sun, which is produced by nuclear fusion of the hydrogen in its interior to form helium and heavier elements.

Solar Energy

1. Fossil energy stocks: coal, oil, natural gas. These fossil energy stocks are made of the organic remnants of ancient plants and animals that grew using the sun's energy (and of course made of the raw element material of the pre-solar nebula). Their biological matter then degraded and transformed into ore grade deposits by geological processes involving deep burial, high pressure, temperature and water, once again driven by plate tectonics.

2. Energy flows

   (a) wind and waves
   (b) solar photovoltaics and solar thermal
   (c) hydroelectric (water transported from sea level to mountains via evaporation and winds but stored to garner the gravitational potential energy via the existence of mountains created by plate tectonics that then flows and the kinetic energy of the falling water is converted into electricity by turbines).
   (d) biomass (wood) and biofuels.

1.1.3   *"The Electricity God"*

Our modern societies depend entirely on continuously available energy for the day-to-day functioning and sustenance of life. In particular, metropolitan areas heavily depend on electrical power-driven facilities, equipment, and appliances, where the

---

sea to be fully compensated by the flux of water from the Atlantic ocean, this would have led to the progressive lowering of the surface of the Mediterranean Sea by up to 200 m, opening large new territories for settlements.

loss of power could severely impair the integrity of the urban societies these metropolitan areas sustain. Transient energy failures bring normal life to a standstill and ensuing nightmare scenarios, as illustrated by Superstorm Sandy that hit the U.S East Coast in October 2012, with an estimated US$65 billion in damages. Power outages occurred for over 8 million customers across 21 states, for days to weeks. A large city without electricity for a few days becomes vulnerable to fraud, theft and exploitation. Social unrest, including riots can result.[24] A blackout may close all places of work, leave traffic jammed, parents isolated from children at school, trucks carrying food isolated, food rotting in supermarket freezers, petrol pumps standing idle, water supply and disposal compromised, bacterial infections thriving.[25]

In particular, electric energy is essential for the smooth operation of public transportation, powering electric trams, subways, and traffic lights. Security systems in many buildings rely on electricity, with access blocked in the presence of an outage. Water purification, waste management, domestic appliances, including electric stoves and induction cooking make our daily life utterly reliant on the "electricity God". Since the generation and distribution of electricity has been organised, the lives of people have been completely reorganized around it. Almost everything is made on the basis of electricity, for refrigerating food,[26] the electricity-powered Internet and digital devices (from laptops to cell phones and many more).

Progress in the last 250 years has been marked by a series of "industrial revolutions" (IR)[27]:

IR1 (1750–1850): coal, steel, steam and railroads;
IR2 (1870–1930): electricity, internal combustion engine, cars, running water, indoor toilets, telephone, wireless telegraphy and radio, movies, petro-chemicals:
IR3 (1960–2000): electronics, computers, the web, the Internet, mobile phones;
IR4 (on-going, from 2000 to the uncharted future)[28]: the progressive fusion of the physical, digital and biological worlds with cloud computing, information storage, the Internet of things, the blockchain, artificial intelligence, robots, self-driving cars, genomics and gene editing, neuro-technological developments.

---

[24]One of the worst example of civil unrest caused by a power outage was the New York City Blackout of 1977, due to a series of lightning strikes on the evening of July 13, 1977, which blew out circuit breakers and caused power lines to overload with electricity and blow out transformers. The loss of power triggered widespread looting, rioting and arson in parts of the New York City, on a backdrop of economic hardships. More than 3700 people were arrested, 1600-plus stores looted, and 550 police officers injured.

[25]Steve Matthewman and Hugh Byrd, *Blackouts: A sociology of electrical power failure*, Social Space Journal, January 2014.

[26]Which we take for granted but is transformative in the way we produce, transport, consume and store food, including great progress in food safety.

[27]Robert J. Gordon, The Rise and Fall of American Growth: The U.S. Standard of Living since the Civil War, The Princeton Economic History of the Western World, Princeton University Press (January 12, 2016).

[28]Klaus Schwab, The Fourth Industrial Revolution, World Economic Forum, Davos, Switzerland (January 12, 2016).

Essentially all the artefacts and processes in IR3 and IR4 rely fundamentally on electricity. In 2013, the main sources of electricity were coal and peat 41%, natural gas 22%, hydroelectric 16%, nuclear power 11%, oil 4%, biomass and waste 2% and wind, geothermal, solar photovoltaic, and solar thermal accounting together for 4%.[29] The trend towards ubiquitous supercomputing and blockchain-based decentralised autonomous organisations requires increasing electric energy resources, with electricity growth expected to be much larger than overall energy growth.

With the development of the Internet economy, direct sales to consumers, decentralized inventories of goods by digital means, the growth of digital informa- tion and entertainment channels, one would think that dramatic reductions in energy consumption and greenhouse gas emissions would follow quickly.[30] However, consider this: the electricity needed just to transmit the trillions of spam e-mails sent every year is equivalent to powering two million homes in the United States and generates the same amount of greenhouse gas emissions as that produced by three million cars.[31] The electricity consumed by cloud computing globally is expected to increase from 632 billion kWh in 2007 to 1963 billion kWh by 2020,[32] i.e. approaching 10% of total electricity production. According to C. Malmo,[33] in June 2015 one Bitcoin transaction required the same amount of electricity as powering 1.57 American households for one day. In June 2015, the Bitcoin net- work[34] was consuming enough energy to power 173,000 American homes while, in January 2017, it was consuming the energy used by more than one million American homes (a sixfold increase). Energy consumption of mining of popular cryptocurrencies has become comparable with the level of energy consumption of Tunisia and Croatia. As of August 23, 2017, energy consumption of mining of bitcoins exceeds 16 TWh per year.[35]

The bold claims of the blockchain revolution to replace a large part of economic transactions by peer-to-peer decentralised globally shared ledgers is presently not

---

[29]OECD Factbook 2015-2016: Economic, Environmental and Social Statistics. 8 April 2016. Retrieved 3 August 2017.

[30]Joseph Romm, Cool Companies: How the Best Businesses Boost Profits and Productivity by Cutting Greenhouse Gas Emissions. New York: Island Press (1999).

[31]The Carbon Footprint of Email Spam Report by Antivirus Company MacAffee: https://resources2.secureforms.mcafee.com/LP=2968

[32]Make IT Green: Cloud computing and its contribution to climate change, Greenpeace report 30 March 2010 (http://www.greenpeace.org/international/en/publications/reports/make-it-green-cloud-computing/).

[33]https://motherboard.vice.com/en_us/article/ae3p7e/bitcoin-is-unsustainable   (accessed 12 May 2018).

[34]The bitcoin network is a peer-to-peer payment network based on a cryptographic protocol, in which users send and receive bitcoins by broadcasting digitally signed messages to the network using bitcoin wallet software (see https://en.wikipedia.org/wiki/Bitcoin_network).

[35]Energy consumption on a mining of bitcoins has exceeded energy consumption of Croatia (http://currentlynews.us/articles/energy-consumption, accessed 24 Aug. 2017).

Table 1.1 Calculation of incremental electricity generation by 2020, based on the hypothesis of an average electricity consumption of 0.17 kWh/km (the present range is from 0.1 to 0.23 kWh/km) and growth of electric vehicles reaching 1.35 million in Europe, less than 0.9 million in the US and more than 5 million in China. The corresponding needed electricity energy is 34.4 TWh per year

Implied electricity generation requirements estimate	EU	US	China
Energy efficiency per EV in 2020 (kWh/km)	0.17	0.17	0.17
Average vehicle distance travelled per annum (km)	11,500	22,000	21,000
Base annual electricity requirements per vehicle (kWh)	1955	3740	3570
Charge/recharge efficiency	85%	85%	85%
Self-discharge efficiency	90%	90%	90%
Grid transmission efficiency	94%	94%	94%
Actual generation per vehicle (kWh)	2719	5201	4965
2020 EV requirements (vehicle)	1,350,000	867,000	5,275,000
Total 2020 incremental generation requirements (TWh)	3.7	4.5	26.2

Source: e-mobility: closing the emissions gap, World Energy Perspectives, E-Mobility (2016) (https://www.worldenergy.org/wp-content/uploads/2016/06/E-Mobility-Closing-the-emissions-gap_full-report_FINAL_2016.06.20.pdf, accessed 24 Aug 2017)

scalable, both in terms of memory requirements, computational resources and electric energy needs. Novel protocols are likely to break the existing bottlenecks and introduce feasible operational distributed blockchain-based ledgers, but one should expect a vigorous increase in electric energy consumption, even if great progress is made to ensure efficiency.

The development of electric vehicles for transportation is expected to accelerate and lead to increasing needs and load on the electric grids. Table 1.1 shows the incremental electricity generation that can be anticipated by 2020, based on a growth of electric vehicles reaching 1.35 million in Europe, slightly less than 0.9 million in the US and more than 5 million in China—a minor fraction of the overall vehicular fleet. The corresponding needed electrical energy is in the range 4–30 TWh, which represents less than 0.5% of 2014 total electricity generation in their respective markets. To compare, a 1 GW nuclear reactor, operating with an average capacity factor[36] of 85%, generates approximately 7.4 TWh per year. Thus, while there are concerns regarding the ability of local grids to support the higher demand for power, utilities and municipalities should be able to absorb this demand with proper planning and balancing capabilities, for the time being. Indeed, compared with the growth in electricity demand associated with digitalization of society, growth in electricity demand for electric vehicles is expected to be much lower and more manageable.

---

[36]The capacity factor is the ratio of the actual electrical energy output over a given period of time to the maximum possible electrical energy output over the same amount of time.

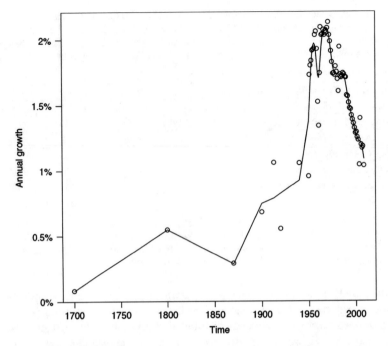

Fig. 1.3 Annualized world population growth rate from year 1700 to 2010, providing a direct diagnostic of a super-exponential growth until the 1960s characterized by a growth of the growth rate. The empty circles are the data points and the continuous line has been obtained by using a smoothing filter. Reproduced from Hüsler and Sornette (2014)

1.2   Mega-Trends and Major Factors

1.2.1   Human Population

As mentioned before, the growth of energy needs is intimately linked to (a) the increase of human population and (b) the level of economic development. Let us discuss these two components in turn.

Human population growth at preindustrial levels is generally estimated to have been below 0.5% per year. Starting with the first industrial revolution, the human population growth rate grew rapidly as can be observed in Fig. 1.3, until its peak in 1964 at about 2.4% per year, from which a slow decrease of the growth rate can be observed. Over the time period 1850–1960, the human population has grown faster-than-exponential.[37] At the time of the coronation of Napoleon as the new French emperor in 1804, the World human population reached its first billion. The second

---

[37] A. Johansen and D. Sornette, Finite-time singularity in the dynamics of the world population and economic indices, Physica A: Statistical Mechanics and its Applications 294 (3-4), 465-502 (15 May 2001).

billion mark occurred around 1927 shortly before the great financial market crash and the onset of the Great Depression, the third billion was reached in 1957, the year of the first artificial satellite (Sputnik 1 launched by the Soviet Union) and the formation of the European Economic Community with the Treaty of Rome signed on March 25. We stand at more than 7.6 billion at the time of writing[38] and grow at the rate of 1% per year, adding the population of one Germany every year to the Planet. The US census bureau estimates that human growth rates should continue to decrease more or less linearly and fall below 0.5% per year at around 2050.[39]

However, the last decade trend suggests, at least for a while, a plateauing of the growth rate at the present level of 1.0%.[40] Human population data is actually well represented by two exponential regimes since 1960: the first one from 1960 to 1990 with a constant growth rate of 1.7% per year, and the second one from 1990 to 2010 with a constant growth rate of 1.0% per year. At present, it is too early to determine whether the plateau of the annual growth rate at 1% per year will continue or will transition to a resumed decrease. A disaggregated analysis is necessary at the level of countries, or even better, regions, taking into account the large heterogeneities in the birth and death rates of the thousands of ethnicities populating the planet. Particularly important for developed countries is the strong trend towards low birth rates below renewal, while developing countries exhibit very strong growth rates.[41] A recent study suggests that there is an 80% probability that world population will increase to between 9.6 billion and 12.3 billion by 2100.[42]

1.2.2   Economic Development

Let us turn to the evolution of economic development. It is well-established that there is a significant positive relationship between energy use per capita and GDP per capita. The more developed a country, the more its citizens consume energy to enjoy a higher lifestyle (more leisure travels, more artefacts such as cars, digital products, more meat and sophisticated and processed foods, etc.) as can be seen in Fig. 1.4. This finding has been for instance quantified for an inclusive set of 171 countries over the period 1950–2004, both taken together and for all subsets of countries. Once a country reaches a high level of development, there is also some indication that

---

[38]see http://www.worldometers.info/world-population/ for a real-time update.

[39]https://www.census.gov/population/international/data/worldpop/table_population.php (accessed 3 Aug 2017).

[40]Andreas D. Hüsler and Didier Sornette, Human population and atmospheric carbon dioxide growth dynamics: diagnostics for the future, Eur. Phys. J. Special Topics 223, 2065–2085 (2014).

[41]The Changing Global Religious Landscape, Pew Research Center (5 April 2017).

[42]Patrick Gerland et al., World population stabilization unlikely this century, Science 346, 234-237 (2014).

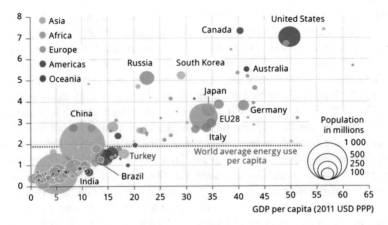

Fig. 1.4 Energy consumption per capita (vertical axis unit is tonnes of oil equivalent) as a function of GDP per capita, PPP (PPP stands for purchasing power parity, and is used to compare countries with different costs of living and to adjust for exchange rates) (horizontal axis unit is thousands of current international U.S. dollars). The size of the bubbles denotes total population per country. All values refer to the year 2011. Reproduced from European Environment Agency (https://www.eea.europa.eu/data-and-maps/figures/correlation-of-per-capita-energy, accessed 3 Aug 2017)

growth in energy use may then decline as GDP per capita continues to rise, as low energy intensity high value services play a growing role.[43]

The countries positioned in the upper right (so-called developed countries) in Fig. 1.4 amount to about 15% of the World population. This means that the remaining 85% of the World population, which aspires to catch the high energy based consumption levels of developed countries, are in a race to multiply energy consumption, with possibly severe environment impacts. This growth is dramatically illustrated by the nightlight (mostly electric lights) maps for three regions of strong development shown in Fig. 1.5. Per capita in 2013, the average U.S citizen consumed about twice the energy of the average German citizen, about three times the energy of the average Chinese citizen and about 10 times the energy of the average Indian citizen.[44,45]

---

[43]Rögnvaldur Hannesson, (2009) "Energy and GDP growth", International Journal of Energy Sector Management, Vol. 3 Issue: 2, pp.157-170, https://doi.org/10.1108/17506220910970560

[44]This counts all the energy needed as input to produce fuel and electricity for end-users. It is known as Total Primary Energy Supply (TPES), a term used to indicate the sum of production and imports subtracting exports and storage changes.

[45]http://data.worldbank.org/data-catalog/world-development-indicators (accessed 3 Aug. 2017).

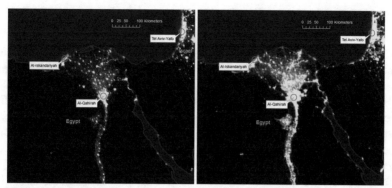

Nile delta in 1992 (left) and 2009 (right). Nearly the totality of light emission of Egypt comes from this area and the banks of the Nile. Note coalescence along interconnections between the settlements.

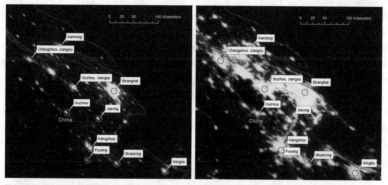

Significant growth of lights in China around Shanghai from 1992 (left) to 2009 (right).

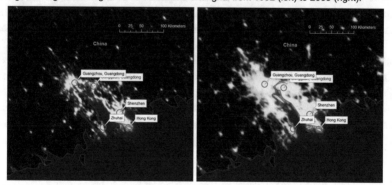

Light growth for the agglomeration of Shenzhen-Guangzhou from 1992 (left) to 2009 (right).

Fig. 1.5 The Nile delta, Shanghai and Shenzhen-Guan agglomeration growth, visualized via the spatial distribution of nighttime light using the open source data of the Defense Meteorological Satellite Program of the U.S. Air Force. Reproduced from Peter Cauwels, Nicola Pestalozzi and Didier Sornette, Dynamics and Spatial Distribution of Global Nighttime Lights, EPJ Data Science 3:2, 10.1186/epjds19 (2014)

1.2.3   Urbanisation

At present and looking decades ahead, a fundamental trend of humanity's development is urbanization. Urbanization may go hand in hand with making more room for
nature but it is crucial for the cities of the future to be designed accordingly. We saw
that electricity demand would increase dramatically in the near future, especially as
developing countries catch up on our modern energy consumption standards. Electricity demand is expected to skyrocket indeed as more and more of the world's
population agglomerates to live in cities. where they may gain access to electricity-
based services and infrastructure and higher levels of prosperity.

In 1950, 30% of the world's population was urban, and by 2050, 66% of the
world's population is projected to be urban, while 54% of the world's population
was residing in urban areas in 2014.[46] While the most urbanized regions are
Northern America (82% in urban areas in 2014), Latin America and the Caribbean
(80%), and Europe (73%), Africa and Asia have just 40% and 48% of their
respective populations living in urban areas. Over the coming decades, Africa and
Asia are expected to urbanize faster than the other regions and are projected to
become 56% and 64% urban, respectively, by 2050.[46]

The urban population of the world has grown rapidly since 1950, from 746 million
to 3.9 billion in 2014 and the trend is expected to accelerate with economic
development and the quest for higher living standards and better lifestyles in
developing countries. This is exemplified by the massive exodus from the rural
areas to the large cities in China in the last three decades that accompanied and
fostered the Chinese industrial revolution: by the end of 2015, 56% of the total
Chinese population lived in urban areas, compared with 26% in 1990. Continuing
population growth and urbanization are projected to add 2.5 billion people to the
world's urban population by 2050, with nearly 90% of the increase concentrated in
Asia and Africa. In 2014, there were 28 mega-cities with more than 10 million
inhabitants (Tokyo with an agglomeration of 38 million inhabitants, followed by
Delhi with 25 million, Shanghai with 23 million, and Mexico City, Mumbai and São
Paulo, each with around 21 million inhabitants). By 2030, the world is projected to
have 41 mega-cities with more than 10 million inhabitants. Hundreds of cities of
more than 1 million inhabitants will appear in the next few decades.[47] Megacities are
large centers of economic activity and creativity[48,49] and thus the biggest energy

---

[46]United Nations, Department of Economic and Social Affairs, Population Division (2014). World
Urbanization Prospects: The 2014 Revision, Highlights (ST/ESA/SER.A/352).

[47]The World's Cities in 2016, United Nations, Department of Economic and Social Affairs, World
Urbanization   Prospects   (http://www.un.org/en/development/desa/population/publications/pdf/
urbanization/the_worlds_cities_in_2016_data_booklet.pdf, accessed 3 Aug. 2017).

[48]Bettencourt, L.M.A., J. Lobo, D. Helbing, C. Kühnert, and G. B. West, Growth, innovation,
scaling, and the pace of life in cities, Proc. Natl. Acad. Sci. USA 104 (17), 7301-7306 (2007).

[49]Schläpfer, M, Bettencourt, L.M.A., Grauwin, S, Raschke, M., Claxton. R., Smoreda, Z., West, G.
B., Ratti, C., The scaling of human interactions with city size, J. R. Soc. Interface 11: 20130789
(2014).

demanders, impacting global emissions the most (urban areas today account for almost 70% of CO_2 emissions and have 55% of population). Megacities are considered areas of high risk (economic, environmental, geopolitical, societal, etc.), and of concentrated wealth and poverty at the same time. Megacities can lead to improved energy efficiency where families live in well-insulated apartments and nearby to where they work. But this improvement can be lost by their greater consumption of material goods and services leading overall to higher energy consumption. This contradiction is sometimes known as the rebound effect or Jevons' paradox[50] and is one of the most widely known paradoxes in environmental economics

These vast concentrations of population require facing a number of development challenges, among them the need to power, feed and manage the wastes generated by these urban areas. While large metropolises tend to produce proportionally more patents, innovations and creative artefacts than smaller ones,[51] they also produce proportionally more crimes. And all residents need to be fully fed and powered. Adopting a cynical metaphor, one can visualise a large metropolitan agglomeration as a kind of "cancerous tumor"[52] in the geography of a region, drawing on and redirecting the resources of a large hinterland to survive and strive. Food must be brought daily from agricultural production areas via a complex transportation network and delivery systems. Electric energy lines feed the city to ensure the functioning of its inner transport system, and most of its functionalities. Wastes need to be removed via a complex sewage system as well as an army of dump trucks carrying their loads away from the city to landfill or recycling and waste processing facilities.

As the existence of megacities depends greatly on the management of their energy resources, it is reasonable to ask oneself whether such dense energy demands can be satisfied by non-dense energy sources like wind and solar? It seems clear that current and future megacities will rely on energy deliveries from outside of their boundaries to meet their total energy needs. From this, it is difficult to determine the ideal mix of

---

[50]Blake Alcott, Jevons' paradox, Ecological Economics. 54 (1), 9-21 (2005).

[51]Geoffrey West, Scale: The Universal Laws of Growth, Innovation, Sustainability, and the Pace of Life in Organisms, Cities, Economies, and Companies, Penguin Press (May 16, 2017).

[52]Cities and malignant cancer tumors share four major characteristics: (i) rapid, uncontrolled growth; (ii) invasion and destruction of adjacent normal tissues (ecosystems); (iii) metastasis (distant colonization); and (iv) de-differentiation. Indeed, many urban forms are almost identical in general appearance, a characteristic that can qualify as "de-differentiation." Large urban settlements display "rapid, uncontrolled growth" expanding in population and area occupied at rates of from 5 to 13% per year. See for instance Fig. 1.4. See Warren M. Hern, Has the human species become a cancer on the planet?; A theoretical view of population growth as a sign of pathology, Biography & News (Speeches and Reports Issue) 36 (6), 1089-1124 (December 1993); Is human culture carcinogenic for uncontrolled population growth and ecological destruction? BioScience 43 (11), 768-773 (December 1993); Is human culture oncogenic for uncontrolled population growth and ecological destruction? Human Evolution 12 (1-2), 97-105 (1997); How Many Times Has the Human Population Doubled? Comparisons with Cancer, Population and Environment 21 (1), 59-80 (1999); Urban Malignancy: Similarity in the Fractal Dimensions of Urban Morphology and Malignant Neoplasms, International Journal of Anthropology 23 (1-2), 1-19 (2008).

centralized base load energy supply (energy from remote plants, including nuclear) and often decentralized energy supply from renewables and related distribution systems. Indeed, the hope that low-density renewable energy sources such as wind and solar could power the concentrated needs of urban centers seems to face a fundamental physical limit as well as huge operational hurdles. A first-order response to the concentration of human population and its needs seems to require concentrated sources of energy with low carbon and pollution impact. This inevitable, and largely beneficial form of urbanization, motivates the research performed in the writing of this book—to articulate the pros and cons of an at least partial nuclear energy solution—since nuclear energy is by far the most concentrated source of energy available to us. As will be described in later chapters, a few medium-sized modular nuclear reactors, the size of a house, preferably built off-site and transported on-site could power a large city for decades.[53]

However, renewable energy supply does not need to be delocalised; some reports argue, "the ideal city is one that consists of multiple self-contained, meaning self-sustainable, centers".[54] The idea of self-contained centers revolves around maximizing compactness of cities and especially compactness of centers in which public transport becomes quasi non-existent and energy cycles are closed due to self-sufficiency. The very dense energy demand of megacities calls for the integration of different energy networks in compact spaces. The report of the International Energy Agency (IEA)[55] focuses on urban energy systems and specifically rooftop solar photovoltaic, municipal solid waste, sewage gas, waste water gas and industrial excess heat as an integrated energy network. The efforts to develop sustainable fully self-sufficient cities are illustrated by the Singaporean Future Cities Laboratories pioneer.[56] The vision is to grow food on the walls of buildings, or in special compounds with optimised LED illumination in multi-shelved agriculture compounds,[57] optimise vegetation to filter out pollution and generate oxygen, recycle water and waste, and produce energy via solar panels distributed on all roofs and via other innovative ideas. This futuristic vision should not be discarded but its thermodynamic realities must surely be questioned and analysed. This requires a new understanding of the physical stocks, resource flows, social institutions, and cultural

---

[53]Taking 10 kWh as a generous daily consumption per capita, this leads to about 3.6 MWh per capita per year, and thus about 3.6 TWh of energy consumed per year for a city of one million people. A medium-sized nuclear reactor of 100 MWe, with average capacity factor of 85% produces 0.75 TWh of electric energy each year. Thus, about 5 medium-sized modules can power a city of one million people.

[54]Allianz SE, Allianz Risk Pulse: "The Megacity State", (2015), https://www.allianz.com/en/press/news/studies/151130_the-megacity-of-the-future-is-smart/ (accessed June 2016).

[55]OECD/IEA Paris, *Energy Technology Perspectives 2016*, http://www.iea.org/etp/ (accessed June 2016).

[56]http://www.sec.ethz.ch/research/fcl.html and https://www.create.edu.sg/about-create/research-centres/sec

[57]Toyoki Kozai, Kazuhiro Fujiwara, Erik S. Runkle, Editors, LED Lighting for Urban Agriculture, Springer Science, Singapore (2016).

catalysts of cities, whose interactions generate a quantifiable 'metabolism'. In this vision, a city transitions from being a parasitic tumor to becoming a super-organism, and its improved design and maintenance should foster its survival and resilience. Thus, the thinking of energy provision should ideally be combined with the design of the whole city.

This idealistic vision should however be tempered by the reality of the diluted nature of solar energy. In Singapore, being very sunny and less than two degrees north of the equator, with a typical daily consumption of about 24 kWh per inhabitant,[58] the modern city-state consumes 140 GWh/day of electricity. Given that a typical 150 W photo-voltaic module is about 1 m^2 in size and produces approximately 0.75 kWh every day on average (for an insolation of 5 sun hours/day),[59] Singapore would need almost 190 km^2 of solar panels, being significantly more than 25% of the total area of the country.

1.3 Current Situation and Looming Dilemmas

1.3.1 Energy Consumption Make-Up and Scenario-Based Outlook

As discussed above, energy production and consumption are key factors to the development of today's societies.[60] The technologies used to provide energy are continuously evolving while the demand is expected to increase in the future to support growing populations and industries, especially in the fastest growing emerging economies (see annexed Fig. 1.18). Electricity demand will be an especially important contributor to this increase, and even more so with the observed fast urbanization trends (see Sect. 1.2.3) and growing use of electric cars. It is commonly agreed that anthropogenic CO_2 emissions, after a significant increase during the last decades, must be reduced to stabilize temperature increase (see annexed Fig. 1.19). Thus, the decarbonisation of our fuel mix through low or non-CO_2 emitting technologies is attractive and is projected by some to become possible (see Fig. 1.6). However, how an electric grid based on the mix shown in Fig. 1.6 is going to be balanced and backed up remains an open problem (see Box 1.1). In general, there is a growing understanding and realisation that humanity must, over the next years to decades, shrink its impacts on the environment to make more room for nature and ensure sustainability.[61]

---

[58]https://data.worldbank.org/indicator/EG.USE.ELEC.KH.PC

[59]Tilak K. Doshi, Neil S. D'Souza, Nguyen Phuong Linh and Teo Han Guan, The Economics of Solar PV in Singapore, Discussion Paper EE/11-01, (August 2011) (http://esi.nus.edu.sg/docs/event/the-economics-of-solar-pv_2011nov28_final.pdf?sfvrsn=0).

[60]Smalley, R.E., *Future Global Energy Prosperity: The Terawatt Challenge*, MRS Bulletin 30(6), pp. 412-417 (2005).

[61]Asafu-Adjaye, J., et al., *An Ecomodernist Manifesto* (2015).

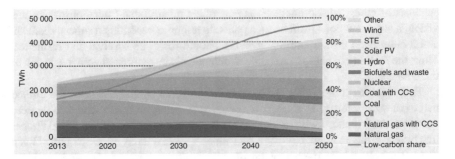

Fig. 1.6 Global electricity generating mix in the 2 °C scenario 2013–2050, produced by the International Energy Agency (IEA),[55] with possible technologies listed on the right. The scenario shows the low-carbon share (orange line) continuously increasing from 2013 to 2050 (right hand scale). CCS means carbon capture and storage. STE stands for solar thermal energy, and solar PV is solar photovoltaic

Box 1.1   Short Discussion of Fig. 1.6 Produced by the International Energy Agency (IEA), and Some Limitations and Costs of Running a High Penetration Renewable Electricity Grid

Future electricity scenarios like the one shown in Fig. 1.6 rely heavily on massive expansion of wind and solar power at the expense of fossil fuels. Scenarios like this often tend to ignore fundamental engineering, economic and political realities. They are problematic because they are produced by respected organisations and may convey a message to politicians that decarbonisation is easy. To be credible, models like this need to be backed up by real grid demand scenarios plotted at high frequency resolution. Some other important concerns are:

1. Is it really feasible to multiply nuclear two-to-three-fold in the 32 year time frame to 2050 given that the majority of reactors currently operational will need to have their life extended beyond the original design life time?
2. Burning biofuel (biomass)releases CO_2, and may be using virgin forest. Is a fourfold increase in biomass use desirable?
3. Hydroelectric power is projected to doubling by 2050. Given that readily available large sites are already developed, is it this feasible?
4. Solar PV has low ERoEI (energy return on energy invested) at high latitudes and produces zero electricity at the time of early evening peak demand in winter when the sun has set. Battery storage can cover the diurnal cycle at low latitudes but, at high cloudy latitudes, there is basically not enough sun in winter to charge a battery during the day for evening use. Similarly, solar thermal plants today are all underperforming and, for

(continued)

Box 1.1 (continued)

example in Spain, on a cloudy day, they produce nothing at all. Can we base a reliable grid on technology like this?

5. Wind turbines produce wildly erratic power output and on certain days they produce nothing at all over large geographic areas. For instance, data shows that the wind on occasions drops across the whole of Europe. At those times, the grid needs to be backed up by fossil fuels which by 2050 are shown to be virtually phased out.

6. The large distributed footprint of wind and PV in the environment, including a massively expanded grid and storage infrastructures, seems at odds with environmental protection.

7. Relatively little coal and gas with carbon capture and storage are assumed to be introduced. It is highly unlikely that they could operate in flexible load following mode. We then need to ask if the small amount of remaining dispatchable sources (unabated flexible fossil, biomass and hydro) will be sufficient to balance and backup the large amount of variable renewable energy that is projected to exist.

8. Already, the utilisation of some fossil fuel plant has become so low that they can no longer make a profit. And they are being paid by the consumer to stay open and do nothing most of the time since they are essential backup for when the wind does not blow and sun does not shine. Do future energy scenarios sufficiently include likely rising electricity costs and balance options accounting for possible spreading energy poverty?

Decarbonisation will inevitably be accompanied by a decoupling (*happy divorce*) of economic growth from CO_2 emissions. We can already see this in countries like France, Sweden and Finland that have high levels of nuclear and hydroelectric power and in countries like Norway (hydro) and Iceland (hydro + geothermal) that have very high penetration of renewable energy. The years 2014/2015 have shown encouraging signs of progress towards this decoupling and more improvements are expected, as shown in the projections of Fig. 1.7.

However, a significant part of the so-called decoupling between GDP and energy consumption in the OECD is simply down to globalization and exporting emissions to China and other developing countries, via outsourcing of the CO_2 intensive manufacturing. Moreover, historic CO_2 and GDP data need to be viewed with some caution. In 2005, the world got 87% of its total energy from fossil fuels and in 2015 that figure had dropped by only 1%. Somewhat perversely, the growth in new renewables has been partly offset by decline in nuclear, rather than fossil fuels, in e.g., Germany and Japan. Growth in the Chinese economy has slowed and the structure of GDP in that country could be evolving towards a higher level of service industries that are less CO_2 intensive. Construction has also slowed in China, reducing cement related CO_2 emissions.

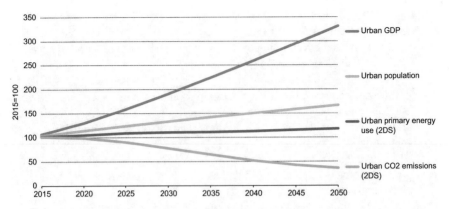

Fig. 1.7 Decoupling of economic growth (GDP) and population growth from CO_2 emissions (OECD/ IEA Paris, *Energy Technology Perspectives 2016*, http://www.iea.org/etp/ (accessed June 2016))

Figure 1.7 is produced by the International Energy Agency (IEA),[55] for which we make some critical observations. Urban population growth is projected to grow faster than global population growth, owing to increasing urbanisation of rural populations. However, a picture of falling urban per capita energy use (rising number of people, flat energy consumption) distorts reality where global per capita energy consumption is rising, and is likely to continue to do so—in particular as the developing world aspires to reach OECD standards. Food, shelter, heat (or cool) and security are the first priorities, usually followed by education, healthcare, large screen TVs and cars. In 1965, global average per capita energy consumption was ~ 1.1 tonnes oil equivalent (toe) and by 2015 this had grown to ~1.8 toe, a rise of 64%. It is therefore difficult to reconcile this projection of the IEA with reality. Said differently, rural peasants, who produce very little CO_2 from combusting fossil fuels, move to cities to get jobs where they begin to consume electricity and processed food that have embedded CO_2 emissions in them and then they get a job in a factory, which consumes a lot of energy making goods that power economic growth and provide them with prosperity.

Figure 1.7 imparts a simple but possibly deceptive message to the political classes that we are on course to decarbonise energy and have strong economic growth at the same time. The projection is also underpinned by the decarbonised energy projections shown in Fig. 1.6, the reality of which are also questionable. An alternative scenario is that seeking de-carbonisation on the renewables route leads to much higher electricity prices and an unstable with regular blackouts (Box 1.1). CO_2 emissions continue to rise while GDP falls as countries like India, Bangladesh and Indonesia (combined population ~ 1.75 billion) spurn climate action and prioritise the welfare of their citizens instead and burn lots of coal to generate cheap electricity.

As primary energy and electricity demands increase, the future of consumption makeups, CO_2 emissions, and other factors have been studied in detail for the next 15, 20 and more years using scenarios. In addition to the IEA outlook discussed above, here we add two more of the most prominent existing scenario-based outlooks to inform our views on the future energy technologies (see Table 1.2, see also

Table 1.2 Comparison of existing scenario-based outlooks: the BP Energy Outlook 2016 considers a base case where world GDP more than doubles but energy demand grows by a third due to energy efficiency improvements; the IEA Energy Technology Perspectives 2016 considers three scenarios with different maximum temperature limits; the MIT 2015 Outlook shows a projection for primary energy use until 2050 for 5-year time intervals

	BP Energy Outlook 2016[a] (see Fig. 1.21, 1.22, and 1.23)	IEA Energy Technology Perspectives 2016[b]	MIT 2015 outlook[c]
Timeline	2035	2050	2050
Assumptions	– Base case (most likely path for energy demand + uncertainty)[b] – Energy demand grows to sustain human activities and living standards – Paris COP21 pledges are met and retained	– ETP model[d] (see Fig. 1.20) – Three scenarios[e] – Energy technology and innovation important in keeping global temperature increase below 2 °C and even 1.5 °C – Cities at the heart of de-carbonization effort[f]	– Emissions are stable and fall in developed countries – Emissions will come from the other G20 and developing countries – Paris COP21 pledges are met and retained
Results/ trends	– World GDP will more than double – CO_2 emissions more than halve – Fossil fuels represent 80% of total energy supply in 2035 – Power gen: non fossil fuels represent 45% – Nuclear grows 1.9% p.a. – Gas is the fastest growing fossil fuel – Renewables (incl. bio-fuels) rise to 9% in share of primary energy	2DS scenario[e]: – World GDP will more than triple between 2011 and 2050 – Energy-related CO_2 emissions already in decline by 2020[f] – Fuel mix (see Fig. 1.24): renewables 44%, fossil fuels 45%, nuclear 11% – Power generation almost completely decarbonized by 2050 (renewables 67%) – Electricity largest final energy carrier	– World GDP increases by 2.7% p.a. over time period 2010–2050 – CO_2 emissions increase by 68% – Emissions from electricity and transportation together account for 51% of CO_2 emissions – Fossil fuels still provide 75% of primary energy (see Fig. 1.25) – Rapid growth in renewables and nuclear (see Fig. 1.26 for projections)

[a]BP, Energy Outlook 2016, https://www.bp.com/energyoutlook (accessed June 2016)

[b]BP considers three alternative cases to the base case: (1) slower global GDP growth, (2) faster transition to a lower carbon world and (3) shale and oil gas having even greater potential

[c]Reilly, J., et al., Energy & Climate Outlook, Perspectives from 2015, http://globalchange.mit.edu/files/2015%20Energy%20%26%20Climate%20Outlook.pdf (accessed June 2016)

[d]The ETP (Energy Technology Perspectives) model comprises four interlinked technology-rich models, one for each of four sectors: energy supply, buildings, industry and transport. Depending on the sector, this modeling framework covers 28–39 world regions or countries, over the period 2013–2050. The scenarios are constructed using a combination of forecasting to reflect known trends in the near term and "back casting" to develop plausible pathways to a desired long-term outcome

[e]The 6DS (6 °C) scenario is an extension of current trends. Primary energy demand and CO_2 emissions would grow by about 60% from 2013 to 2050. The 4DS (4 °C) scenario takes into account recent pledges by countries to limit emissions and improve energy efficiency; it would require significant changes in policy compared with 6DS and yet would cause serious climate impacts. The 2DS scenario (2 °C) requires at least a 50% chance of limiting the average global temperature increase to 2 °C; it includes cutting CO_2 emissions by 60% by 2050. The focus is on the energy sector but the 2DS recognizes that greenhouse gas emission reductions concern the non-energy sectors, too (e.g., agriculture, industry, passenger aviation)

[f]This ignores the explosion in CO_2 emissions in China that accompanied urbanization

annexed Table 1.4 for a more comprehensive overview). In particular, we summarize the view of three leading energy institutions, namely British Petroleum (BP), the International Energy Agency (IEA) and the Massachusetts Institute of Technology (MIT). BP, despite their technical excellence and global energy overview underpinned by their annual statistical review, are beholden to the political will of the countries where they operate. The IEA is a political organ of and funded by the OECD. Note that the forecast dates differ, BP is looking ahead to 2035 and the IEA and MIT look ahead to 2050.

These three studies contain similarities and differences. The **main features** to retain are the following:

1. **Strong global economic growth:** BP global GDP more than doubles by 2035; IEA GDP more than triples by 2050 and MIT GDP increases by a factor of 2.7 by 2050. A tripling of GDP over 32 years corresponds to a 3.5% annual growth rate. This value is essentially the same as the average GDP growth rate observed since 1975.
2. **Expectation that decarbonisation of the energy supply of the more advanced economies will be visible in the near future. Moreover, the scenarios assume a progressive shift of developing countries to a low carbon development path.** The IEA report and other reports[62] (e.g. Energy [R]evolution Scenario by Greenpeace[63] and Roadmap 2050 by the ECF[64]) all agree on this trend. Both MIT and BP see that Paris COP21 commitments will be largely met. However, if this route causes hardship, countries may quickly rescind, like the USA has indicated its intention to do under the Trump administration.
3. **Extensive deployment of low carbon electricity generation options.** The IEA report agrees on this point by mentioning a decarbonized energy production in 2050. Greenpeace, among others, advocates for a 100% renewable energy sourced power supply. The BP report[65] puts more emphasis on gas and a rebound of fossil fuels.
4. **Growth in energy demand concerns mainly electric power generation.**
5. **Declining share of fossil fuels.** No consensus on the extent of this decline, some reports are more optimistic than others. Looking at the individual institutional cases. BP's is perhaps the most extraordinary. They see fossil fuel as 80% of

---

[62] The Greenpeace Energy [R]evolution scenario presents a widely decarbonized energy system by 2050, accompanied by a quick short-term phase-out of fossil fuels and nuclear. The ECF Roadmap 2050 presents a decarbonized power sector by 2030 for a low carbon European economy.

[63] Greenpeace, *The Energy [R]evolution*, (2015), http://www.greenpeace.org/international/en/cam paigns/climate-change/energyrevolution/ (accessed June 2016).

[64] European Climate Foundation (ECF), *Roadmap 2050: Power Perspectives 2030*, (2011), http://www.roadmap2050.eu/attachments/files/PowerPerspectives2030_FullReport.pdf   (accessed June 2016).

[65] The BP report is an industry report, therefore subject to industry subjectivity. One should not be, however, naïve and think that NGO, academic-based and IEA reports are necessarily less subjective; they could be subjective in different ways. For other industry reports and a more complete comparison of studies, see Prognos 2011 (Table 1.4).

global total energy by 2035, which seems reasonable considering 2015 was 86% and 2005 was 87%. But they then go on to suggest that CO_2 emissions more than halve.[66]

6. **Increasing share of renewables.** The consensus on a growing renewables share is argued on the basis that renewables are becoming cheaper, with better technologies, with all this supported by international agreements on the need for decarbonisation.

7. **Ambiguous share of nuclear energy.** Most studies argue for a larger share of nuclear in agreement with the decarbonisation trend, others for a rapid phase out of nuclear (i.e. Greenpeace) due to its relatively minor role played in the global energy mix and supposedly major risks.

The IEA see global power generation almost totally decarbonized by 2050 with renewables at 67% of electricity with no engineering or economic plan on how this can be achieved at a cost that also enables global GDP to more than triple. Very high energy prices in the past have caused recession and not growth (recall e.g. the oil shocks in the 1970s). The MIT projections are the only ones that seem to us to be grounded in reality: CO_2 emissions stable or falling in the OECD but growing in the rest of the G20 and the developing nations; CO_2 emissions up by 68% in 2050. In 2050, the world still gets 75% of its total primary energy from fossil fuels (compare with BP estimate of 80% in 2030).

The importance of energy technology, and especially low-carbon energy technology, calls for a closer examination in terms of which criteria are considered important in energy makeup scenarios. As we saw above, possible decarbonisation strategies include switching to a renewable energy-base or to a nuclear-based or to a mix of both.[67] Consensus on a stable future energy production includes the following points:

1. Abundant, affordable, clean and sustainable energy (see Sect. 1.4.1 on sustainability);
2. Reduced import dependencies creating a greater balance in the global economy, reducing exposures to price spikes and geopolitical risks;
3. Improved security of supply and diversification of the energy mix;
4. Regulating new countries that wish to harness nuclear power.

As an addition to this, the Paul Scherrer Institut (PSI) in Switzerland has proposed an *Energy Trilemma*, whereby one will have to make trade-offs when maximizing

---

[66]BP probably envisions massive deployment of carbon capture and storage (CCS)—which may have strong impact on the cost of electricity and on the growth potential of the economy, and is unlikely to be accepted by the public.

[67]PwC, *Decarbonisation and the Economy*, (2013), https://www.pwc.nl/nl/assets/documents/pwc-decarbonisation-and-the-economy.pdf (accessed June 2016).

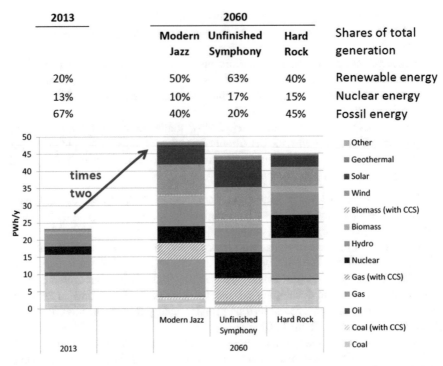

2013		2060			
		Modern Jazz	Unfinished Symphony	Hard Rock	Shares of total generation
20%		50%	63%	40%	Renewable energy
13%		10%	17%	15%	Nuclear energy
67%		40%	20%	45%	Fossil energy

Fig. 1.8 Global electricity generation scenarios from the report of the Paul Scherrer Institut (PSI) in Switzerland (S. Hirschberg and P. Burgherr, Sustainability Assessment for Energy Technologies, Handbook of Clean Energy Systems (2015)). For more details see annexed Fig. 1.27

energy security, energy equity (accessibility and affordability) and environmental sustainability.[68] Conclusions—which are in line with main trends mentioned above—include:

1. Doubling of demand for electricity up to 2060 (see Fig. 1.8);
2. Unprecedented growth rate for wind and solar technologies;
3. Global transport transition will be the most challenging and determining factor for our success in decarbonizing our energy systems;
4. Global cooperation, sustainable economic growth, and technology innovation will be key in addressing this transition.

While each of these scenarios sees the share of fossil fuel falling in the electricity mix, CO_2 emissions will still rise under their "Modern Jazz" and "Hard Rock" scenarios since the quantity will have doubled (let us not confuse share or fraction with absolute quantity), in contradiction with the IEA scenario shown in Fig. 1.7.

---

[68]PSI, *World Energy Scenarios: The Grand Transition*, October 2016, https://www.worldenergy. org/work-programme/strategic-insight/global-energy-scenarios/

Under the scenario they dub "unfinished symphony", as we have already discussed, it is doubtful that 20% fossil fuel will be sufficient to balance any grid.

To conclude this section, we see a risk that going the high renewables route may in fact lock the world into long-term dependency on fossil fuel, mainly natural gas, that is required to back up and balance renewables on the grid. This is the exact opposite of the often-stated policy goals. In contrast, nuclear power is the only scalable low carbon option already proven to be able to power a modern society, as evidenced by France.

1.3.2 Decoupling Human Development from Impact on Nature

Centralized and decentralized energy supplies call for different demands on land. How can a future energy mix be designed to be in line with reducing land exploitation/usage and reduce the coupling of human development with nature (i.e. decreasing the impacts on nature)? Table 1.3 shows the energy densities of the most common fuels for comparison. Clearly, uranium has by far the greatest energy density of the fuels listed above. Second is hydrogen, but six orders of magnitude below uranium. This means approximately a factor of the order of one million in favor of nuclear energy.[69] Although a more careful comprehensive analysis must be made (a comparative sustainability assessment, below), nuclear energy is attractive on the basis of its very large energy density and almost zero CO_2 emissions while other "zero carbon energy sources" are mostly very dilute in comparison (e.g., wind and solar).

Decarbonisation also requires commitment to technological progress: this means technological progress in dilute renewable energies and in energy-dense nuclear, too. As energy demand grows in the next decades, notably for electricity, and as access to modern energy and especially electricity in big cities develops—decarbonisation seems to be reachable but requires a portfolio involving sufficiently energy dense technologies as well as a smart design of our cities/energy supply networks.

At the policy level, decarbonisation is believed to require some form of carbon taxes, which introduce a tariff on CO_2 emissions by making polluters pay. From economic theory, levying such a Pigovian tax is meant to correct for inefficient market outcomes in activities that generate negative externalities, here the fact that

---

[69]This is the result of the physics of the binding between nucleons (protons and neutrons) within atom cores, which involves binding energies in the range of 1 to several MeV (Mega-electron-volts), compared with the binding of atoms within molecules (gas, oil) or solids (coal) with energies in the range of 1 eV (electron-volt). In other words, when burning coal, gas, oil or hydrogen, the reconfigurations of the atoms from the reactant to the product molecules lead to rupturing and producing covalent bonds between atoms with typical energy of 1 eV. During nuclear fission of the core of an atom of Uranium 235, say, the rupture and building of bonds between nucleons involve energies millions times larger, per bond.

Table 1.3 Specific energy density values for different fuels and energy carriers. The energy content is the potential energy stored in the fuel. Specific energy density refers to energy per unit mass. Hydrogen and electrochemical batteries are put at the end of the list because they are not primary energy sources but just energy storage. The energy density for Uranium corresponds to the composition enriched with Plutonium that is used in breeder reactors

Energy carriers	Lower (=net) heating value or specific energy density (MJ/kg) (wind and solar units are W/m^2)
Uranium in breeder reactors (238U–239Pu mix)	80,620,000
Natural gas	47.1
Crude oil	42.7
Gasoline	43.4
Diesel	42.8
Black Coal	22.7
Petroleum coke	29.5
Biomass (sugarcane, coffee, coconut)	16–18
Wind (1 Watt = 1 J/s)	Proportional to cubic law of wind speed: poor (<150 W/m^2), fair ($150 \sim 250$ W/m^2), good ($250 \sim 350$ W/m^2), or excellent (>350 W/m^2)
Solar	5 W/m^2 in Europe, 20 W/m^2 in desert solar PV farms (due to 10–15% efficiency only)
Hydrogen (energy storage)	120.2
Electrochemical batteries (energy storage)	0–1

Natural uranium consists of roughly 99.3% of U-238 and 0.7% of U-235, which is the fissile component. In contrast, U-238 is not fissile but fertile, i.e. it becomes fissile when converted into fissile Pu-239 via neutron capture and two following beta-minus decays. Therefore, the energy density value given for U-238/Pu-239 of 80,620,000 MJ/kg is almost representative of natural uranium but could be slightly modified regarding the small (0.7%) fraction of U-235 and slightly smaller energy density, resulting in 80,611,290 MJ/kg

the costs to society of CO_2 emissions are not internalized in the market price since they are not supported directly by the emitters. In this respect, the latest 2015 Paris Accord with its 195 national signatories asks each nation to specify its contributions to reducing CO_2 emissions and to increase those contributions over time, but without any legal enforcement or deadlines. Such voluntary actions with no imposed time-line is meant to minimize the economic impact and costs of decarbonisation for the many countries that are still struggling to recover from the Great Recession of 2009, which has been followed by a "new normal" of reduced economic growth and high unemployment in many countries like Greece, Italy and Spain.[70] However, Kotlikoff[71] notes that this may actually backfire, since the only real "achievement"

---

[70]Youth unemployment still stands at 40% or more in these countries at the end of 2017, while aggregate unemployment stands at 20.7% in Greece and 16.4% in Spain.

[71]Kotlikoff, L., *Write me in but don't send me a penny*, Kotlikoff for President, 2016 (www.kotlikoff2016.com).

of the Paris COP 21 agreement is to notify the dirty-energy producers that their days are numbered. He thus concludes: "this greatly incentivizes them to accelerate their extraction of fossil fuels and, thereby, increase the planet's temperature. The recent extremely low price of oil can be partly explained by the "use it or lose it" calculus underlying today's oil production".[72] The standard economic theory of extractable resources shows that a rising rate of carbon taxation will accelerate emissions, that a constant rate of carbon taxation will have no impact on emissions and that a declining rate of carbon taxation will slow emissions.[73] Kotlikoff advocates to treat fossil fuel producers harshly now but far better in the future. This is to keep them from focusing on "use it or lose it", as they seem to presently do. Such large-to-low tax policy aims at making more attractive delayed extractions and the diversification towards alternative energy sources. This fundamental insight needs to be considered seriously in the policy actions that are needed to encourage the transition to decarbonisation.[74]

At the time of finalizing this book version, the COP23 held in Bonn ended in Nov. 27, 2017, with none of the fanfare of the Paris COP 21 of December 2015: with a glaring current lack of momentum (and with still increasing CO_2 emissions globally), the needed cohesive climate actions are in jeopardy. Without the development, adoption and scaling of technologies to significantly counter greenhouse gas emissions, the forecasts are bleak. Fighting climate change is likely to be replaced by the reality of adapting to a hotter world, with the dangers of significant impacts on the global order with climate change induced conflicts, and no guarantee that the current world order will be maintained.[75]

1.4 Comprehensive Assessments

1.4.1 Sustainability

To overcome the limitations of technology assessments based on single factors, a more comprehensive approach is needed. With the publication of the Brundtland-

---

[72]Another contribution for the low oil price is the attempt by OPEC to dump the oil frackers in the US and maintain market share. Moreover, given the development time of the order of 5 year for new large oil projects, true oil price equilibrium is never reached, with supply either overshooting or undershooting demand.

[73]Hotelling, H., The Economics of Exhaustible Resources, Journal of Political Economy 39 (2), 137-175 (April 1931).

[74]There is also U.S. President Trump's pledge to revive the beleaguered American coal industry and put miners back to work. One year into the new U.S. administration, coal production and exports increased slightly in 2017, but mining employment barely changed and coal-fired power plants continued to close.

[75]A gloomy forecast for climate change, Assessments by Stratfor, 6 Dec. 2017 (https://worldview.stratfor.com/article/gloomy-forecast-climate-change).

Report,[76] "sustainability" has become the dominant model of societal development and a standard of evaluation in many disciplines, including technologies in the energy sector.

There is no single, generally, accepted definition of sustainability. Rather, the concept evolved over a long time. The EU-Project NEEDS[77,78] represents the current state of the art in the establishment of the framework for indicator-based technology assessment with application to prospective energy generation technologies in the year 2030 and even 2050, reflecting expected future developments based on three scenarios: "pessimistic", "realistic-optimistic", and "very optimistic" and using current technologies as a benchmark. Overall, 26 future advanced electricity generation technologies were analyzed, including fossil (coal, lignite, natural gas), nuclear (European pressurized water reactor (EPR) and sodium-cooled fast reactor (EFR), 1600 MW electric each) and a wide range of renewables (biomass, solar, on- and offshore wind), all with relevant upstream and downstream processes within the respective energy chain.

In the project, a full set of 36 technology-specific evaluation criteria and indicators,[79] covering the environmental, economic, and social dimension of sustainability, was established, namely:

- 11 environmental criteria covering energy and mineral resources, climate change, ecosystem impacts from normal operation and severe accidents, and special chemical and medium and high level radioactive wastes;
- 9 economic criteria including impacts on customers (electricity price), overall economy (employment and autonomy of electricity generation), and utility (financial risk and operation);
- 16 social criteria addressing security/reliability of energy provision, political stability and legitimacy, quality of residential environment (landscape and noise) and social and individual risks (normal operation and accidents) based on experts' assessments and social and individual risks as they are often perceived by the public, such as terrorist threat.

Numerous approaches have been used for quantification, including 'life cycle assessment (LCA)', 'impact pathway assessment', 'severe accident risk assessment' based on adopted historical experience with energy-related accidents and probabilistic safety analysis (PSA, see Chap. 4) for the nuclear options, and finally internal

---

[76]WCED, Our Common Future, Oxford, Oxford University Press (1987).

[77]New Energy Externalities Developments for Sustainability (NEEDS), *NEEDS Documents* [Online], 20.10.2016, http://www.needs-project.org

[78]S. Hirschberg and P. Burgherr, Sustainability Assessment for Energy Technologies, Handbook of Clean Energy Systems (2015).

[79]See annexed Tables 1.5, 1.6, and 1.7 for detailed descriptions and units.

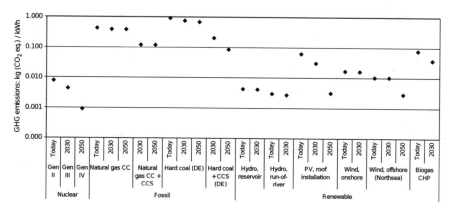

Fig. 1.9 Greenhouse gas emissions of selected technologies at different time periods, from S. Hirschberg and P. Burgherr, Sustainability Assessment for Energy Technologies, Handbook of Clean Energy Systems (2015). Note the logarithmic y-axis scale

and external cost assessment. Most of the social indicators were quantified based on a survey and use of the Delphi technique among experts in relevant areas.[80]

Some examples of quantitative indicators are provided:

- Renewables and nuclear have greenhouse gas (GHG) emissions that are one to two orders of magnitude lower than those from fossil technologies without carbon capture and sequestration (CCS), with CCS having the potential to reduce the emissions by one order of magnitude (Fig. 1.9).
- The aggregated impacts on ecosystems are the highest for biogas, followed by coal, while hydro impacts are lowest (Fig. 1.10). Indeed, hydro is often claimed to have the lowest ecosystem impact. This may be over-optimistic in situations where the dam totally obliterates all natural life in the flooded valley. As an illustration, the Three Gorges dam in China flooded archaeological and cultural sites and displaced some 1.3 million people, while causing significant ecological changes, including an increased risk of landslides. Radioactive wastes are inherently associated with nuclear energy. The volumes of critical non-radioactive wastes are today highest for solar PV (45 m^3 stored underground/TWh) but expected to be strongly reduced in the future.
- Fossil technologies have the highest expected operational risks due to severe accidents (Fig. 1.11). New generations of nuclear exhibit very substantial reduction of risk compared with the current ones. However, the maximum credible consequences of accidents are clearly highest for nuclear, which leads to a high

---

[80]The Delphi Technique is a method used to estimate the likelihood and outcome of future events. A group of experts exchange views, and each independently gives estimates and assumptions to a facilitator who reviews the data and issues a summary report. See Norman Dalkey and Olaf Helmer, An Experimental Application of the Delphi Method to the use of experts, Management Science. 9 (3), 458-467 (1963); Harold A. Linstone and Murray Turoff, The Delphi Method: Techniques and Applications, Reading, Mass.: Addison-Wesley (1975).

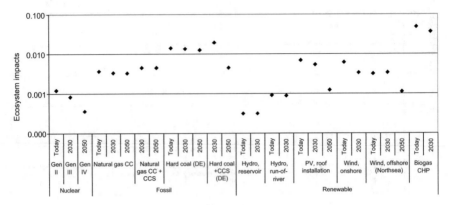

Fig. 1.10 Ecosystem impacts of selected technologies: the "potentially damaged fraction" (PDF) of species quantifies loss of species (flora and fauna) due to land use, ecotoxic substances released to air, water, and soil, and acidification and eutrophication, from Hirschberg and Burgherr (2015)

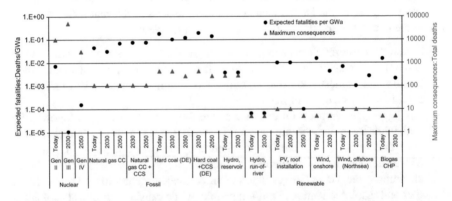

Fig. 1.11 Expected fatality rates due to severe accidents (left hand scale shows the death count per GWa; 1 GWa is the energy that a one GW power plant delivers over one year) and maximum consequences per accident (right hand scale), related to the amount of electricity produced. Source: Hirschberg and Burgherr (2015). See Sects. 4.2 and 4.3 for a critical discussion of the methodology used to estimate the maximal potential deaths due to a nuclear accident

level of risk aversion. Also, large hydro installations, depending on the location, have the potential for very large consequences of accidents, but without the component of very long-term land contamination—a feature unique to nuclear (see Chap. 5).

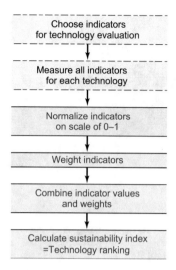

Fig. 1.12 Chart of the multi-criteria analysis process (subjective elements in red, objective steps in blue), from Hirschberg and Burgherr (2015)

If desired, aggregation can be based on the total[81] cost approach or a Multi-Criteria Decision Analysis (MCDA). Not all factors that play a role in the evaluation of a technology are amenable to being expressed in monetary units, and doing so may be controversial—above all concerning subjective aspects like perceived risks, human suffering, or visual disturbances to the landscape. This will be discussed further in terms of the external costs of nuclear power in the section on risk.

Multi-Criteria Decision Analysis (MCDA) has the capability to explicitly reflect subjective social acceptance issues. The approach builds on the steps shown in Fig. 1.12. Single indicators can already be used individually for technology comparisons. And from them, a single comprehensive index value can be calculated, which reflects how sustainable the individual technologies are compared to each other. Then the indicators are each weighted, based on the individual user preferences. Figure 1.13 shows the average indicator weights assigned by individual European stakeholders in the NEEDS project.

The overview of the results based on all stakeholder responses as elucidated in the NEEDS project is shown in Fig. 1.14 along with total costs. While, within the applied cost estimation framework, nuclear energy exhibits the lowest total costs, its ranking in the MCDA framework tends to be lower, mainly due to consideration of a variety of social aspects not reflected in external costs. Thus, largely thanks to predicted improved economic performance of renewables, nuclear energy ranks in MCDA mostly lower than renewables.

---

[81]Internal plus external costs, the latter are not born by the party that causes them, but rather by society as a whole. They include the costs of health damages that result from air pollution, converted to monetary units with high uncertainty.

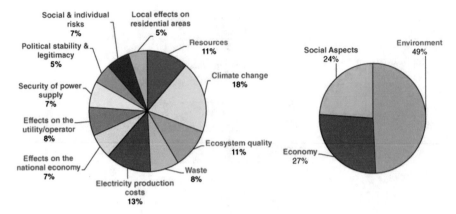

Fig. 1.13 Average indicator weights for technology assessment, obtained via online survey from stakeholders engaged in the European energy sector (not representative of the overall population), from Hirschberg and Burgherr (2015)

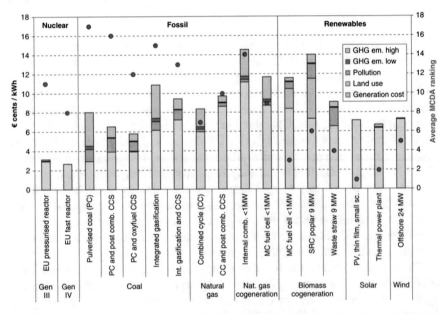

Fig. 1.14 Average Multi-Criteria Decision Analysis (MCDA) ranking (red dot, against right-hand axis) of future (year 2050) technologies compared with total costs (Bachmann, T. M. and Schenler, W., Final report on establishment of economic indicators. EU Project NEEDS on New Energy Externalities Developments for Sustainability", Research Stream 2b "Technology Roadmap and Stakeholder Perspectives". EDF Research and Development, Paris, France and Paul Scherrer Institut, Villigen, Switzerland (2007)). The figure shows a subset of the 26 systems evaluated. GHG low/high values represent low and high estimates of damage costs due to climate change. *CCS* carbon capture and storage, *MC* molten carbonate, *PV* photovoltaic. Note that the ranks on the right-hand axis go from bottom to top: hence, the fact that the red dots for nuclear are higher than renewable means that nuclear ranks behind, as stated above in the text (low ranks are better)

The individual preference profiles have a decisive influence on the Multi-Criteria Decision Analysis (MCDA) ranking of technologies. Given equal weighting of environmental, economic, and social dimensions, and with the emphasis on the protection of climate and ecosystems, together with the minimization of objective risks and the maximization of affordability for customers, the nuclear options are top ranked. On the other side, focusing on radioactive wastes, land contamination due to hypothetical accidents, risk aversion and perception issues, terrorist threat, and conflict potential, the ranking changes to the disadvantage of nuclear energy. The ranking of fossil technologies highly depends on the degree of emphasis put on the environmental performance, which in relative terms remains to be a weakness, more pronounced for coal than for gas. Renewables show mostly a stable very good performance in terms of relatively low sensitivity to changes in preference profiles, based on highly improved economics expected to occur in the next few decades.

Independent from this assessment, an interdisciplinary team of academic researchers in economics, ecology, environment and climate sciences, history, philosophy and public policy and members of several NGOs published a report, entitled "the ECOMODERNIST Manifesto".[82] This report praises nuclear fission as "the only present-day zero-carbon technology with the demonstrated ability to meet most, if not all, of the energy demands of a modern economy. However, a variety of social, economic, and institutional challenges make deployment of present-day nuclear technologies at the scale necessary to achieve significant climate mitigation unlikely. A new generation of nuclear technologies that are safer and cheaper will likely be necessary for nuclear energy to meet its full potential as a critical climate mitigation technology".

1.4.2 Externalities

The term externality, coming from economics, is for a cost (or benefit) accrued by a third party from the actions of others, where the third party did not choose to acquire said costs or benefits. It plays an important role in discussions about what should be included in the market price of electricity, such that the electricity market functions well. The term, augmented with the adjective "negative", and related to "social costs", has been widely adopted to describe negative impacts of energy production systems, such as the harm done by the combustion of fossil fuels, particularly greenhouse gas emissions. The argument goes that the fossil fuel producers and consumers do not currently pay for this harm and should do so, with the sought-after consequence of promoting cleaner substitutions. On the other hand, one can worry that a consequence of adopting these measures, like a carbon tax, will be higher energy prices that spread energy poverty, prevent development of innovative energy

---

[82]Asafu-Adjaye, J. et al., An ecomodernist manifesto, www.ecomodernism.org (April 2015).

systems, and hold back economic growth. There are also externalised benefits (or positive externalities) that the same energy production systems provide.

Furthermore, having much in common with the sustainability framework, the quantification of externalities allows for an "all-inclusive" comparison of energy sources on the basis of their "full cost". As outlined by the OECD/NEA,[83] the price levels for energy production, all defined per unit of electricity generated, are:

1. **Plant-level:** Defined by the LCOE (levelized cost of electricity), which includes capital, operation, and decommissioning costs. At this level, fossil fuels are cheapest, and wind and PV can be competitive with nuclear, which depends heavily on the capital cost.
2. **Grid/system-level:** Various costs incurred by the system due to a source, notably grid and balancing costs associated with integrating variable renewables. Assessment at this level is somewhat uncertain, but indicates a less competitive position of current wind and PV whose costs at this level are at least an order of magnitude greater than dispatchable technologies.
3. **Social costs:** Diverse negative impacts on society associated with the energy system: air pollution (health), land use, as well as extreme and highly uncertain impacts like climate change, and severe accidents. See Box 1.2 for a discussion of uncertainties.

The current market price of electricity is essentially determined by plant-level costs, neglecting system and social costs, which together therefore form the externality—see Fig. 1.15 for a quantification of the external cost by energy source. Further, some make the statement that the current price omits more than it includes. On the basis of morbidity and loss of life alone (see Sect. 4.4), hard coal leads to social cost of around USD 32 Million per TWh.[84] This can be compared with a typical market value of electricity of USD 100 million per TWh. Considering climate change risk and/or energy sources with worse emissions (e.g., lignite) could indeed plausibly eclipse the plant-level costs.

System level costs can also be decisive. In particular, plant-level generation costs of PV and wind have fallen sharply, but often forgotten are the system or "grid-level" costs of integrating these distributed intermittent technologies into the grid. At present, grids are balanced by cycling either gas or coal plants up or down, and with nuclear also contributing to baseload power. These services to the grid incur wear, tear and reduce efficiency at plant-level. At the same time, providers of baseload power are losing market share, as growing amounts of renewable electricity is given priority access to the grid. However, a reliable system featuring a high stake

---

[83] OECD/NEA. "The Full Costs of Electricity Provision", NEA No. 7298 (2018). http://www.oecd-nea.org/ndd/pubs/2018/7298-full-costs-2018.pdf

[84] The disability-adjusted life years lost per TWh generated (DALY/TWh) of hard coal is 320/TWh, and the assumed cost of a DALY is 100,000 USD.

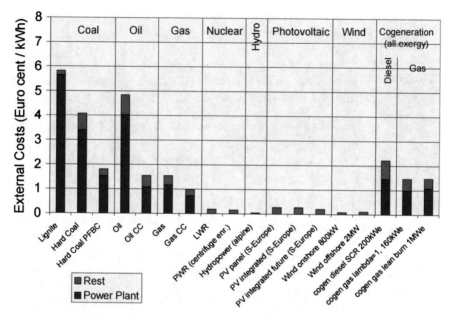

Fig. 1.15 Externalities of a range of energy sources, from the PSI and ExternE, 2005 (Dones, Roberto, et al. "ExternE-Pol Externalities of Energy: Extension of Accounting Framework and Policy Applications." Paul Scherer Institut (2005)). See the NEEDS project for further analysis. Nuclear (PWR and LWR) omits the severe accident externality

of wind generation will require an extensive grid ("it's always windy somewhere"[85]) as well as significant backup infrastructure. For instance, already some coal plants are being paid to stay open and "do nothing" most of the time, to provide backup for when the wind does not blow.

There is the aspiration to store surplus generation of renewables to cover periods of low generation, which has yet to be implemented. Pumped hydro storage (PHS), where mountainous geography allows, offers an economical storage solution, but a rough calculation shows the large scale of costs involved: A small country like Scotland, with 4 GW mean winter demand, would require 672 GWh of storage if the wind did not blow for a week. This is equivalent to 22 large PHS schemes that would cost of the order of 40 billion euros to build and integrate with the grid.

A full price would be substantially higher than the one we currently pay, and a market with a full price would therefore co-ordinate towards lower carbon sources, such as nuclear and PV/wind, with system costs becoming highly relevant. Carbon taxes make a step in this direction, however being around 0.01 USD per kWh, tend to fall short of covering the true social (and grid level) costs.

---

[85]In fact, this is not case when a strong anticyclonic regime installs itself over all Europe, with no wind anywhere.

Box 1.2 Views on Uncertainty of Externalities

Externalities (costs and benefits) typically have vast uncertainties associated with them, which complicates assigning economic cost, and presents a major challenge to their adoption within the markets. For example, three classes of externality that may be associated with coal are: (1) health effects and mortality associated with coal mining, (2) health effects and mortality associated with air pollution and (3) the risks associated with climate change and rising sea levels. To get a feeling of the complexity of the question, consider air pollution. The risks associated with inhaling smoke, normally quantified as the concentration of <2.5 μ particles in the air (pm 2.5) are linked to respiratory illnesses that express themselves mostly at the end of life and a shortening of life. But this should be put in the perspective of ageing populations on the back of medical and social advances supported by food created from fossil fuels and comfort by way of warmth provided largely by fossil fuels until now. How to rationally "trade" the loss of a few months or years at the end of life in exchange for all the benefits acquired? Quantifying the possible impacts of climate change are even more difficult. For example, higher atmospheric CO_2 does lead to enhanced food production. The possible costs of manmade climate change are perhaps best condensed into the impact of rising sea level. If sea level was to rise 10 m by the end of this century, the costs, both human and economic, would be vast. If, on the other hand, the rise is 10cm, then there is probably no cost at all. So where does the truth lie? Thus, the climate externality associated with burning coal and other fossil fuels may range from zero to many trillions of dollars. The benefits of cheap electricity and more food are plain to see. As another example, consider also the rare-earth elements such as Neodymium (Nd) used to make the magnets in the actuators of wind turbines or Cobalt (Co) used in Lithium Ion batteries, and the huge environmental and human labor costs in China where these elements are mined. The social dimension here is that the public often naively sees a solar panel and thinks it is making magic clean electricity without being made to look behind the façade and the dirt being created in someone else's backyard.

1.4.3 The Case of Renewable Energy Sources

Renewable energy sources, i.e. hydro, wind, solar, geothermal and biomass, have been claimed to have the potential to cover the world energy/electricity demand in the future[86] (note that this paper and its assumptions have been severely criticized[87]).

---

[86]Mark Z. Jacobson, Mark A. Delucchi, Mary A. Cameron and Bethany A. Frew, Low-cost solution to the grid reliability problem with 100% penetration of intermittent wind, water, and solar for all purposes, Proc. Acad. Sci. USA 112 (49),15060-15065 (2015).

[87]see the criticism in Christopher T. M. Clack et al., Evaluation of a proposal for reliable low-cost grid power with 100% wind, water, and solar, Proc. Natl. Acad. Sci. USA 114 (26), 6722-6727

At least, there is a consensus towards having a significantly growing share of renewable energy sources on the basis that they are becoming cheaper, with better technologies. However appropriate sites are not available in all countries and output is not growing fast enough to replace fossil fuels, which dominate current power production by still providing roughly 70% of the total. As shown in Sect. 1.1, the history of energy generation has not been generally that of energy replacement but of energy addition. Given the continued strong human population growth and, even more importantly, the fast economic development and energy consumption of developing countries, it remains highly uncertain whether renewable energy sources will be able globally to replace fossil sources, and instead just add to the pool of existing energy contributions.[88] Moreover, the problem of their intermittency (night and clouds for solar, wind variability, water levels for hydro, etc.) is yet to be solved for scenarios using decentralized energy generation, which awaits, among other solutions, new generation large-scale batteries[89] or other storage capabilities such as power to gas. Progress to store solar energy generated in a more centralized way is more advanced, as the natural solution could be pumped hydroelectric energy storage (PHES) whose role would be to provide daily storage and not inter-seasonal storage as is the case with conventional hydroelectric dams.[90] However, the costs of such solution may be prohibitive and needs to be assessed.

The use of and switch to renewable energy sources is supported by international agreements and largely accepted by the public in many countries,[91] while nuclear energy is largely missing such backing. In general, capital costs and construction times for renewable energy sources (except for hydro dams) are low, though expensive infrastructure is mostly needed (transmission grids, storage capacities) and subject to growing opposition.

(2017) and the rebuttal in Mark Z. Jacobson, Line-by-Line Response by M.Z. Jacobson, M.A. Delucchi to Evaluation of a proposal for reliable low-cost grid power with 100% wind, water, and solar (2017) (http://web.stanford.edu/group/efmh/jacobson/Articles/I/CombiningRenew/Line-by-line-Clack.pdf, accessed 4 Aug. 2017) ; Further coverage at https://www.technologyreview.com/s/608126/in-sharp-rebuttal-scientists-squash-hopes-for-100-percent-renewables/

[88]Jason Scott Johnston, Debunking the 100% Renewables Fantasy, Real Clear Energy (July 18, 2017), reproduced at the CATO Institute, (https://www.cato.org/publications/commentary/debunking-100-renewables-fantasy, accessed 6 Aug. 2017).

[89]Batteries have themselves a number of issues. In particular, they contain heavy metals and toxic chemicals that need to be recycle to avoid health effects as well as shortage of some their constituting rare-earth elements.

[90]Denis Bonnelle, Everything about photovoltaics except... semi-conductors science, residential rooftops, self-consumption, start-ups, grid parity, lithium batteries, exponential growth, maximum power point, record-breaking efficiencies and beautiful sunny photographs, The BookEdition.com, Collection développons, 2017.

[91]To be complete, one should note that there is vast opposition to wind in Scotland and in USA for instance. This is still a minority, and the opposition differs widely between city dwellers (who don't care in general: NIMB) and rural dwellers (who support the inconvenience of the spreading wind farms).

There is however a knowledge gap between the idealistic view (and unrealistic expectations) of a large part of the public with respect to solar energy in particular, and the reality. This can be illustrated by the little-known but important concept of "energy return on energy invested" (ERoEI). Taking a solar photovoltaic panel as illustration, ERoEI is simply the ratio of energy generated (over the lifetime of the solar panel) to the amount of energy used to get this energy. Following the methodology pioneered by Murphy and Hall (2010),[92] in order to get energy from a solar panel, energy must be used not only to produce the panels but also to erect and install them. One should also take into account the labor at every stage of the process and the energy spent for mining, manufacture and disposal. Energy consumption is also involved in the utilization of pre-existing infrastructure and energy investment, as well as in the integration of intermittent photovoltaic panels onto the grid.

Taking all these factors into account, Ferroni and Hopkirk (2016) estimated the ERoEI of temperate latitude solar photovoltaic systems to be 0.83 with an error of $\pm 15\%$[93] which if correct would mean that a temperate latitude PV system would never produce the amount of energy used to create it. However, the article by Ferroni and Hopkirk has been strongly criticized by Raugei et al. (2017),[94] who argue that several fundamental mistakes were made, such as the use of outdated information, invalid assumptions on photovoltaic specifications and an incorrect choice of boundaries. Raugei et al. (2017) conclude with an ERoEI of 7–8 for solar photovoltaic, which is supported by other studies like Battisti and Corrado (2005).[95] Of course, at lower latitudes that enjoy stronger insolation, the ERoEI is significantly larger, which has motivated projects such as the desertec initiative.[96] Moreover, the strong pace of innovations in photovoltaic technology suggests that it will

---

[92]Murphy, D.J.R. and Hall, C.A.S., Year in review-EROI or energy return on (energy) invested, Ann. N. Y. Acad. Sci. Spec. Issue Ecol. Econ. Rev. 1185, 102-118 (2010).

[93]Ferruccio Ferroni and Robert J. Hopkirk, Energy Return on Energy Invested (ERoEI) for photovoltaic solar systems in regions of moderate insolation, Energy Policy 94, 336-344 (2016). If correct, the ERoEI of 0.83 estimated in this article would mean that the energy used to make the PV panels will never be recovered from them during their 25 years of estimate lifetime. At European latitudes corresponding to Germany or the UK, photovoltaic panel would produce more CO_2 than if coal or oil were simply used directly to make electricity. A photovoltaic panel would thus not be an energy source but an energy conversion device, transforming the concentrated fossil energy used to produce it into a diluted source of electric energy. This would also mean that all the CO_2 from the production of an existing solar panel is in the atmosphere today, while burning the fossil fuel that has produced it to generate electricity would be spread over the 25-year period of the solar panel lifetime. In this view, more CO_2 emissions occur in China (where solar panels are mainly produced) to make Europeans believe that they contribute to the global reduction of CO_2 emissions.

[94]Marco Raugei et al., Energy Return on Energy Invested (ERoEI) for photovoltaic solar systems in regions of moderate insolation: A comprehensive response. Energy Policy 102, 377-384 (2017).

[95]Riccardo Battisti and Annalisa Corrado, Evaluation of technical improvements of photovoltaic systems through life cycle assessment methodology, Energy 30(7), 952-967(2005).

[96]Desertec was a large-scale project aimed at creating a global renewable energy, mainly solar from concentrated solar farms installed in deserts in North Africa, and transporting it through high-voltage direct current transmission to European consumers such as Germany (see http://www.

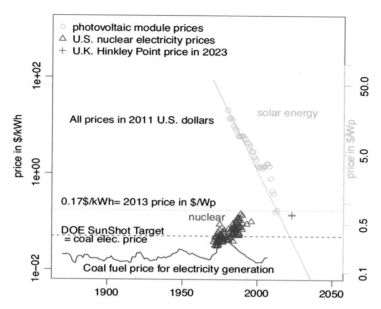

Fig. 1.16 Comparison of long-term price trends for coal, nuclear power and solar photovoltaic modules. Reproduced from Farmer, J. Doyne and Lafond, F., How Predictable Is Technological Progress? Research Policy (45): 647-655 (2016)

outperform nuclear and even coal energy, on a cost basis, within a decade (see Fig. 1.16). However, one should not forget the cost of intermittency, which is usually neglected in such projections and which could add significant cost to mitigate.

By comparison, it is generally estimated that hydroelectric power has an ERoEI typically larger the 50. Thus, despite substantial environmental harm and social disruptions, dispatchable hydroelectric power is a fantastic energy source. Euan Mearns gives an instructive estimate for the Three Gorges Dam on the Yangtze River in China, the largest hydroelectric scheme in the World, with 22.5 GW installed capacity. Taking into account labor and embedded energy of the concrete and steel and assuming a 45% capacity factor and 70-year life, he finds a very large partial ERoEI of 147,[97] Moreover, its advantage lies in its much longer lifetime. However, hydropower may have unintended negative side-effects, with dams causing disruption of natural systems, affecting river environments, fisheries and land,[98,99] as well as

---

desertec.org). The project stalled in 2012 due to enormous costs, the need for very large subsidies, the impact of renewable on existing grids, as well as security concerns.

[97]Euan Mearns, The Energy Return of The Three Gorges Dam, Energy Matters, Posted on June 3, 2016 (http://euanmearns.com/the-energy-return-of-the-three-gorges-dam/, accessed 6 Aug. 2017).

[98]http://www.conserve-energy-future.com/Disadvantages_HydroPower.php, accessed 8 Aug. 2017.

[99]International Rivers https://www.internationalrivers.org/environmental-impacts-of-dams accessed 8 Aug. 2017.

triggering earthquakes.[100] More so, a physical shortage of appropriate locations provides an upper bound for its potential.

Turning briefly to wind energy generation, many more studies have been conducted for the wind industry than for solar. Thus, the ERoEI is better understood for wind energy generation. An in-depth meta-analysis taking into account 119 wind turbines from 50 different analyses from 1977 to 2007 reveals the existence of large differences in the quoted ERoEI, with an average value 25.2 and standard deviation of 22.3.[101] The very large range values (68.2% of wind turbines having their ERoEI between $25.2 - 22.3 = 1.9$ and $25.2 + 22.3 = 47.5$) is mainly due to different assumptions regarding the operating characteristics of wind turbines, for example their assumed lifetime. The ERoEI depends strongly on the rotor diameter and the wind speed, i.e. is very sensitive to the precise conditions of the installation of the wind turbine. In addition to reasonable (but site-dependent) energy production qualities, one can state that, in general, wind turbines have low environmental impacts compared with fossil power sources. They however have some potential effect on property values, aesthetics, and there is some concern about stress and health effect from the very low-frequency radiated sound perceived by some individuals.

Another myth is the efficiency of concentrated solar power systems, which generate electricity by using mirrors or lenses to concentrate a large area of sunlight onto a small area, whose heat is transformed into electricity via a steam turbine connected to an electrical power generator or via thermochemical reactions. Built-in heat storage capacity can be added using molten salts. Roger Andrews writing at the Energy matters blog recently reviewed the evidence showing that the concentrated solar power plants operating in the US are costly, heavily subsidized, generally performing below expectations (with large production shortfalls in electricity generation between 25% and 40%) and no more efficient than utility-scale photovoltaic plants.[102] They often use substantial quantities of natural gas each morning to jump-start them. Clouds and stormy weather have been invoked as the causes of the underperformance, but there seems to be technological issues, such as the continuous delicate synchronization of the 600,000 parabolic mirrors across 7.3 km^2 (for the 250 MW Genesis solar energy project located in Blythe, California).[103] Keeping the hundreds of thousands of mirrors clean in dusty deserts is also a non-trivial issue.

In Europe, matching the positive public view and political support, renewable subsidies are substantial and this perhaps is not factored into the positive perception.

[100]J.R. Grasso and D. Sornette, Testing self-organized criticality by induced seismicity, J. Geophys. Res. 103 (B12), 29965-29987 (1998).

[101]Ida Kubiszewski, Cutler J Cleveland, and Peter K Endres, Meta-analysis of net energy return for wind power systems, Renewable energy 35(1), 218-225 (2010).

[102]Roger Andrews, Concentrated solar power in the USA: a performance review, Energy Matters, Posted on April 24, 2017 (http://euanmearns.com/concentrated-solar-power-in-the-usa-a-performance-review, accessed 6 Aug. 2017).

[103]Concentrating Solar Power, Solar Energy Industry Association (http://www.seia.org/policy/solar-technology/concentrating-solar-power, accessed 6 Aug. 2017).

Fig. 1.17 Correlation between installed wind and solar capacity and electricity price in European countries. Reproduced from Euan Mearns, The High Cost of Renewable Subsidies, Posted on April 16, 2017 (http://euanmearns.com/the-high-cost-of-renewable-subsidies, accessed 6 Aug. 2017)

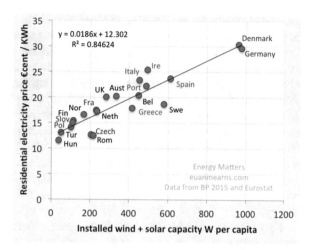

A recent report by the Council of European Energy Regulators[104] shows that the weighted average subsidy paid to renewable generators in EU 26 in 2015 was 110 €/MWh. The maximum was 184 €/MWh in the Czech Republic and the minimum 16.2 €/MWh in Norway. This should be compared with the wholesale price of electricity in Europe, which lies in the range from 40 € to 60 €/MWh, Thus, renewables are costing on average about 3 times as much as conventional power (wholesale ≈ 50, subsidy ≈ 110, total ≈ 160). In fact, this estimation is a lower bound as it ignores the "system costs" to expand the grid itself and to provide a portfolio of balancing and backup services for these very intermittent energy sources. In particular, there is need for an almost 100% backup from conventional energy sources, which face a dwindling market share and thus rising costs. Figure 1.17 shows a significant correlation between electric prices and installed solar and wind energy in European countries, suggestive (but not proof) of a causation. However, we should stress that the high electricity costs of renewables are likely to decrease substantially in the future, with technological progress and economies of scale. Hence, one should not necessarily extrapolate the data in Fig. 1.17 as predicting the future. The transition to renewable energies is, in the end, a political choice, with costs and benefits that have many facets reaching beyond pure economic reasons.

Renewable energies are often presented as ideal pollution-free sources. However, such assertions overlook that many potentially hazardous chemicals are used during the production of solar modules,[105,106] Nitrogen trifluoride (NF3) is a gas used in

---

[104]Status review of renewable support schemes in Europe, Council of European Energy Regulators, Ref: C16-SDE TF-56-03, 11-04-2017.

[105]Dustin Mulvaney, Solar Energy Isn't Always as Green as You Think. IEEE Spectrum, Posted 13 Nov 2014 (http://spectrum.ieee.org/green-tech/solar/solar-energy-isnt-always-as-green-as-you-think, accessed 6 Aug. 2017).

[106]Different PV technologies have different chemical signatures: for instance, thin film solar is much worse than polycrystalline when it comes to generation of harmful chemicals.

process chambers for the cleaning of the silicon contaminants, where silicon is one of the compounds used for solar panels. According to the IPCC (Intergovernmental Panel on Climate Change), this gas has a global warming potential of approximately 16,600 times that of CO_2.[107] Two other similarly undesirable "greenhouse" gases appearing in the manufacture of solar panels are hexafluoroethane (C2F6) and sulphur hexafluoride (SF6). However, overall, lifecycle inventories and assessments of photovoltaic systems demonstrate that photovoltaic technology achieve significant reductions, over carbon-based sources, of harmful air emissions on a life cycle basis. For instance, emissions of cadmium are over 100 times lower than those from coal-fired generation on a life cycle basis.[108] There are however concerns about the enforcement of recycling and clean environmental practices, and the occurrence of a number of pollution scandals has somewhat tainted the green credentials of solar energy.[109]

Let us also mention biomass, which appears to be a favored source of 'renewable' energy in Europe,[110] with some of its supply coming from American and Canadian forests that are cut to create wood pellets.[111,112] The negative side is that plantations for biofuels displace ecologically diverse ecosystems that could have absorbed carbon more efficiently, or compete with land that could have been used for food production. There is also the concern that, with the biofuels targets set by the U.S. and Europe, the amount of land used to create fuel rather than food will increase dramatically. Then, food prices could rise as occurred in 2010–2011 (which

---

[107]Arnold, T., Harth, C.M., Mühle, J., Manning, A.J., Salameh, P.K., Kim, J., Ivy, D.J., Steele, L.P., Petrenko, V.V., Severinghaus, J.P., Baggenstos, D., Weiss, R.F., Nitrogen trifluoride global emissions estimated from updated atmospheric measurements, Proceedings of the National Academy of Sciences USA 110 (6), 2029-2034 (2013).

[108]Vasilis M. Fthenakis, Hyung Chul Kim, and Erik Alsema, Emissions from Photovoltaic Life Cycles, Environmental Science & Technology 42 (6), 2168-2174 (2008); Hsu, D., O'Donoughue, P., Fthenakis, V., Heath, G., Kim, H., Sawyer, P., Choi, J. and Turney, D., Life Cycle Greenhouse Gas Emissions of Crystalline Silicon Photovoltaic Electricity Generation: Systematic Review and Harmonization, Journal of Industrial Ecology (16:S1), S122-S135 (2012); Kim, H., Fthenakis, V., Choi, J and Turney, D., Life Cycle Greenhouse Gas Emissions of Thin-film Photovoltaic Electricity Generation: Systematic Review and Harmonization, Journal of Industrial Ecology (16:S1), S110-S121 (2012); R. Frischknecht, R. Itten, P. Sinha, M. de Wild-Scholten, J. Zhang, V. Fthenakis, H.

[109]Dustin Mulvaney, Solar Energy Isn't Always as Green as You Think, IEEE Spectrum, Posted 13 Nov 2014 (http://spectrum.ieee.org/green-tech/solar/solar-energy-isnt-always-as-green-as-you-think, accessed 6 Aug. 2017).

[110]https://ec.europa.eu/energy/en/topics/renewable-energy/biomass, accessed 8 Aug. 2017.

[111]The Economit, Wood: The fuel of the future. Environmental lunacy in Europe http://www.economist.com/news/business/21575771-environmental-lunacy-europe-fuel-future, accessed 8 Aug. 2017.

[112]https://www.dogwoodalliance.org/our-work/our-forests-arent-fuel/

catalysed the so-called 'Arab spring'[113]), pushing hundreds of millions of people into hunger,[114] with the potential to trigger social instabilities. While cultivation of biofuels that displaces food production is normally prohibited, much of the production moves to natural land such as forests or grasslands,[115] leading to loss of biodiversity,[116] deforestation and the actual net increase of emissions[117] in Europe and beyond. Moreover, biomass can only be a minor contributor to our energy needs: cutting all the hard wood in the US would provide enough energy to power the US for just about 4 months.[118]

Finally, most present forms of renewable energy raise growing concerns regarding the consumption of scarce raw materials as well as the scale of land use and other environmental impacts. For instance, if we assume that 10% of the windiest part of the UK was covered with windmills, only 20 kWh/day per person of power would be generated on average,[119] which is half of the power used by driving an average fossil-fuel car 50 km per day. This can also be compared with the average of approximately 100–200 kWh/day of power needed for the consumption of a Western European.[120] Similarly, if the UK had 5% of its territory covered with 10%-efficient photovoltaic solar panels, this would generate about 50 kWh/day/person.[119] The main message here is the desperately dilute nature of these sources of energy. This is where nuclear energy excels in contrast, as already mentioned, by having energy density millions of times larger than any other sources. To make the point vividly, Sir David MacKay points out that a typical Brit creates 30 kg of waste carbon dioxide per day for his/her total energy consumption. A standard nuclear fission reactor produces the same energy with just 2 g of natural uranium, which generates just 0.25 g of waste per day with Generation II nuclear technology.[121] This amounts to a factor 120 million difference (less waste) in favour of nuclear power.

---

[113]Lagi, M., K.Z. Bertrand and Y. Bar-Yam, The food crises and political instability in North Africa and the Middle East (https://arxiv.org/abs/1108.2455).

[114]https://www.actionaid.org.uk/sites/default/files/publications/biofuels_fuelling_hunger.pdf, accessed 8 Aug. 2017.

[115]http://ec.europa.eu/energy/en/topics/renewable-energy/biofuels, accessed 8 Aug. 2017.

[116]F. Stuart Chapin III, Erika S. Zavaleta, Valerie T. Eviner, Rosamond L. Naylor, Peter M. Vitousek, Heather L. Reynolds, David U. Hoope, Sandra Lavorel, Osvaldo E. Sala, Sarah E. Hobbie, Michelle C. Mack and Sandra Díaz, review article: Consequences of changing biodiversity, Nature 405, 234-242 (2000).

[117]Walsh, B. Even Advanced Biofuels May Not Be So Green, Time (2014), http://time.com/70110/biofuels-advanced-environment-energy, accessed 8 Aug. 2017; Steer, A. and C. Hanson, Biofuels are not a green alternative to fossil fuels, The Guardian (29 July 2015), http://www.theguardian.com/environment/2015/jan/29/biofuels-are-not-the-green-alternative-to-fossil-fuels-they-are-sold-as, accessed 8 Aug. 2017.

[118]Private communication from Euan Mearns.

[119]David JC MacKay, Sustainable Energy – Without the Hot Air, UIT Cambridge Ltd.; 1 edition (February 20, 2009).

[120]Which is the same as saying that the energy of 200 kWh is consumed each day by a typical well-off western European.

[121]Farmer, J. Doyne and Lafond, F., How Predictable Is Technological Progress? Research Policy (45): 647-655 (2016).

Annexes

	GDP growth rate	Population (millions)	Pop. Growth rate
Myanmar	8.60%	54	0.90%
Ivory Coast	8.50%	24	2.50%
Bhutan	8.40%	0.8	1.20%
India	7.50%	1340	1.10%
Laos	7.40%	6.8	1.50%
Iraq	7.20%	38	2.90%
Cambodia	7.00%	16	1.50%
Tanzania	6.90%	57	3.10%
Bangladesh	6.60%	165	1.10%
Senegal	6.60%	15.8	2.80%

Fig. 1.18 World's fastest growing economies (Source: IMF World Economic Outlook, April 2016)

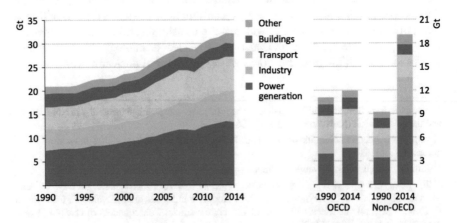

Fig. 1.19 CO_2 emissions makeup by sectors and regions in Gt (Giga tons) over the time period 1990–2014. The "Other" category includes agriculture, non-energy use (such as cement), oil and gas extraction and energy transformation (such as refining) (Source: International Energy Agency (2016), World Energy Outlook Special Report, OECD/IEA, Paris)

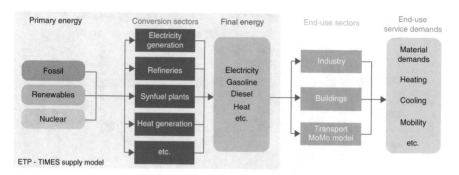

Fig. 1.20 ETP model: technology, bottom-up analysis of the global energy system (Source: International Energy Agency (2016), Energy Technology Perspectives 2016, OECD/IEA, Paris)

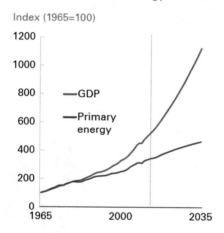

Fig. 1.21 Economic growth drives energy demand but not proportionately due to improvements in energy efficiency, production of high value low energy goods like iPhones, and low energy high value services like Netflix: Global GDP grows by 107% while energy demand grows by 34%. (Source: BP 2016 Energy Outlook)

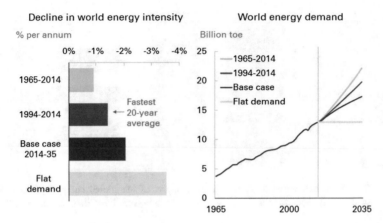

Fig. 1.22 Decline in world energy intensity and world energy demand in the three alternative scenarios. Toe stands for ton of oil equivalent (Source: BP 2016 Energy Outlook). The % decline in energy intensities for different scenarios shown in the left bar graph are used to project different energy demand scenarios into the future

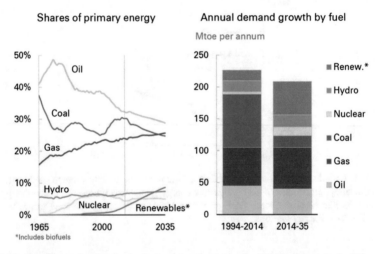

Fig. 1.23 Significant changes in the fuel mix projected in the BP Energy outlook: Oil and coal shares are seen to decrease and renewables and gas shares are seen to increase. The largest projected growth is expected to be that of renewables. (Source: BP 2016 Energy Outlook)

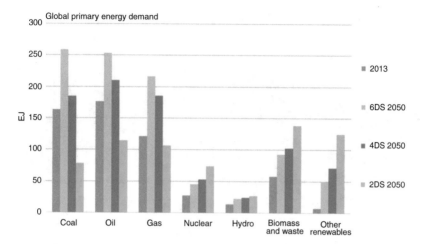

Fig. 1.24 Shares in primary energy demand in the three different scenarios for 2050 (6DS, 4DS, 2DS) and the situation in 2013 (Source: International Energy Agency, Energy Technology Perspectives 2016)

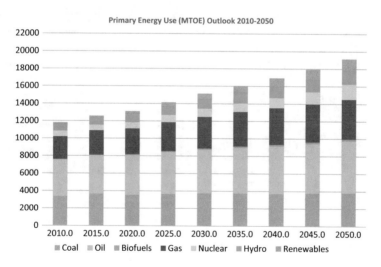

Fig. 1.25 Energy makeup projections for time period 2010–2050, reproduced from MIT Energy and Climate Outlook (Reilly, J., Paltsev, S., Monier, E., Chen, H., Sokolov, A., Huang, J., & Schlosser, A. (2015). Energy & climate outlook: perspectives from 2015. *MIT Joint Program on the Science and Policy of Global Change.* Available at: http://globalchange.mit.edu/files/2015%20Energy,20,26). These projections are at odds with Figs. 1.6 and 1.7, illustrating some of the differences between the BP, IEA and MIT views on the energy future

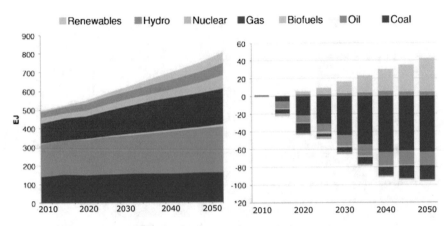

Fig. 1.26 Left: Scenario of global primary energy use (in (exajoules)); Right: Changes in outlook from 2014 to 2017 also in exajoules. 1 EJ (exajoules) corresponds to 23.88 Mtoe. Reproduced from MIT Energy and Climate Outlook (2015)

Critical uncertainties	MODERN JAZZ	UNFINISHED SYMPHONY	HARD ROCK
Productivity and Economic Growth	• Open economies • Digital boost	• Intelligent growth • Circular economies	• Domestic growth and expertise • Local content emphasis
Tools for Action	• Free markets • Enabling policies • New business models	• Climate focused policy • Global policy convergence	• Security focused policy action
Climate Challenge	• Consumer driven technology adoption • Technology support	• Local support • Global mandates • Unified action	• Lower GDP growth • Energy security drives renewables
International Governance	• Complex globalisation • Shifting hubs • Growing global connections	• Strong global cooperation • Regional integration	• Fragmented political and economic systems • Power balancing alliances

Fig. 1.27 Critical uncertainties in the PSI's Energy Trilemma (PSI, *World Energy Scenarios: The Grand Transition*, October 2016, https://www.worldenergy.org/work-programme/strategic-insight/global-energy-scenarios/). PSI is the Paul Scherrer Institut in Switzerland, a Swiss National Laboratory funded within the ETH domain (the Swiss Federal Institute of Technology)

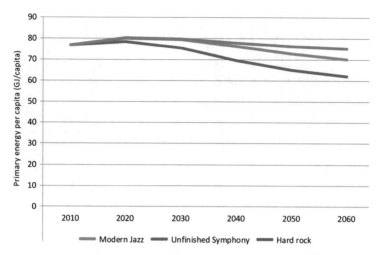

Fig. 1.28 Global per capita annual primary energy consumption peaking before 2030; *reproduced from PSI, World Energy Scenarios: The Grand Transition, October 2016* (PSI, *World Energy Scenarios: The Grand Transition*, October 2016, https://www.worldenergy.org/work-programme/ strategic-insight/global-energy-scenarios/). The peak of 80 GJ can be expressed in toe (ton of oil equivalent) units as equal to 0.570 toe annual consumption per capita on average for the World. In comparison, Brits consume ~ 5 toe while Canadians consume about 10 toe. This comparison suggests that energy efficiency needs to accelerate so much that it is able to compensate for the increasing need for energy of the 85% of the world population who are trying to catch up with the developed world in terms of quality of life

Table 1.4 Overview of outcomes of studies and short-listed studies (Source: Prognos, Analysis and comparison of relevant mid- and long-term energy scenarios for EU and their key underlying assumptions, 2011: Basel/Berlin)

Nr.	Study	Year of publication	Time horizon	Amount of scenarios (sensitivities)	World	Europe	EU-27	Quantifiability	Residential	Commercial / Services	Industrial	Transportation	Power	Other conversion sectors	Amount of sectors covered	Level of detail (rating Prognos)	Bottom-up	Top-down	None	(International) Relevance (rating Prognos)	Short list for selection
	Inter-/Governmental institutions:																				
1	EA (2010). International Energy Outlook 2010	2010	2035	1 (+4)	x	(OECD)	(EU-19)	x	x	x	x	x	x	x	6	+++	x			+++	x
2	EU DG TREN / ENER (2008, 2010). Energy trends to 2030, Update 2007 and Update 2009	2008 / 2010	2030	1 / 2		(OECD)		x	x	x	x	x	x	x	6	+++	x	x		+++	x
3	IEA (2010). Energy Technology Perspectives	2010	2050	2 (+4)	x	x	(x)	x	x	x	x	x	x	x	5	+++	x			+++	
4	European Parliament / DG Internal Policies (2009). Future Energy Systems in Europe	2009	2050	3		x	x	x	x	x	x	x	x	x	3	+	x			+	
5	EA/NEA (2010). Technology Roadmap: Nuclear Energy	2010	2050	3	Based on IEA (2010)			x	Based on IEA (2010)						0	+				++	
6	NEA (2009). Energy, Electricity and Nuclear Power Estimates for the Period up to 2030	2009	2030	2	x	(OECD)							(Nuclear)		0	+		(x)		++++	x
7	IEA (2009) World Energy Outlook 2009	2009	2030	2	x	(OECD)	x	x	x	x	x	x	x	x	5	+++	x			++++	
8	NEA (2008). Nuclear Energy Outlook 2008	2008	2050	2	x	(OECD)							(Nuclear)		0	+			x	+	
9	EU DG Research (2006). World Energy Technology Outlook. WETO - H2	2006	2050	3	x	x	x	x	x	x	x	x	x	x	4	++	x			++	x
10	EU DG TREN (2006). Scenarios on energy efficiency and renewables	2006	2030	3 (+2)			(EU-25)	x	x	x	x	x	x	x	6	++	x			++	
11	EEA (2005). European environment outlook	2005	2030	2		x		x	x	x	x	x	x	x	5	+	x			+	
	Non-governmental organisations:																				
12	EGF (2010). Roadmap 2050	2010	2050	3 (+1)		x	x	x	x	x	x	x	x	?	6	+++	x			++	x
13	EREC (2010). RE-Thinking 2050	2010	2050	1 (+1)		x	x	x					x		2	++	x			++	
14	Greenpeace/EREC (2010). Energy [r]evolution	2010	2050	3	x	(OECD)	x	x	x	x (Renewable heat)	x	x	x (RES-E)	x	5	+++	x			++++	x
15	WEC (2007). Deciding the Future: Energy Policy Scenarios to 2050	2007	2050	4	x	x		(x)	x		x	x	x		0	+				++	
	Industry:																				
16	ExxonMobil (2009). Outlook for Energy: A View to 2030	2009	2030	1	x			x	x	x	x	x	x		4	+++	x (?)			++	
17	IHS Global Insight (2008). European Energy and Environmental Outlook	2008	2030	1		x	x	x	x	x	x	x	x	x	6	++	x (?)			++	
18	Shell (2008). Shell energy scenarios to 2050	2008	2050	2	x			x	x	x	x	x	x		4	+	x (?)			++	
19	PWC (2006). The World in 2050	2006	2050	6	x			x							0	+	x (?)			++	
	Industry associations:																				
20	Eurelectric (2009). Power Choices	2009	2050	1 (+ base)			x	x	x	x	x	x	x	x	5	++	x	x		++	x
21	Euracoal (2007). The future role of coal in Europe	2007	2030/2050	5				x					x		1	+	x	x		+	
22	Eurelectric (2007). The Role of Electricity	2007	2030/2050	4			(EU-25)	x	x	x	x	x	x	x	5	+	x			++	
	Research / academic consortia:																				
23	FEEM et al. (2010). Probabilistic Long-term Assessment of New Energy Technology Scenarios	2010	2100	10	x	x	x	x	x	x	x	x	x	x	6	+++	(x)	x		++	x
24	Capros et al. - EU DG ENV (2008). Model-based Analysis of the 2008 EU Policy Package	2008	2030	9		x	x	x	x	x	x	x	x	x	6	++	x	x		++	x
25	Energy Watch Group (2008). Renewable Energy Outlook 2030	2008	2030	2	(OECD)			x	x	x	x	x	x	x	4	+	x			+	
26	Oko-Institut (2006). The Vision Scenario for the European Union	2006	2030	2		x	(EU-25)	x	x	x	x	x	x	x	6	+	x			++	
27	ECN (2005). The next 50 years: Four European energy futures	2005	2050	4		x								x	0	+			x	+	
28	ISIS et al. (2005-2009). NEEDS New Energy Externalities Development for Sustainability	2006*/2009	2050	7		x		x	x	x	x	x	x	x		+	x			++	

*Year of scenario analysis

Table 1.5 Environmental criteria and indicators established in the NEEDS project[a]

Criteria/Indicator	Description	Unit
ENVIRONMENT	Environment-related criteria	
RESOURCES	Resource use (nonrenewable)	
Energy	Energy resource use in whole lifecycle	
Fossil fuels	This criterion measures the total primary energy in the fossil resources used for the production of 1 kWh of electricity. It includes the total coal, natural gas, and crude oil used for each completed electricity generation technology chain	MJ/kWh
Uranium	This criterion quantifies the primary energy from uranium resources used to produce 1 kWh of electricity. It includes the total use of uranium for each complete electricity generation technology chain	MJ/kWh
Minerals	Mineral resource use in whole lifecycle	
Metal ore	This criterion quantifies the use of selected scarce metals used to produce 1 kWh of electricity. The use of all single metals is expressed in antimony-equivalents, based on the scarcity of their ores relative to antimony	kg (Sb-eq.)/ kWh
CLIMATE	Potential impacts on the climate	
CO_2 emissions	This criterion includes the total for all greenhouse gases expressed in kg of CO_2	kg(CO_2-eq.)/kWh
ECOSYSTEMS	Potential impacts to ecosystems	
Normal operation	Ecosystem impacts from normal operation	
Biodiversity	This criterion quantifies the loss of species (flora and fauna) due to the land used to produce 1 kWh of electricity. The "potentially damaged fraction" (PDF) of species is multiplied by land area and years	PDF*m2*a/kWh
Ecotoxicity	This criterion quantifies the loss of species (flora fauna) due to ecotoxic substances released to air, water, and soil to produce 1 kWh of electricity. The "potentially damaged fraction" (PDF) of species is multiplied by land area and years	PDF*m2*a/kWh
Air pollution	This criterion quantifies the loss of species (fora and fauna) due to acidification and eutrophication caused from production of 1 kWh of electricity. The "potentially damaged fraction" (PDF) of species is multiplied by land area and years	PDF*m2*a/kWh
Severe accidents	Ecosystem impacts in the event of severe accidents	
Hydrocarbons	This criterion quantifies large accidental spills of hydrocarbons (at least 10,000 tones) which can potentially damage ecosystems	t/kWh
Land contamination	This criterion quantifies land contaminated due to accidents releasing radioactive isotopes. The land area contaminated is estimated using probabilistic safety analysis (PSA). Note: Only for nuclear electricity generation technology chain	km^2/kWh
WASTE	Potential impacts due to waste	
Chemical waste	This criterion quantifies the total mass of special chemical wastes stored in underground repositories due to the production of 1 kWh of electricity. It does not reflect the confinement time required for each repository	kg/kWh
Radioactive waste	This criterion quantifies the volume of medium and high level radioactive wastes storied in underground repositories due to the production of 1 kWh of electricity. It does not reflect the confinement time required for the repository	M^3/kWh

[a]P. Burgherr, S. Hirschberg and E. Cazzoli, "Final set of sustainability criteria and indicators for assessment of electricity supply options, NEEDS Deliverable no. D3.2 – Research Stream 2b, NEEDS project", Brussels, Belgium (2008)

Table 1.6 *Economic criteria and indicators established in the NEEDS project*[a]

Criteria/ indicator	Description	Unit
ECONOMY	Economy-related criteria	
CUSTOMERS	Economic effects on customers	
Generation cost	This criterion gives the average generation cost per kilowatt-hour (kWh). It includes the capital cost of the plant (fuel) and operation and maintenance costs. It is not the end price	€/MWh
SOCIETY	Economic effects on society	
Direct jobs	This criterion gives the amount of employment directly related to building and operating the generating technology, including the direct labor involved in extracting or harvesting and transporting fuels (when applicable). Indirect labor is not included. Measured in terms of person-years/GWh	Person-years/GWh
Fuel autonomy	Electricity output may be vulnerable to interruptions in service if imported fuels are unavailable due to economic or political problems related to energy resource availability. This measure of vulnerability is based on expert	Ordinal
UTILITY	Economic effects on utility company	
Financial	Financial impacts on utility	
Financing risk	Utility companies can face a considerable financial risk if the total cost of a new electricity generating plant is very large compared with the size of the company. It may be necessary to form partnerships with other utilities or raise capital through financial markets	€
Fuel sensitivity	The fraction of fuel cost to overall generation cost can range from zero (solar PV) to low (nuclear power) to high (gas turbines). This fraction therefore indicates how sensitive the generation costs would be to a change in fuel prices	Factor
Construction time	Once a utility has started building a plant, it is vulnerable to public opposition, resulting in delays and other problems. This indicator therefore gives the expected plant construction time in years. Planning and approval time is not included	Years
Operation	Factors related to a utility company's operation of a technology	
Marginal cost	Generating companies "dispatch" or order their plants into operation according to their variable cost, starting with the lowest cost base-load plants up to the highest cost plants at peek load periods. This variable (or dispatch) cost is the cost to run the plant	€ cents/ kWh
Flexibility	Utilities need forecasts of generation they cannot control (renewable resources such as wind and solar), and the necessary start-up and shut-down times required for the plants they can control. This indicator combines these two measures of planning flexibility, based on expert judgment	Ordinal
Availability	All technologies can have plant outages or partial outages (less than full generation) due to either equipment failures (forced outages) or maintenance (unforced or planned outages). This indicator tells the fraction of the time that the generating plant is available to generate power	Factor

This table includes a measure for fuel security, which is important. One should also be concerned with the cost of dispatch. The "Flexibility" should incorporate a "controllability factor", for instance we estimate Gas = 3, nuclear = 2, RE = 1/3, meaning that gas is much more flexible and controllable than renewable energy (RE), nuclear being in between

[a]P. Burgherr, S. Hirschberg and E. Cazzoli, "Final set of sustainability criteria and indicator s for assessment of electricity supply options, NEEDS Deliverable no. D3.2 – Research Stream 2b, NEEDS project", Brussels, Belgium (2008)

Table 1.7 Social criteria and indicators established in the NEEDS project[a]

Criteria/Indicator	Description	Unit
SOCIAL	Social-related criteria source: NEEDS Research Stream 2b survey of social experts. Quantitative risk based on PSI risk database	
SECURITY	Social security	
Political continuity	Political continuity	
Secure supply	Market concentration of energy suppliers in each primary energy sector that could lead to economic or political disruption	Ordinal scale
Waste repository	The possibility that storage facilities will not be available in time to take deliveries of waste materials from whole life cycle	Ordinal scale
Adaptability	Technical characteristic of each technology that may make it flexible in implementing technical progress and innovations	Ordinal scale
POLITICAL LEGITIMACY	Political legitimacy	
Conflict	Refers to conflicts that are based on historical evidence. It is related to the characteristics of energy systems that trigger conflicts	Ordinal scale
Participation	Certain types of technologies required public, participative decision-making processes, especially for construction or operating permits	Ordinal scale
RISK	Risk	
Normal risk	Normal operation risk	
Mortality	Years of life lost (YOLL) by the entire population due to normal operation compared with/without the technology	YOLL/kWh
Morbidity	Disability adjusted life years (DALY) suffered by the entire population due to normal operation compared with/without the technology	DALY/kWh
Severe accidents	Risk from severe accidents source: NEEDS Research Stream 2b for severe accident data	
Accident mortality	Number of fatalities expected for each kWh of electricity that occurs in severe accidents with five or more deaths per accident	Fatalities/kWh
Maximum fatalities	On the basis of the reasonably credible maximum number of fatalities for a single accident for an electricity generation technology chain	Fatalities/ accident

(continued)

Table 1.7 (continued)

Criteria/Indicator	Description	Unit
Perceived risk	Perceived risk	
Normal operation	Citizens' fear of negative health effects due to normal operation of the electricity generation technology	Ordinal scale
Perceived acc	Citizens' perception of risk characteristics, personal control over it, scale of potential damage, and their familiarity with the risk	Ordinal scale
Terrorism	Risk of terrorism	
Terror potential	Potential for a successful terrorist attack on a technology. On the basis of its vulnerability, potential damage and public perception of risk	Ordinal scale
Terror effects	Potential maximum consequences of a successful terrorist attack. Specifically for low probability, high consequence accidents	Expected fatalities
Proliferation	Potential for misuses of technologies or substances present in the nuclear electricity generation technology chain	Ordinal scale
RESIDENTIAL ENVIRONMENT	Quality of the residential environment	
Landscape	Overall functional and aesthetic impact on the landscape of the entire technology and fuel chain. Note: Excludes traffic	Ordinal scale
Noise	This criterion is based on the amount of noise caused by the generation plant, as well as transport of materials to and from the plant	Ordinal scale

[a]P. Burgherr, S. Hirschberg and E. Cazzoli, "Final set of sustainability criteria and indicator s for assessment of electricity supply options, NEEDS Deliverable no. D3.2—Research Stream 2b, NEEDS project", Brussels, Belgium (2008)

Chapter 2
Basics of Civilian Nuclear Fission

Abstract Uranium is by far the most concentrated available energy source, but with the downside that the physical process of fission generates a surplus of neutrons and radioactive fission products. Objectives exist to ensure the control of reactivity, confinement of radioactive substances and decay heat removal as well as long-term waste management, which are supported by mature methods and stringent safety requirements. Operators and regulators claim that the risk of well-designed and operated power plants with light-water reactors, which dominate the current worldwide fleet of 449 units, is justifiably small. The operating experience has accumulated to more than fifteen thousand years, with typical capacity factor now around 80%. Another 60 facilities are under construction in 15 countries; at the same time, some countries are phasing out nuclear while promoting renewables. New and future generation designs aim at further—in some cases radical—reduction of the risk of core melt accidents and minimize proliferation risks. New designs may also use advanced fuel cycles including thorium that extend and use fuel resources more efficiently and reduce reliance on husbandry of long-lived waste from millennia to centuries.

Most current fuel cycles end with long term storage, relying on the deep geological repository concept, subject to a range of geo-scientific and social uncertainties. Human doses due to potential long-term releases from a repository are calculated to be extremely small, much below natural radioactivity. Although no operating civil disposal facility exists yet, plans are advanced in some countries, especially Finland, where construction has begun in 2016.

2.1 Physics and Characteristics

The **fundamentals of nuclear fission** are depicted in Fig. 2.1 for Uranium-235, the only natural material being fissionable by neutrons of low energy and allowing a self-sustaining chain reaction, which means that each fission event causes, on average, exactly one additional such event in a continued chain.[1] When Uranium-

---

[1]The average number of neutrons that cause new fission events is called effective neutron multiplication, usually denoted by k-effective. If k-effective is equal to 1, the assembly is called critical,

© Springer Nature Switzerland AG 2019
D. Sornette et al., *New Ways and Needs for Exploiting Nuclear Energy*,
https://doi.org/10.1007/978-3-319-97652-5_2

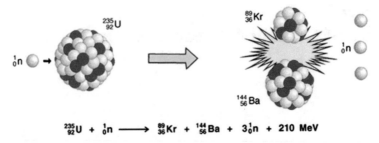

$$^{235}_{92}\text{U} + {}^{1}_{0}\text{n} \longrightarrow {}^{89}_{36}\text{Kr} + {}^{144}_{56}\text{Ba} + 3{}^{1}_{0}\text{n} + 210 \text{ MeV}$$

Fig. 2.1 Exemplary nuclear uranium fission equation for uranium-235, 3.1×10^{10} fissions/second are needed to generate 1 Watt). MeV: million electron volts (W. Koelzer, Lexikon zur Kernenergie, 7/2015. https://www.kit.edu/downloads/Lexikon_zur_Kernenergie_2011.pdf)

235 undergoes nuclear fission, it typically releases between one and seven neutrons (with an average of 2.4). There is a 1/2.4 probability of causing another fission as opposed to either being absorbed by a non-fission capture or escaping from the fissile core. Most neutrons are emitted directly from the fission within a very short time of about 10^{-14} s, while a small fraction (less than 1% on average) comes from radioactive fission products with short half-lives after a decay of up to several minutes. These delayed neutrons allow the power level to be controlled via movement of control rods which contain neutron poison, such as boron or hafnium.

Resulting challenges to safety include:

- Control of reactivity because of neutron surplus,
- High inventory of radioactive substances (10^{18}–10^{19} Becquerel, Bq[2]) in a commercial reactor, requiring confinement and core cooling despite stoppage of chain reaction.

Typically, the fission energy appears as kinetic energy of the nuclei flying apart; nuclear fuel contains at least ten million times more usable energy per mass unit than does chemical fuel making nuclear fission a very dense energy source. By contrast, the physical process includes a potential of power excursions (as outlined before) and most fission products are radioactive at a level far higher than the heavy elements of the raw material (uranium core) over a long period of time, giving rise to decay heat removal and waste management problems, in general. Indeed, heat production due to radioactive decay is in the range of 6% one second, 4% one minute and still 1% one day of the original power after reactor shutdown, calling for sufficient continuous core cooling.

---

and if less than or greater than 1, subcritical and supercritical, respectively. An assembly is called prompt critical (resp. prompt super-critical) if critical (resp. super-critical) without contribution from delayed neutrons. In a supercritical assembly the power increases exponentially with time, with the rate depending on the average time it takes for neutrons to be released in a fission event to cause another fission, and is worse (power excursion) in the case of prompt supercriticality.

[2]One Becquerel is the SI unit of radioactivity corresponding to one disintegration per second.

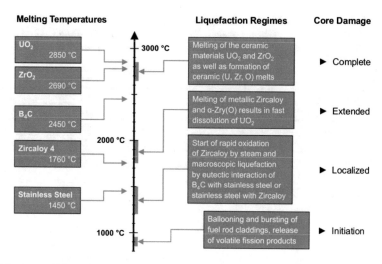

Fig. 2.2 Susceptibility of light water reactors' (LWR), notably fuel, fuel cladding, and structural material, to temperature rise and resulting core damage

Most current power reactors use ordinary light water as coolant, to which the fission energy is transferred, and as a moderator to lower the energy of fission neutrons to "thermal" level. To fuel such thermal Light Water Reactors (LWR), natural uranium that contains 0.71% U-235 needs to be enriched to 3–5%, before conversion to uranium dioxide and insertion into the reactor core as stacks of cylindrical pellets encapsulated in metal tubes made of Zircaloy. These characteristics place current large size LWR (up to 1600 MWe) at risk of loss of cooling accidents that may lead to the core overheating, resulting in meltdown in less than one hour (see Fig. 2.2).

Absorption of neutrons by U-238 (99.28% of natural uranium) leads to the production of "man-made" fissionable material, that is Plutonium-239,

$$_{92}U^{238} + {_0}n^1 \rightarrow {_{92}}U^{239} \xrightarrow[\text{(23 min)}]{\beta^-} {_{93}}Np^{23} \xrightarrow[\text{(2.3 days)}]{\beta^-} {_{94}}Pu^{239},$$

which significantly contributes to energy production by in-site conversion.

Table 2.1 demonstrates the low materials flow due to the high-power density of the physical process, slightly depending in the burn-up rate: on average, a large-sized LWR produces per year about 30 tons of heavy metal (HM) including 1400 kg of fission products and 350 kg of recyclable plutonium; the volume totals to 15 m^3. Plutonium can be separated and reused as mixed-oxide (MOX) fuels in commercial LWR.[3]

---

[3]Replacement of 1/3 of enriched UO$_2$ fuel by MOX (U, Pu)O$_2$ fuel would avoid net plutonium production. Source: R. Brogli, K. Foskolos, C. Goetzman, W. Kröger, A. Stanculescu, P. Wydler,

Table 2.1 Exemplary impact of increasing burn-up on the back-end of the fuel cycle (GWd: gigawatt-days; tHM: tons of heavy metal; TWhe: terawatt-hours of electricity) (M. Debes, Increased fuel burn-up and fuel cycle equilibrium. 9 international conference on nuclear engineering, France (2001))

Impact areas			
Burn-up (GWd/tHM)	33	45	60
Fission products (kg/TWhe)	140	140	140
Cladding and structural material (kg/TWhe)	1210	890	660
Recyclable uranium (kg/TWhe)	3830	2810	2100
Minor actinides (kg/TWhe)	4.3	4.5	4.7
Recyclable plutonium (kg/TWhe)	37	32	27
Total (1300 MWe LWR, 10 TWhe annually)	52 tHM	38 tHM	29 tHM

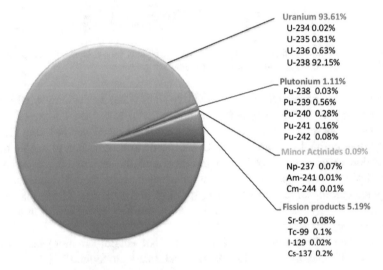

Uranium 93.61%
U-234 0.02%
U-235 0.81%
U-236 0.63%
U-238 92.15%

Plutonium 1.11%
Pu-238 0.03%
Pu-239 0.56%
Pu-240 0.28%
Pu-241 0.16%
Pu-242 0.08%

Minor Actinides 0.09%
Np-237 0.07%
Am-241 0.01%
Cm-244 0.01%

Fission products 5.19%
Sr-90 0.08%
Tc-99 0.1%
I-129 0.02%
Cs-137 0.2%

Fig. 2.3 Discharge of isotopic composition of a Westinghouse 17 × 17 assembly with initial en-richment of 4.5 wt% that has accumulated 45 GWd/MTU burnup

Power reactors operate in a steady state. The composition of the discharge fuel depends on the initial enrichment and fuel burn-up. With an initial enrichment of about 3.5% and average burn-up of about 33 GWd/t HM (gigawatt-days/metric ton of heavy metal), the fraction of U-235 will be down to almost 1% at the end of the fuel cycle. See also Fig. 2.3 for discharge composition for a selected assembly.

As opposed to thermal LWR, in fast reactors, the fission chain is sustained by fast neutrons and without moderating fission neutrons. They could, in principle, extract almost all of the energy contained in uranium (or thorium as an alternative fuel) by

Fortgeschrittene Nukleare Systeme im Vergleich, PSI Bericht Nr 96–17, Paul Scherrer Institut, Villigen (1996).

making use of significantly increased fission probabilities[4,5] Fast reactors can also be used for fission/transmutation of selected plutonium-isotopes and other actinides. Fast reactors with conversion factors[6] larger than one and producing more fuel than they consume, are called fast breeder reactors (FBR), potentially extending nuclear fuel reserves and nullifying the fuel shortage argument against uranium fueled reactors. Current FBR designs[7] use liquid metal, i.e. sodium, as coolant, while lead or helium as coolants are currently being evaluated. Fast reactors are more complex, and costly to build and operate. Needed enrichment is much higher than in thermal reactors, which raises greater nuclear proliferation and security concerns.

2.2 Safety Principles and Requirements

To protect people and the public against harmful effects of ionizing radiation, three basic safety objectives must be ensured by the design and operation of a nuclear facility for all stages of its lifetime:

– Controlling the reactivity and power,
– sufficiently cooling the fuel and core at all times,
– confining radioactive substances.

Under the umbrella of the International Atomic Energy Agency (IAEA), and according to its status, a design philosophy was developed and ten **fundamental safety principles** were internationally agreed,[8] see Table 2.2.[9] They constitute a basis on which safety requirements were deduced and safety measures prescribed; to demonstrate that they are achieved, comprehensive deterministic and probabilistic safety assessment of any design is required to be carried out in order to examine normal operation of the plant, its performance in anticipated operational occurrences

---

[4]So-called cross section that qualifies the intrinsic likelihood of a physical event like scattering or absorption/fission when a (neutron) beam strikes a target object (atom), typically denoted sigma with units of area (barn).

[5]For example, the cross section of U-238 for thermal fission is almost zero (non-fissile: 11.77 micro barn) but sufficiently high for fast fission (1136 barn).

[6]Ratio of fissile atoms created compared to consumed fissile material; LWR have a conversion factor of approximately 0.6.

[7]Prototype/commercial fast breeder reactors are operating in Russia (BN-600 (since 1981), BN-800 (since 2015)) or commissioning in India (FBTR) while test and prototype reactors have been operated and shutdown in many countries, including France (Phénix, (1975–2010); Superphénix (1985–98)) and Japan (Monju (1995-dormant)).

[8]IAEA Safety Standards, Safety Fundamentals, No. SF-1, IAEA Vienna (2006).

[9]States have an obligation of diligence and duty of care and are expected to fulfill their national and international undertakings and obligations. International safety standards provide support for States in meeting their obligations under general principles of international law, such as those relating to environmental protection.

Table 2.2 IAEA fundamental safety principles (IAEA Safety Standards (2006))

Principle 1: *Responsibility for safety*	The prime responsibility for safety must rest with the person or organization responsible for facilities/ activities that give rise to radiation risks.
Principle 2: *Role of government*	An effective legal and governmental framework for safety, including an independent regulatory body, must be established and sustained.
Principle 3: *Leadership and management for safety*	Effective leadership and management for safety must be established and sustained in organizations concerned with, and facilities/activities that give rise to, radiation risks.
Principle 4: *Justification of facilities and activities*	Facilities/activities that give rise to radiation risks must yield an overall benefit.
Principle 5: *Optimization of protection*	Protection must be optimized to provide the highest level of safety that can reasonably be achieved.
Principle 6: *Limitation of risks to individuals*	Measures for controlling radiation risks must ensure that no individual bears an unacceptable risk of harm.
Principle 7: *Protection of present and future generations*	People and the environment, present and future, must be protected against radiation risks.
Principle 8: *Prevention of accidents*	All practical efforts must be made to prevent and mitigate nuclear or radiation accidents.
Principle 9: *Emergency preparedness and response*	Arrangements must be made for emergency preparedness and response for nuclear or radiation incidents.
Principle 10: *Protective actions to reduce existing or unregulated radiation risks*	Protective actions to reduce existing or unregulated radiation risks must be justified and optimized.

(and disturbances) and accident conditions.[10] The primary means of preventing and mitigating the consequences of accidents (principle 8) is the application of the concept of **defense in depth**, which requires the combination of a number of consecutive and independent levels of protection that would all have to fail before harmful effects could be caused to people or to the environment. The strategy ensures that no single technical, human or organizational failure could become harmful, and that the combinations of failures giving rise to significant harmful effects are of very low probability. This is achieved through redundancy and diversity; plants are not allowed to operate unless a minimum number of diverse systems is available[11,12]

---

[10]IAEA Safety Standards, Specific Design Requirements No.SSR-2/1 (Rev. 1), p.17, IAEA, Vienna (2016).

[11]OECD, Comparing Nuclear Accident Risks with Those from Other Energy Sources, OECD Publishing, Paris (2010).

[12]IAEA, Basic Safety Principles for Nuclear Power Plants 75-INSAG-3 Rev. 1, INSAG Series No. 12, Vienna (1999).

Protection against radiological hazards is regarded as effective, if the overall risk associated with the entire set of potential adverse internal and external events—defined as the arithmetic product of their frequencies and consequences—is less than the risk of any other industrial activity, already accepted, or competing energy source. To achieve that general goal, the technical measures against accidents apply to precaution, management and mitigation in a balanced way.

Many inherent and engineered safety features have to be incorporated into the design of the plant to ensure that, for all accidents taken into account, radiological consequences would be so minor, and the likelihood of severe accidents with serious radiological consequences would be extremely small, "practically eliminated" in more recent terms.[13] The design shall be conservative so as to provide assurance that accidents are prevented as far as practicable and that a small deviation in a plant parameter does not lead to an instance of severely abnormal behavior ("cliff edge effect"); the need for operator actions in an early phase of deviations from normal operation is minimized, e.g., by automation.

A set of "design base accidents" (DBA) and required deterministic analyses have been established to demonstrate compliance with qualitative protective goals. Probabilistic safety analysis (PSA, see Sect. 4.2) complements this by demonstrating that protective measures are balanced against each other, and are sufficient to meet suggested or prescribed target values. For most countries, accepted targets for licensing requirement are as follows (for new/future plants, lower by a factor of 10):

- Total core damage frequency (CDF) less than 10^{-4}/reactor year;
- Adequate precautions against accidents for CDF between 10^{-4} and 10^{-5}/reactor year[14];
- Frequency of large early release of radioactive substances (LERF) significantly less than CDF, e.g. by a factor of 10;
- Proof of sufficient protection against natural events for hazards with size beyond the largest that would be expected with frequency 10^{-4}/year, e.g., the ten-thousand-year earthquake;
- Protection against aircraft crash for the types of planes in operation when applying for a construction license.

The concept of DBA inevitably suggests the possibility of "beyond design base accidents" (BDBA), see Textbox 2.1 for definitions, supported by the occurrence of recent major accidents. Therefore, the basic safety concepts have been further developed, blurring the clear distinction between types of accidents and putting protection levels and related safety measures into perspective (Fig. 2.4); fundamental principles relating to management responsibilities including safety culture, technical issues and defense in depth have been strengthened.

---

[13]IAEA Safety Standards, Safety of Nuclear Power Plants: Design. SSR-2/1 (Rev. 1), IAEA, Vienna (2016).

[14]For the EPR in Finland, a CDF of 2×10^{-6} per reactor year has been estimated and accepted by the Finnish regulator, STUK.

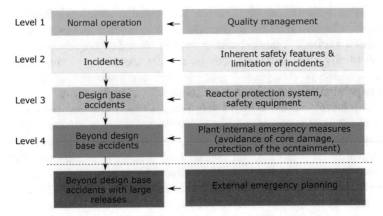

Fig. 2.4 "Historical" levels of possible state of a nuclear power plant, corresponding protection levels and related safety measures (level 1–3 are part of the 'traditional' licensing process; level 4 was added to the 'traditional' licensing process, level 5 goes beyond, both are subject of PSA)

Design Base Accidents (DBA)

- Selection of infrequent events, which are expected to occur during the lifetime of a nuclear power plant or cannot be excluded following human discretion and which require the most extreme design parameters; their mean frequency is in the range between 10^{-4} and 10^{-2} per reactor-year.
- Design of the plant in such a manner that the occurrence of such an event does not lead to core damage and unacceptable consequences for people and the environment; dose limits apply.
- For the verification, both initiating event and the unavailability of an independent safety system, needed to handle accidents, are assumed (this results in redundant systems design, and no need to assume multiple system failures).

Beyond Base Design Accidents (BDBA)

- Accidents are beyond design base, when they can be characterized by multiple failures of systems designed to handle accidents or if they are instantiated by very rare severe events. The occurrence of such accidents is understood, based on the experience and analysis, as very unlikely.
- In contrast to DBA, it cannot be excluded that radioactive substances in a harmful amount are released to the environment; no dose limits around the site are defined. They are grouped into (a) accidents which do not directly lead to core damage and which may be mitigated due to existing safety margins not credited in the design base, and (b) accidents leading directly to core damage and calling for accident management including confinement protection; their mean frequencies are in the range (a) between 10^{-6} and 10^{-4}; (b) smaller than 10^{-6} per reactor-year.

Textbox 2.1 "Historical" Definitions

More recently, after the accident at the Fukushima Daiichi nuclear power plant, the IAEA safety standard SSR-2/1 has been revised, mainly relating to severe accidents, i.e. their prevention by strengthening the design base, the prevention of unacceptable radiological consequences and mitigation of consequences to avoid or minimize radioactive contamination—the latter two being off-site consequences. The term "design extension conditions" (DEC), including accidents with core melting, was introduced to replace the term BDBA. "The main technical objective of considering design extension conditions is to provide assurance that the design of the plant is such as to prevent accident conditions that are not considered design basis accident conditions, or to mitigate their consequences, as far as is reasonably practicable". A set of design extension conditions shall be derived on the basis of engineering judgment, as well as on deterministic and probabilistic assessments. This aims at further enhancing the plant's capabilities to withstand, without unacceptable radiological consequences, accidents that are either more severe than design basis accidents or that involve additional failures or adverse conditions. This might require additional safety features, notably to maintain the integrity of the containment, i.e., the containment and its safety features shall be able to withstand extreme scenarios, among other things, loads from melting of the reactor core.

The defense in depth concept is centered on given levels of protection (Fig. 2.5) including a series of successive physical barriers, as well as a combination of active, passive and inherent safety features that contribute to the effectiveness of the physical barriers in confining radioactive substances and preventing their release, respectively, or to mitigate the consequences. As mentioned before, the application of the concept ensures that, if one level/barrier were to fail, the subsequent one would come into play, and so on. Therefore, special attention must be paid to hazards or plant states that could impair several levels of defense at the same time such as fire, flooding, earthquake or aircraft crash as well—or propagate into subsequent levels—such as a core meltdown may do. Precautions can be taken and systems can be designed to cope with such hazards depending on the quality of the overall safety concept and understanding of severe accident mechanisms.[15] However, these so-called common cause failures or common triggering events are at least possible and require special attention.

---

[15]While standard second generation LWR are not designed against consequences of the core melt accidents and physical barriers including the confinement structure might be subsequently get lost, most generation three designs cope with such extreme events.

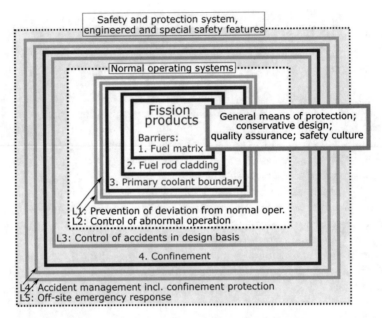

Fig. 2.5 Defense in depth: the four physical barriers (blue) and five levels of protection (brown), split into normal (white inner) and safety/protection (grey outer) operation layers. These components form a series of obstacles between the radioactive material in its normal state and its potential release due to an accident, adapted from IAEA (1999)

In summary, current partially revised safety objectives, concepts and principles ensure, if properly implemented, good performance and a remarkably high safety level of nuclear power plants. As the intrinsic features, e.g., of LWR, are not sufficient to cope with all possible accidents, sophisticated engineered systems and human actions are necessary, which potentially may fail, although precautions against such secondary failure scenarios have been further enhanced, following the Fukushima Daiichi disaster in particular.

When combining probability and consequences, over the whole spectrum of possible adverse events, the resulting risk of well designed, maintained and operated nuclear power plants is justifiably small in absolute and relative terms. However, severe accidents are still the matter of some disagreements. For some, such 'low-probability-high-consequence events', typical for risk estimates of today's large nuclear facilities, and still associated with large uncertainties, are cause for great concern. They further state that consequences endangering large areas, up to whole societies, should not be "excused" by their rare occurrence. Others point to the risk of not having reliable and affordable energy, which does not seem to be factored into the analysis.

More than a decade ago, more stringent safety requirements have been prepared by utility organizations[16] to support the development of evolutionary designs and to downplay the problem mentioned above inter alia by[17]:

- Increased margins to reduce sensitivity to disturbances and the number of challenges to plant safety,
- Provision of more time for the operator to act in accident situations, and
- Design measures to cope with severe accidents.

The European utilities[12] set lower release targets than required by the US to limit emergency protection and delayed/long-term actions as well as restrictions of foodstuff and crops in terms of distance "to a minimum"[12] from the site of the accident. The IAEA has implemented an international project to provide guidance for the evolution of innovative reactors and fuel cycles, suggesting to enhance the defense in depth strategy by more extensive use of inherent safety characteristics and greater separation of redundant systems.[18] It is a general aim to design future reactors in such a way so that emergency measures are no longer needed from a technical point of view to protect the public and reduce their risk significantly in case of "severe accidents". However, some basic problems remain, leaving room for "revolutionary" (rather than evolutionary) approaches and designs.

2.3   Fleet of Commercial Nuclear Reactors

As of the end of 2017, there were 449 Nuclear Power Reactors/Plants (NPP) in operation, distributed throughout 31 countries (see Fig. 2.6), including 16 countries in Europe (Fig. 2.7), amounting to a total net installed capacity of about 392.5 GWe (Gigawatt electric).[19] In 2016, nuclear energy supplied 2441 TWh of electricity, which corresponds to a share of about 11% of worldwide electricity production. The power comes from reactors of various types; the most common ones will be described in greater detail in Sect. 2.4.

Currently 82% of the nuclear power is produced by different variations of the Light Water Reactor (LWR) technology, particularly the Pressurized Water Reactor

---

[16]Electric Power Research Institute (EPRI), Japanese and Korean Utilities (JURD, KURD), European Utilities (EUR); the EUR documents (www.europeanutilityrequirement.org) formed the basis for the European Pressurized-Water Reactor (EPR, see annexed Fig. 2.11 for Containment Heat Removal System).

[17]IAEA, Implementation of Defense in Depth for Next Generation Light Water Reactors, IAEA-TECDOC-986, IAEA, Vienna (1997).

[18]IAEA, Guidance for the Evaluation of Innovative Nuclear Reactors and Fuel Cycles Report of Phase 1A of the International Project on Innovative Nuclear Reactors and Fuel Cycles (INPRO), IAEA-TECDOC-1362, IAEA, Vienna (2003).

[19]The IAEA offers an online tool that presents the up-to-date information online: http://www.iaea.org/PRIS

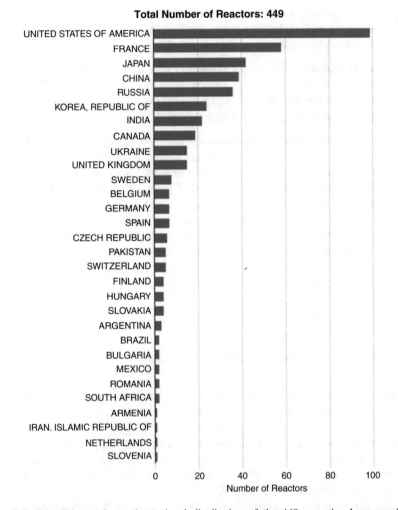

Fig. 2.6 This diagram shows the regional distribution of the 449 operational power plants throughout the world at the end of 2017, from the IAEA PRIS database. The total also includes 6 reactors in Taiwan, included in China

(PWR) and Boiling Water Reactor (BWR). Reactors of the PWR type represent 69.6% of the total net installed capacity mentioned above and, with 291 reactors, they account for a major share of currently operated NPP, followed by 78 BWR. Other designs to be mentioned are the Pressurized Heavy Water (moderated and cooled) Reactor (PHWR, CANDU type) with 49 operating reactors, the Gas Cooled (graphite-moderated) Reactor (GCR) with 14 facilities in operation, 11 commercial Light Water cooled, Graphite-moderated Reactors (LWGR or better known as RBMK) and three sodium-cooled Fast Breeder Reactors (FBRs). It is worth mentioning that the RBMK is an early Generation II reactor (see Fig. 2.12) and the oldest

Fig. 2.7 Map showing the distribution of nuclear power plants throughout Europe in 2017 (updated based on Nuklearforum Schweiz, Kernkraftwerke der Welt, 2014)

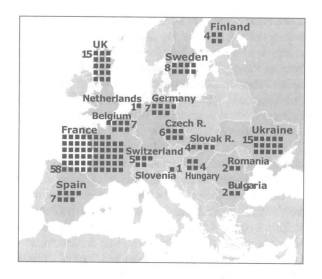

commercial design still used, although only in Russia, with its development roots in the former Soviet Union, for military purpose.[20] The bulk of the currently operated plants were put into operation in the time period between 1980 and 1990; the thermal gross-efficiencies range from a minimum of 32% (RBMK) up to 38% (modern PWR), meaning that the thermal energy gets converted to electrical energy with those efficiencies.

In order to obtain a broader perspective on the situation of nuclear power plants globally, it is useful to take a step back and look not only at the operational plants, but also include suspended and permanently shut-down reactors. With this extension, the number of reactors ever built worldwide reaches 615. Combined, they have a total of about 16,500 reactor-years of experience, which, discounted by an assumed long-term average load factor of 0.75,[21] leads to almost 13,200 actual reactor-years.

Moreover, there are another 56 facilities under construction in 15 different countries as of the end of 2017 and more than 140 planned. With a power output of 57 GWe (net), these projects under construction have a significant capacity. Concerning the facilities that are currently under construction, the most favored type by far is the PWR, closely followed by the BWR. These projects are a result of a number of ambitious nuclear programs in China, India, United Arab Emirates and some others, seemingly indicating a rebirth of nuclear power. In more detail, China has 39 operating units, 18 units under construction and another 38 units planned,

---

[20]RBMK reactors are an enlarged version of previously existing military reactor developed for production of plutonium; so, their construction required minimal restructuring of existing machinery plants, and RBMKs could use cheap natural uranium, while western analogs require more expensive enriched uranium (Source: Nikolay Dollezhal: At the root the man-made world, Moscow, 2010, fourth edition, p.160–162); new builds are not expected.

[21]The load factor is the percentage of time when the reactor is in operation, the rest of the time being used for maintenance, re-fueling, addressing incidents and safety improvements.

announcing plans to build 6–8 new reactors per year. If these goals were realized, it would make China the top nuclear energy supplier by 2030. Interest in nuclear energy is also seen in the United Kingdom, which partnered with EDF and China Nuclear Power (CNP) for the construction of the Hinckley Point C, with large-scale twin-EPR-units, and envisages building twin-units at least two other existing sites. A "renaissance of nuclear power" can also be found in the USA, where four AP1000 (advanced PWR) reactors are under construction, two of which should be commissioned as soon as 2018, and small and medium sized reactors are gaining attraction.

Financing of huge capital costs and long construction times are seen as major obstacles for new builds, notably for large units. Thus, lifetime-extension of operating plants is favored in many countries for economic reasons.[22]

These extensive plans for future nuclear expansion are accompanied by a rising number of permanent shutdowns, which are motivated by a variety of causes. From a technical point of view, aging is the primary reason, as there are several forms of wear specific to nuclear facilities, like embrittlement of the reactor vessel due to neutron bombardment, mechanical and thermal stress transients experienced by the primary cooling system, corrosion of steam generator tubes, etc. Nonetheless, economic and political issues also play a major role in determining the future of a plant. Some countries have decided to phase out nuclear energy, have forbidden to build new plants, or have established a shutdown plan for existing plants, although operating them safely and economically while promoting renewable energy sources such as wind and solar. The most prominent example for such a strategy ("Energiewende") is Germany. This transition has been accelerated since Chancellor Angela Merkel's decision in 2011 to shut down all 17 of Germany's nuclear power stations by 2022 in the wake of Fukushima's disaster. However, this has led to a significant ramping up of coal use to compensate for the resulting gap in electricity production,[23] with alleged harmful consequences for health in Europe as a result of the increased pollution by particulates.[24]

More generally, with the advent of the recent "Fukushima" catastrophe in mind, it seems that a large portion of the public in some parts of the world is feeling an increasing unease about nuclear energy, demanding the shutdown of facilities and the switch towards "new" renewable energies—putting constant pressure on politicians and leaders in the energy sector, while certain anti-nuclear advocates became pro-nuclear after Fukushima for environmental reasons.[25]

---

[22]87 out of 99 nuclear power plants were granted life time extension to 60 years.

[23]Andresen, T., Coal Returns to German Utilities Replacing Lost Nuclear, Bloomberg, April 15, 2014 (http://www.bloomberg.com/news/articles/2014-04-14/coal-rises-vampire-like-as-german-utilities-seek-survival)

[24]*Guerreiro, Cristina, et al. Air Quality in Europe-2016 Report. Publications Office of the European Union, 2016.*

[25]Such as George Monbiot, the British writer known for his environmental and political activism (https://en.wikipedia.org/wiki/George_Monbiot)

2.4   Present Reactor Designs and Fuel Cycle Concepts

A nuclear power plant is a thermal power station where the heat source is the reactor core. The heat is used to generate steam, which drives a steam turbine and a generator to produce electricity. As mentioned before, roughly 80% of currently operating[26] nuclear power reactors use light water as coolant and neutron moderator. Less than one quarter of these LWRs are boiling water reactors (BWR, see Fig. 2.8) where the primary coolant undergoes phase transitions to steam inside the reactor. More than three quarters are pressurized water reactors (PWR, see Fig. 2.9) that operate at higher pressure (155 bar instead of 80 bar) on the primary cooling circuit, separated from the conventional water-steam-cycle by a heat exchanger. LWR use uranium fuel with U-235 enriched to 3–5%, either following an open fuel cycle with direct disposal of spent fuel or a partially closed fuel cycle with reprocessing and recycling of uranium as mixed-oxide (MOX) fuel.

Pressurized heavy water is used in CANDU-type reactors (See Fig. 2.10), mainly used in Canada and India. These reactors are fueled by natural uranium (U-235 enrichment 0.71%), that is feasible because of much lower absorption probability of deuterium compared to protium.[27] Another interesting reactor design worth mentioning here is the RBMK. The RBMK is also water-cooled but graphite-moderated, originally designed and developed for military purposes and scaled-up for

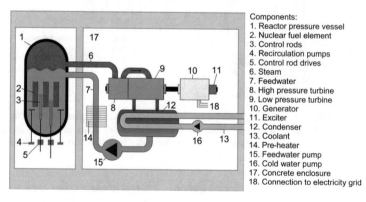

Components:
1. Reactor pressure vessel
2. Nuclear fuel element
3. Control rods
4. Recirculation pumps
5. Control rod drives
6. Steam
7. Feedwater
8. High pressure turbine
9. Low pressure turbine
10. Generator
11. Exciter
12. Condenser
13. Coolant
14. Pre-heater
15. Feedwater pump
16. Cold water pump
17. Concrete enclosure
18. Connection to electricity grid

Fig. 2.8 Schematic view of a simple boiling water reactor (https://upload.wikimedia.org/wikipedia/commons/3/38/Schema_Siedewasserreaktor.svg). Unlike the more common pressurized water reactor (Fig. 2.9), a steam generator is not needed, as steam is formed inside the reactor and then led to the turbine-generator, all housed in a safety-grade building

---

[26]Total of 442 (including 40 units in Japan suspended after Fukushima Daiichi nuclear accident) in 31 countries by end of 2015.

[27]Hydrogen with one proton but without any neutron is protium (H), with one neutron is deuterium (D), and with two neutrons is tritium (T). A (light) water molecule is made of one atom of oxygen and two atoms of protium (H_2O). So-called heavy water is made by replacing the two protium atoms by two deuterium atoms (D_2O). Tritiated water contains two atoms of tritium instead of protium (T_2O).

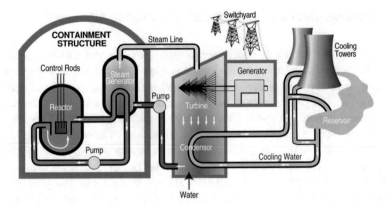

Fig. 2.9 A schematic of a Pressurized Water Reactor (https://commons.wikimedia.org/wiki/File%
3APWR_nuclear_power_plant_diagram.svg)

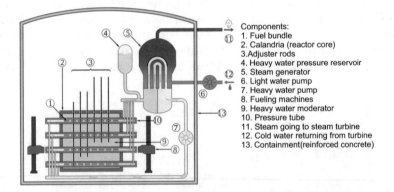

Fig. 2.10 A CANDU-type reactor, with horizontally inserted pressure tubes into the heavy water
filled vessel; the coolant is also heavy water, which flows into the steam generator after passing
through the pressure tubes and afterwards returning to the reactor (https://commons.wikimedia.org/
wiki/File%3ACANDU_Reactor_Schematic.svg)

commercial use in the former Soviet Union. Fuel and coolant tubes pass vertically
through the graphite block and, as with the CANDU design, the reactor supports
on-line refueling, allowing a high availability.

The LWR development line is basically rooted in the USA military naval sector.
The first unit for power generation, devoted exclusively to peacetime uses, was the
PWR Shippingport station, Pennsylvania, USA. It reached criticality in December
1957 and, aside from stoppages for three core changes,[28] remained in operation until

---

[28]The first core used highly enriched uranium (U-235 93%) as seed, surrounded by a blanket of
natural U-238; the second core was similarly designed but more powerful while the third core kept
the same seed- and blanket design now using U-233 as seed and thorium as blanket material
('thermal breeder reactor').

October 1982. As a government-sponsored prototype in the range of 60 MWe, it was not built on a commercial basis. The first purely commercial PWR was Yankee Rowe (185 MWe, 1960–92), the first BWR was Dresden-1 (210 MWe, 1960–78), both being so called Generation I (short Gen I) prototypes (see Fig. 2.12 for development lines), which were striving for commercial use of nuclear power and for paving the ground for its wider use.

These early Gen I reactors were equipped with safety systems to stop the chain reaction and remove decay heat in case of emergency, particularly of loss-of-coolant accidents; redundancy and diversity served as design principles.

The prototype and large demonstration plants in the range of roughly 200–400 MWe, built in many countries, which embarked on civilian use of nuclear power in the late 1960s to early 1970s, were followed by more powerful designs in the mid/late 1970s. These Generation II plants in the range of 400 to 800 MWe were based on the same basic safety concept but the reliability of safety systems was further improved, the barriers against release of radioactive substances, including the containment was strengthened and the spectrum of design base accidents extended by including external events: The final goal was to make severe core meltdown accidents unlikely, to the point of becoming 'hypothetical'; offsite emergency measures were planned as a precaution.

Results of probabilistic safety analyses (PSA) and reviews of severe accidents, notably the Three Mile Island accident (see Sect. 5.1), all amplified by anti-nuclear pressure groups, caused the authorities to question this safety concept (see also Sect. 5.2). Subsequently, measures were taken to prevent core meltdown accidents after ordinary safety systems would fail, to mitigate potential consequences and to enable operators to do so by planned emergency management measures (e.g., to prevent containment failure). Thus, extensive back-fitting of safety systems (i.e., retroactive improvements) were agreed upon and severe accident management guidelines were established, although not implemented adequately in all parts of the world. Furthermore, external accident initiating events were integrated consistently into the safety concept, leading to a new set of independent emergency systems.

Generation III designs, in the range of up to 1600 MWe, go one step further, aiming at further reduction of the frequency of core damage accidents and at mitigation of their consequences. That is, in the extremely rare case of core damage, offsite emergency measures like evacuation, relocation, ban on foodstuff, etc., should no longer be necessary to reduce the public risk due to the design of safety systems. The European Pressurized Reactor (EPR) is a prominent example (Fig. 2.11).

Here we wish to differentiate between (a) the evolutionary concept such as the EPR, having advanced safety features, from (b) the revolutionary concept, proposing a change from active to passive systems such as the Westinghouse AP-1000/1400 design. The latter are sometimes called Generation III+ with deployment times in 2010–2025.

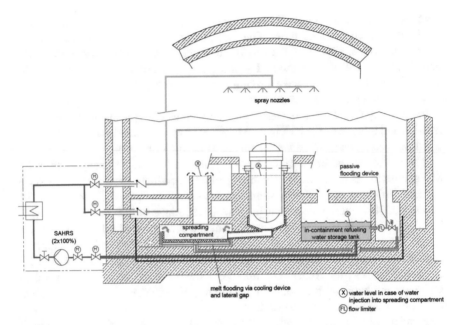

spray nozzles

passive
flooding device

SAHRS
(2x100%)

spreading
compartment

in-containment refueling
water storage tank

melt flooding via cooling device
and lateral gap

Ⓧ water level in case of water
injection into spreading compartment

Ⓕ flow limiter

Fig. 2.11 Containment Heat Removal System of the European Pressurized-Water-Reactor (EPR), designed to control severe core meltdown accidents (Wikipedia, Areva EPR, https://upload.wikimedia. org/wikipedia/commons/4/45/EPR_CHRS.jpg). All EPR are equipped with a core catcher, in case of vessel melt-through. The containment is designed to withstand a global hydrogen combustion taking into account the implementation of passive hydrogen recombiners that limit the hydrogen accumulation. The severe accident heat removal system (SAHRS) is the primary means of controlling heat and pressure in the containment under severe accident conditions, and has its own support systems

Designs that go even further with deployment times (2030 and beyond) are called Generation IV (see Fig. 2.12 for development line). The Generation IV International Forum (GIF)[29] provides the framework for the development of six selected nuclear systems including reactor and fuel cycle designs, centered around potential neutron spectra (thermal to fast), purpose (fuel and waste burner, breeder) and coolant (supercritical water, gas, lead, sodium, molten salt), see Fig. 2.13.

Development goals were commonly agreed as compromising between:

1. *Sustainability*: Systems will provide energy generation that meets clean air objectives and promotes long-term availability and effective use of fuel. They will minimize and manage their nuclear waste and notably reduce the long-term stewardship burden in the future, thereby improving protection for public health and the environment.

---

[29]With 10 active members (Canada, China, EU, France, Japan, South Korea, Russia, Switzerland, USA), three of the 13 signatories are inactive (Brazil, South Africa, UK), www.gen4.org

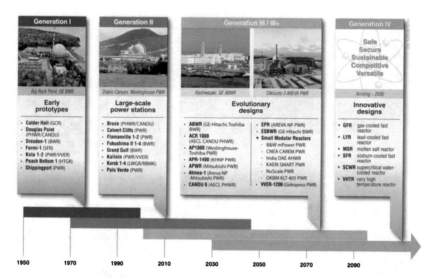

Fig. 2.12 Development line of nuclear power reactors, from generation I to IV, and related deployment times (Generation IV International Forum, A Technology Roadmap for Generation IV Nuclear Energy Systems (2002))

2. *Economics*: Systems will have a clear lifecycle cost advantage over other energy sources. They will have a level of financial risk comparable to other energy projects.
3. *Safety & Reliability*: Systems operations will excel in safety, and reliability will have a very low likelihood and degree of reactor core damage, and they will eliminate the need for offsite emergency response.
4. *Proliferation Resistance and Physical Protection*: Systems will increase the assurance that they are a very unattractive and the least desirable route for diversion or theft of weapon-usable materials and provide increased physical protection against acts of terrorism.

The expectations were set very high in order to drive the research agenda; most probably no single approach will dominate all four categories.[30] The safety goal does not exceed current requirements of Gen III/III+ plants which, however, do not meet the sustainability criterion. Nevertheless, these goals provide a challenging background for our own thoughts, reminding us to follow a system approach and to take several, in part contradictory, criteria into consideration, e.g. low power rate and power density of the reactor may satisfy safety criteria best but infringe upon economics.[31]

---

[30]https://commons.wikimedia.org/wiki/File%3ACANDU_Reactor_Schematic.svg

[31]The selected Gen IV concepts, their fundamental characteristics and development lines in various as well as related topics, including safety and non-proliferation, are explained in detail in: I.L. Piori (editor) et al., Handbook of Generation IV Nuclear Reactors, Woodhead Publishing Series in Energy, ISBN: 978-0-08-100,162-2 (online), (2016).

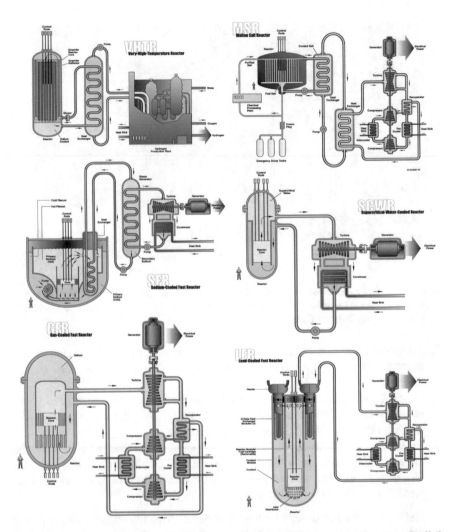

Fig. 2.13 Reactor concept under development within the GIF-framework (for more detailed descriptions of these and other concepts see Sect. 6.2) (Generation IV International Forum, A Technology Roadmap for Generation IV Nuclear Energy Systems (2002))

2.5   Fuel-Cycle Characteristics and Strategies

Following the fundamentals of nuclear fission based on present light water reactor (LWR) technology using uranium-235 as fissile material, three fuel cycle options can be distinguished: "once-through", "partially closed" and "fully closed".

As depicted in Fig. 2.14, all cycles start with natural uranium mining, refining and conversion, followed by enrichment of uranium-235 from 0.71% (fraction in natural uranium) to the 3–5% needed to fuel reactors with a thermal neutron spectrum like

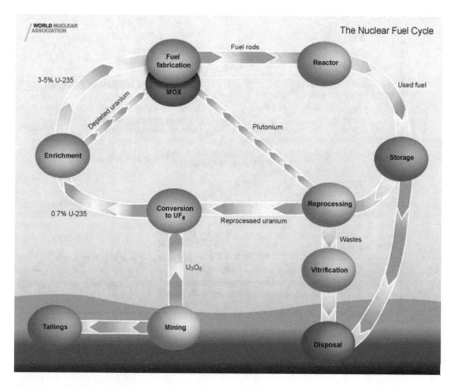

Fig. 2.14 The nuclear fuel cycle and options, partially with fractions of fissile material (U-235, Pu-239), minor actinides and fission products. MOX stands for mixed oxide nuclear fuel. U3O8, which is the first stage of processing at the mine is commonly known as yellowcake

LWR and fabrication of uranium dioxide fuel. This level of enrichment of uranium-235 is unsuitable for weapons use, however, the technology to enrich to weapons grade is the same, and a country that possesses the enrichment technology can easily make weapons grade uranium-235. After its useful life of 3 to 7 years in an LWR, the spent fuel is unloaded and stored in actively cooled water ponds, either inside the containment building or in its immediate proximity, at least for several months until the need for cooling drops sufficiently. During this period, short-lived isotopes decay away to almost nothing and once they are gone, the risk of high temperatures and the fuel melting diminishes.

The irradiated spent fuel can then be sent to a facility for extended storage, pending conditioning and emplacement in long-term storage facilities or, mostly favored in principle, to a deep geological storage facility.

With this process, called the *once-through cycle*, about one third of the fissile material remains in the spent fuel[32] and is not separated, which is sometimes

---

[32]With an enrichment of 3.5% and average burn-up of 33 GWd/tHM, the fraction of U-235 will be down to almost 1% (Source: [42]); see also Fig 2.3.

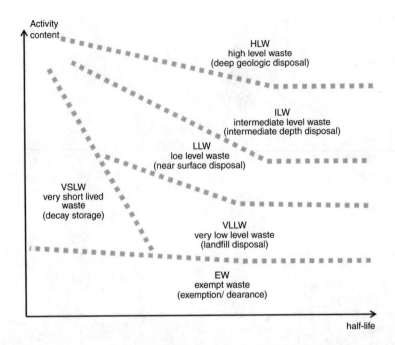

Fig. 2.15 IAEA waste classification scheme as currently established (based upon IAEA, Geological Repositories, IAEA, Vienna (2009))

considered favorable in terms of non-proliferation. As no removal of the more radioactive components takes place, this results in a relatively large volume of high-level waste (HLW)—see Fig. 2.15 for the classification of waste. The radiotoxicity[33] of the waste is particularly high as a result of the presence of plutonium and the other actinides (see Table 2.7 for inventories of safety-relevant radionuclides in a reference canister). Increase of fuel burn-up would help to reduce the fractions of remaining fissile material and the amount of highly radioactive cladding and structural material (per electricity unit). However, the amount of long lived, heat generating minor actinides and therefore the pre-cooling times, would rise slightly.

Using technologies originally driven by military/naval applications, thermal reactors with uranium-based fuel cycles were developed to generate electricity and to accumulate plutonium for the start-up of advanced fuel cycles. Fast breeder reactors were also developed to support a large-scale growth of nuclear power. Usually, the once-through cycle is selected by countries with a small nuclear power program, with some notable exceptions. For example, Belgium, the

---

[33]Measure of toxicity (Sv/TWeh), i.e. health effects after assumed incorporation of a radionuclide including the effects of radioactive daughter products. Radiotoxicity is dependent on radiation quality (type and energy of ionizing radiation) and the bio-kinetics of the radionuclide in the human body.

Table 2.3 Current fuel cycle options in selected countries (source: OECD Nuclear Energy Agency, 2008)

Country	Number of reactors	Material	Destination
United States	104 in operation	Spent fuel	Deep geological disposal
France	59 operation 1 under construction	Vitrified HLW	Deep geological disposal
		Separated U and Pu	Recycling
Japan	55 in operation 3 under construction	Vitrified HLW	Deep geological disposal
		Separated U and Pu	Recycling
Russian Federation	31 in operation 7 under construction	Vitrified HLW	Deep geological disposal
		Separated U and Pu	Recycling
Korea	20 in operation 6 under construction	Spent fuel	Deep geological disposal
United Kingdom	19 in operation	Vitrified HLW	Deep geological disposal
		Separated U and Pu	To be determined
Canada	18 in operation	Spent fuel	Deep geological disposal
Germany	17 in operation	Vitrified HLW (past)	Deep geological disposal
		Separated U and Pu (past)	Recycling
		Spent fuel (future)	Deep geological disposal

HLW High level radioactive waste

Netherlands, and Switzerland have opted for reprocessing contracts with other countries' facilities and for recycling of the fissile material as mixed oxide fuel. Nonetheless, the country with the largest number of reactors, the USA, has opted for a once-through fuel cycle since 1977 to eliminate the spread of technologies that were considered to increase the risk of nuclear proliferation (see Table 2.3 for an overview of fuel cycle options in selected countries).

Indeed, an important option for spent fuel management is reprocessing, i.e. the extraction of usable material (uranium and plutonium) by wet chemical treatment (PUREX process) and its recycling in current reactors, mainly LWR. A "western country" wishing to select this *partially closed fuel cycle option* and to reprocess its spent nuclear fuel will usually need to use either the AREVA plant at Cap de la Hague in northern France or the BNFL plant at Sellafield in the UK. The wastes need to be returned to the country from which the spent nuclear fuel originated. One of the first short-term advantages of this option is the reduction in the volume of spent fuel to be stored on site or in centralized facilities. One of the major short-term disadvantages is the increased number and volume of transports of radioactive materials that are potentially hazardous and subject to strict international regulation[34] and also at risk from terrorist attacks.

Another even more important disadvantage of the partially closed fuel cycle, in general, is the separation of fissile and radiotoxic material (in particular plutonium) and its storage on site; this is done for a certain period of time depending on the amount being recycled. Typically, about two-thirds of the discharged plutonium consist of the fissile isotopes 239 and 241. To recycle this plutonium, it is possible to fabricate mixed oxide (*MOX*) *fuel*, containing about 7% to 9% plutonium mixed with depleted uranium (uranium depleted in the isotope 235), which is equivalent to enriched uranium fuel at a level of about 4.5% U-235. MOX fuel can be used in a conventional LWR.

In the past, the output of reprocessing plants has exceeded the rate of plutonium usage in MOX fuel, resulting in the build-up of inventories of civil plutonium in several countries. Due to limited capability to fabricate and deploy MOX fuel,[35] these stocks are expected to exceed 250 tons before they start to decline once MOX fuel use increases. MOX fuel is expected to supply 5% of world nuclear fuel in the near future. In France, opposite to the UK, the MOX fabrication plant is not sited adjacent to the reprocessing facility but in the south of France, necessitating the transport of depleted plutonium across the length of the country.

Advanced fuel cycles strive for better meeting sustainability goals, i.e. by designing a *closed fuel cycle*, coupled with further developed reprocessing technology and various types of advanced reactors, either with thermal or fast neutron spectrum. In a closed fuel cycle, most of the fissile material is "burned". In particular, the following goals are pursued:

–   Extension of fuel reserves by better exploiting its energy content and converting uranium-238 or thorium-232 to new fuel (so called breeding process);
–   Reduction of radiotoxicity of the waste and decreased requirements of stewardship by more selective separation ("partitioning") of the various long-lived fission products (iodine-129, technetium-92) and minor actinides (such as neptunium-237, americium-241, -243, curium-244) and by subsequent "transmutation" into shorter-lived elements.

According to the current state-of-knowledge, these fully closed fuel cycles recycling both plutonium and minor actinides (long-lived fission products not yet included) will make use of fast neutron spectrum systems for transmutation and multiple recycling schemes with very low losses (0.1% for plutonium and 0.5% for minor actinides). This could significantly reduce, typically a hundred-fold, the transuranic isotopes and long-term radiotoxicity inventories finally disposed of; the energy content of the final waste could also be reduced dramatically.

---

[34]NIREX (Nuclear Industry Radioactive Waste Executive), What is the Nirex Phased Disposal Concept? Harwell, Dicot, 2002.

[35]Streffer C., Gethmann C., Kamp G., Kröger W., Rehbinder E., & Renn, O. (2011). *Radioactive Waste: Technical and Normative Aspects of its Disposal*. Berlin: Springer, 2011 p. 122.

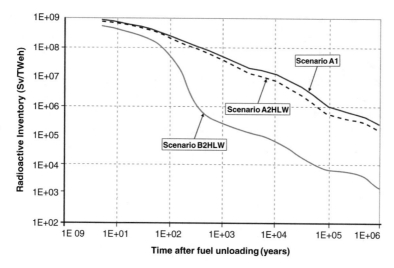

Fig. 2.16 Development of radiotoxicity inventory with time for various fuel cycle Schemes (A1: "open", A2: "Pu burning", B2: "Pu plus minor actinides (MA) burning, assuming realistic losses 0.1% for Pu, 0.5% for MA); results confirm the attractiveness of fully closed fuel cycles with burning/transmutation of plutonium and minor actinides with a sharp decrease of the radioactive inventory after 100 years and finally its reduction by factor of >100 (Gonzales EM (2008) Impact of Partitioning and Transmutation on Nuclear Waste Management and the Associated Geological Repositories. In: Euradwaste '08. Seventh European Commission Conference on the Management and Disposal of Radioactive Waste. Conference Presentations, Papers and Panel Reports, cordis. europa.eu/fp7/euratom-fission/euradwaste2008-presentations_en.html) after 1000 years. The unit Sievert (Sv) represents the equivalent biological effect from the exposure of one joule of radiation energy per kilogram of human tissue. HLW: High-level radioactive waste

This potential has been confirmed by numerous studies comparing various fuel cycle scenarios, as illustrated by Fig. 2.16, which shows the development of the normalized radioactive inventory over time. Mostly, the open fuel cycle with direct disposal of spent uranium dioxide (UOX) fuel serves as the reference scenario (A1); a plutonium burning Scheme (A2), with multiple recycling of plutonium only and using LWR and conventional fast reactors, presents a partially closed fuel cycle with mainly disposal of MOX spent fuel. Transmutation schemes may for instance include burning of plutonium in LWR and of minor actinides (and plutonium after first recycling) in accelerator-driven systems (ADS)[36] (scenario B2).

The effect of fuel cycle schemes including separation and burning/transmutation of plutonium and minor actinides depends proportionally on the removal factors or losses, respectively, which are the subject of ongoing research.[37] In this way, there is

---

[36] An accelerator-driven system provides neutrons to allow for a subcritical reactor to become critical (see Sect. 7.2.5).

[37] Pilot-scale continuous testing at Marcoule, France, confirmed the feasibility of adapting the-state-of-technology PUREX process to recover 99% of neptunium (1% losses); for americium and curium, new extractants need to be developed that would allow recovery factors of 99.9%.

a high potential to bring husbandry times down to more manageable scales: Complete removal of uranium, plutonium and minor actinides would reduce the radiotoxicity level of remaining fission products to the level of natural uranium ore after 300 years.[38]

Thorium is three to four times more abundant than uranium in the Earth's crust,[39] may also serve as fuel for nuclear reactors although the technology is not yet considered to be mature.[40] All uranium fuel cycle options apply to Thorium in principle. Thorium, in metallic and oxide states, has some favorable physical properties compared to respective uranium states, such as higher melting points, better thermal conductivity, and smaller expansion coefficient.

Thorium-232 does not undergo fission itself but, on capturing a neutron, it leads to uranium-233, which is fissile.

$$_{90}Th^{232} + _{0}n^{1} \rightarrow \ _{90}Th^{233} \ \overset{\beta^-}{\underset{(22.3 \ min)_{91}}{\rightarrow}} \ Pa^{233} \ \overset{\beta^-}{\underset{(26.97 \ days)_{92}}{\rightarrow}} \ U^{233}$$

Since thorium is fertile but has no naturally occurring fissile isotope,[41] it needs another fissile material, either uranium-235 or plutonium-239, as "seed material" or for neutrons to be generated externally. Due to the long reaction chain, i.e. differences of atomic masses from thorium-232/uranium-232 to plutonium isotopes and minor actinides (see Fig. 2.17), the probability of generating, highly toxic elements is drastically reduced, compared to the uranium cycle, alleviating the requirements (husbandry time) for underground repositories. However, the radiotoxicity of the fission fragments waste (fission products) is roughly the same for the uranium and thorium cycles to begin with, hence "the differences will only become visible after a few centuries have passed, say 500 years".[42] As uranium-233 can be misused for weapon production[43] and its forerunner palladium-233 can be separated effectively, the thorium fuel cycle is basically not proliferation proof,[44] but offers other significant features and barriers to open the way for much more proliferation resistant future reactor systems.

---

[38]R. Cashmore et al., *Fuel Cycle Stewardship in a Nuclear Renaissance*, The Royal Society Science Policy Centre report 10/11, Royal Society, London, Figure 6.

[39]Bagla, P., Thorium seen as nuclear's new frontier, Science 350 (6262), 726–727 (2015).

[40]R. Cashmore et al., *Fuel Cycle Stewardship in a Nuclear Renaissance*, The Royal Society Science Policy Centre report 10/11, Royal Society, London, p. 29 (2011).

[41]Fissile material is capable of sustaining a nuclear fission chain. Fertile material itself is not fissionable, but can be converted into fissile material by neutron absorption and subsequent nuclei conversion.

[42]E. Mearns, *Molten Salt Fast Reactor Technology—An Overview*, [Online] http://euanmearns.com/molten-salt-fast-reactor-technology-an-overview/

[43]The IAEA considers 8 kg of U-233 enough to construct a nuclear weapon, double compared to Pu-239.

[44]Ashley, S.F., G.T. Parks, W.J. Nuttall, C. Boxall and R.W. Grimes, Nuclear energy: Thorium fuel has risks, Nature 492, 31–33 (2012).

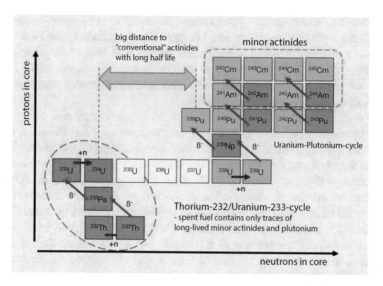

Fig. 2.17 Excerpt of the periodic table illustrating and comparing physical transitions of the thorium- 232/uranium-233 and the uranium/plutonium fuel cycle; as the fertile material of the thorium fuel cycle has an atomic weight of 232 compared to 238 of the uranium fuel cycle, the resulting mass difference means that thorium-232 requires six times more neutrons before transuranic elements/minor actinides can be produced (represented by the term "big distance" in the figure) (source: H.-M. Prasser, Electro Suisse Bulletin 11/2014)

2.6   Radioactive Waste Management Concepts

2.6.1   General Framework

Long-term radioactive[45] waste management usually focuses on deep underground storage where geological factors come into play. It involves compromises between technical and social dimensions: It must not only be technically achievable, but also acceptable to the public. The technical solutions have to demonstrate "beyond reasonable doubt" that a method/concept exists to ensure safe and secure containment of long-lived highly radioactive waste for the indefinite/distant future and that undue burdens on future generations are avoided (adapted from the Flowers report[46]). This section will focus on deep underground storage, as the most favored option, where geological factors come into play (see [47] for further details and other options).

---

[45] According to the German Atomic Energy Act, a material is considered radioactive if it contains "one or more radionuclides and whose activity or specific activity in conjunction with nuclear energy or radiation protection cannot be disregarded". The Act refers to the possibility of clearance or exemption in the case that certain levels of activity or specific activity are not exceeded.

[46] Flowers, Sir Brian (September 1976). Nuclear Power and the Environment (6th ed.). London: Royal Commission on Environmental Pollution, ISBN 0 10 166,180 0. Retrieved October 30, 2016.

[47] IAEA (International Atomic Energy Agency) (2009a) Geological Repositories. IAEA, Vienna.

Addressing uncertainties appropriately is key for confidence in both the geo-scientific/technical solutions and societal/cultural attitude (preferences, trust). In general, inherent safety features (based on physical laws) and passive safety (no energy, no triggering needed) and security features (barriers) might be more secure and preferred to active safety and security means (monitoring and control system) that may be more prone to failure.

Geo-scientific uncertainty issues entail the inventory for disposal, and the safety case is based on detailed knowledge of the local geology and technical features coupled with the understanding, predictability, and modeling of mechanical, hydraulic, and chemical processes relevant for safety (e.g., transport by groundwater). Questions regarding the robustness of the repository concept as a whole, the validity of research results and whether or not basic scientific premises will change, belong to the spectrum of geo-scientific uncertainties as well, and are hard to answer.

Further it is uncertain whether irradiated fuel will still be considered solely as waste—and not at least partially as a resource—in the far future. The stability of societies over extreme long periods of time is mostly assumed but by no means ensured, thus representing a major source of uncertainty.[48]

Radioactive waste from nuclear power production includes, in general,

- Spent nuclear fuel[49] and low-level operational waste;
- Waste arising during the production of nuclear fuel such as mine and mill tailings, enriched and depleted uranium;
- High-level waste (HLW) from reprocessing, including hulls and end-cups from fuel assemblies, extracted plutonium and uranium, and associated operational wastes;
- Waste from the decommissioning of Nuclear Power Plants.

The amount of long-lived radionuclides contained in the waste defines the timeframes over which the waste needs to be isolated and/or monitored while the heat production of the waste defines cooling requirements and the layout of transport means, storage, and disposal facilities. Thus, waste categorization schemes focus on radioactivity (Becquerel, Bq, or number of disintegration per second), activity content and heat production (kW/m^3) and some—like the IAEA—propose safe disposal options like geological disposal facilities for high level and long lived waste, in particular (see Table 2.4 as an example[50]).

Waste comes in a large variety of types in terms of radionuclide content, chemical composition, and physical conditions. Beyond dispute, spent nuclear fuel and—if reprocessing is applied—vitrified high level waste place the strongest requirements

---

[48]Notably, low level radioactive waste is also a by-product of use of nuclear fission or technology in other fields such as research and medicine.

[49]For typical inventories of safety-relevant radionuclides in canister of spent fuel of nine UO_2 fueled BWR, after decay of 40 years, see annexed Table 2.7.

[50]See also IAEA waste classification scheme as currently established.

Table 2.4 Typical waste characteristics (based upon IAEA, Classification of Radioactive Waste Safety Guide, IAEA Safety Series No. 111-G-1.1, Vienna (1994))

Waste classes	Typical characteristics	Disposal options
1. Exempt waste (EW)	Activity levels at or below clearance levels based on an annual dose to members of the public of less than 0.01 mSv	No radiological restrictions
2. Low and intermediate level waste (LILW)	Activity levels above clearance levels and thermal power below about 2 kW/m^3	
2.1. Short lived waste (LILW-SL)	Restricted long lived radionuclide concentrations (limitation of long lived alpha emitting radionuclides to 4000 Bq/g in individual waste packages and to an overall average of 400 Bq/g per waste package)	Near surface or geological disposal facility
2.2. Long lived waste (LILW-LL)	Long lived radionuclide concentrations exceeding limitations for short lived waste	Geological disposal facility
3. High level waste (HLW)	Thermal power above about 2 kW/m^3 and long lived radionuclide concentrations exceeding limitations for short lived waste	Geological disposal facility

on radioactive waste management.[51] Waste management strategies include a series of steps with safe disposal as the ultimate goal. In contrast to storage, the term (final) disposal is characterized by the intention of not retrieving the radioactive materials ("close and walk away"). In many countries with a large number of reactors in operation, deep geological disposal is favored/prescribed for highly radioactive, heat generating waste exceeding clearance levels.[52]

Analyses have been made with regard to potential effective radiological doses to members of a critical group—within a self-sustaining community, in a stable society—living in an area where the radionuclides released from the repository might reach the biosphere in the very long run. The analyses also demonstrate the effect of fuel cycle options on potential doses. It has been assumed that engineered and geological barriers will function as expected, with container lifetime of 2000 years or longer, slow release of radionuclides due to highly delayed matrix degradation and solubility limitations as well as very slow transport through the host rock due to diffusion and sorption processes. In recent scenario-type simulations, the calculated annual effective doses—shown up to ten million years—indicate that (see Fig. 2.18):

– Releases and thereby annual effective doses depend on host rock formation and are caused by long-lived fission products, mainly iodine-129;

---

[51]See annexed Table 2.7 for inventories of safety-relevant nuclides in a canister of PWR fuel assemblies after 40 years decay.

[52]For Germany, the amount of 30,000 m^3, or about 17,000 tons, of heavy metal has been estimated for a phase-out according to the Atomic Act 2002.

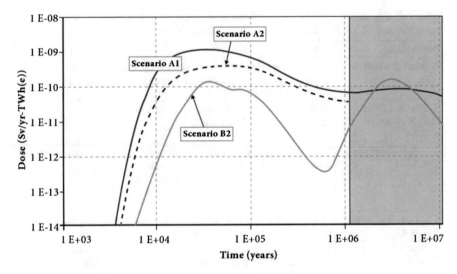

Fig. 2.18 High-level waste in granite: Calculated annual effective doses to the members of a critical group after release to biosphere for various fuel cycle Schemes (A1: "open", A2: "Pu burning", B2: "Pu plus minor actinides burning" assuming realistic losses of 0.1% for Pu, and 0.5% for minor actinides (Bagla, P., Thorium seen as nuclear's new frontier, Science 350 (6262), 726–727 (2015).). For assessment timeframes beyond one million years, dose calculations become increasingly meaningless due to the decreasing confidence in the underlying assumptions and in comparison with natural releases and natural background radiation of 2.4 mSv/year as global average

- no significant improvement is expected with closed fuel cycles with separation and plutonium-burning exclusively (scenario A2);
- separation and plutonium-burning and transmutation of minor actinides (scenario B2) delay and reduce potential doses significantly (up to a factor of 100) over an extremely long period of time (hundreds of thousands of years) but may increase afterwards.

All these theoretical analyses assumed a "normal evolution of the repository", i.e. intrusion scenarios and/or extreme events, either socio-political or natural have been excluded. The accuracy of the results should not be overestimated. However, there are clear indications that potential doses are extremely small, both in absolute and relative terms.[53]

The "safety community" and most countries follow the "close and walk away" approach that requires designs of the deep geological disposal to be passively safe

---

[53]Taking the inventory of 25 plants, operated over 40 years with an annual power production of 10 TWe, as basis, the hypothetical annual dose would be 0.001 mSv; the average value of natural radiation dose adds up to 1 to 2 mSv; the Swiss regulatory body (ENSI) claims to prove less than 0.01 mSv/a for a final repository.

and environmentally sound, needing no active institutional oversight after closure.[54] As commonly agreed, the safety case for such a disposal should not be dependent on the potential retrieval, while some argue flexibility should be built into the decision-making process so that earlier decisions including site selection can be re-evaluated and reversed where appropriate, leading to the concept of retrievability. It is important to note that, currently, there is no operating civil geological disposal facility, although disposal plans are well advanced in some countries, in particular in Finland, where construction works have begun in 2016 and operations are expected to begin in 2023.

2.6.2 Implementation of Waste Management Programs

Since dealing with nuclear waste and its management remains one of the primary issues for nuclear power, we can observe several approaches to the implementation of safeguard clauses, regulatory guidelines, and management programs including final storage.

The ongoing globalization, which reached the energy sector in general and the nuclear sector in particular, shifts the focus from a national approach, as during the Non-Proliferation Treaty Negotiations, towards a multi-national and global perspective. Organizations such as the World Institute of Nuclear Security (WINS[55]) and the World Association of Nuclear Operators (WANO[56]) were founded, which serve as platforms to share, promote and exchange best practices on safety and operational procedures (WANO) and on preventing misuse of nuclear and radioactive material (WINS). While the WINS is a non-profit NGO and the WANO a voluntary member based organization, global governance is also in place, notably the *Convention on the Safety of Spent Fuel Management* and *Safety of Radioactive Waste Management*,[57] established by the IAEA. A peer review process of national reports submitted by committed member states then shows whether the corresponding countries have complied with the Joint Convention and fulfilled the standards. A more detailed outlook on existing international and European organizations and their cooperation is provided in Chap. 3, including a tabular overview of responsible risk communication frameworks.

A step back at this point shows that the accumulation of stockpiles started in the early 1970s, after industrial scale power plants were built in several countries,

---

[54]R. Cashmore et al., *Fuel Cycle Stewardship in a Nuclear Renaissance*, The Royal Society Science Policy Centre report 10/11, Royal Society, London, p. 37.

[55]http://www.wins.org

[56]http://www.wano.info

[57]The Joint Convention is the first legal document globally addressing these issues and entering into force in June 2001. It involves 72 parties as of 2016 (http://www.iaea.org/Publications/Documents/Conventions/jointconv_status.pdf) and derives the obligations to the contracting parties from the IAEA fundamentals document "The Principles of Radioactive Waste Management", 1995.

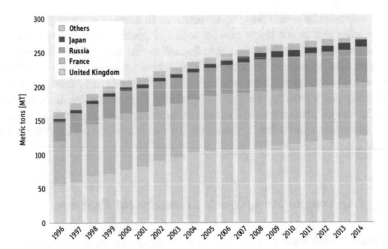

Fig. 2.19 Civilian stockpile of separated plutonium from 1996 to 2014 based on the yearly INFCIRC/549 declaration to the IAEA; stocks are listed by their location, not ownership (IPFM, Global Fissile Material Report 2015: Nuclear Weapon and Fissile Material Stockpiles and Production. Eighth annual report of the International Panel on Fissile Materials, 2015.). A simple implosion bomb requires about 6 kg of plutonium (2 to 4 kg for a more sophisticated implosion device) (Union of Concerned Scientists, Preventing Nuclear Terrorism, www.ucsusa.org (April 2004)) to obtain a yield of 10–30 kt (kilo-tonnes of TNT equivalent energy). For comparison, the Hiroshima bomb had a yield of 13 kt. Thus, 280 MT is enough plutonium to make 50 million weapons with each having the same yield as the Hiroshima bomb. Clearly, we should burn these 280 MT of plutonium in nuclear reactors for civil energy generation

including Germany, France, Russia, the UK, Japan, and the USA. Since the processing and fabrication method of separated plutonium to MOX was known, the recycling of spent fuel and of plutonium, in particular, increased rapidly leading to the construction of large and commercial MOX fabrication plants. Some of the biggest are in Europe. France has La Hague with reprocessing capability/annual throughput of 1400 tonnes heavy metal (tHM) and the UK has Sellafield with 1500 tHM. These plants even undertake reprocessing for utilities in other countries. Considering the amounts of stored separated plutonium as seen in Fig. 2.19, this step has a high significance. The storage of large, still increasing amounts of separated weapon-grade plutonium raises major safety concerns.

The waste management process including reprocessing, if the closed fuel cycle strategy is applied, has to follow international guidelines or to be oriented on global recommendations that then flow into national policies and legislation. A current effort in this area is the Radioactive Waste Safety Standards (RADWASS) Program by the IAEA, which provides guidance to member states for producing their own regulations for the safe management of radioactive waste including final disposal.

In addition, a directive "for the disposal of used nuclear fuel and radioactive waste"[58] was adopted by the European Commission (EC) and became effective in

---

[58]https://ec.europa.eu/energy/topics/nuclear-energy/radioactive-waste-and-spent-fuel

August 2011. The "waste directive" required all member countries to develop national waste management plans for the EC review by 2015, including cost assessments, financing schemes and timetables for the construction of disposal facilities, which can be operated jointly, but ultimate waste has to return to the originating EU country.[59] Moreover, the IAEA standards would become legally binding within the EU policy framework. With 143 facilities in 14 EU's member states generating used fuel, the impact of the directive has a big potential and it acknowledges two main branches: disposing of spent nuclear fuel as waste or reprocessing plutonium and uranium, leaving only the remainder of the reprocessing as waste to be disposed of.

On a more local scale, a successful example of Waste and Spent Fuel Management is Sweden and Finland, who combined operate 14 reactors, generating roughly 11.7GWe, with nuclear power having a 30–40% share in each country's electricity mix. Both countries share the policy of direct spent fuel disposal in deep geological facilities/bedrocks. In Finland, two near surface disposal facilities have operated for low and intermediate level operational waste since 1992 and 1998 respectively (Olkiluoto and Loviisa). By 25 November 2015, a construction license was granted to the company Posiva Oy for a 400 m final deep geological disposal repository for used fuel at the Olkiluoto site, expanding the existing facilities and aiming for an overall repository capacity of 6500 tons.[60] In Sweden, planning for its own repository near Forsmark is ongoing, with future capacity of 12,000 tons of spent fuel at 500 m depth, which should start construction work in 2020.

Another notable example of the current development of deep geological storage facilities and its planning and execution is the Swiss "Sectorial Plan for Deep Geological Repositories"[61] (SGT). Developed by the Swiss Federal Office of Energy and approved by the Federal Council in 2008, it specifies three steps for the selection of a deep geological waste repository site within Switzerland. With the first step, the selection of geological siting areas, already completed, two steps are left: the selection of at least two sites and the selection of a final site, since the final goal of the SGT is to determine either two sites for the separate storage of high and intermediate radioactive waste (HLW, ILW) or one which does both. The plan is considered exemplary and was highlighted by the Joint Convention as a model for participation and involvement of the public and neighboring countries.

---

[59]Following the compliance check with these obligations the Commission has opened infringement cases against several Member States; two of them have been closed while Romania has been taken to the EU Court of Justice.

[60]By Finnish nuclear energy law, such a license has to be approved by the Radiation and Nuclear Safety Authority (STUK), which it did, confirming the license on November 25, 2016.

[61]A full presentation of the plan including relevant documents and official statements can be found at https://ensi.ch/en/waste-disposal/deep-geological-repository/sectoral-plan-for-deep-geological-repositories-sgt/

On the one hand, this seems to be a proof of concept for local communities applying to the government as a qualified place for a final storage point and the successful approach for the search of deep geological repositories. On the other hand, however, the example of Yucca Mountain in the USA shows that there can be a lot of complication with the development and construction of long term storage facilities. In 2008, the US Department of Energy first submitted an application for the construction of the site in the Nevada desert. But since then, several problems in terms of potential groundwater contamination, earthquake activities and problems with transportation of nuclear material to the site arose and the project experienced major delays that led to the project being abandoned by the Obama administration, also due to political reasons.[62]

Thus, while positive developments, in the sense of a substantial regulation with acknowledgement of appropriate guidelines on the national level, are crucial for a successful implementation of waste management programs, the international perspective in terms of directives, guidelines, treaties and synergies has to be included as well. The consensus for solving the problem of the disposal of the nuclear wastes, which remain after all reprocessing efforts, is the use of deep geological disposal facilities to store HLW and ILW. This is already planned in countries including but not limited to France, Sweden, Finland and the USA.

Textbox 2.2 Remarks on Retrievability (According to the Royal Society[63])

Flexibility could involve reversing decisions about site selection or choice of disposal concept; or, at later stages, decisions about the construction, operation and date of closure of the Geological Disposal Facility (GDF). Retrievability is a special case of reversibility, namely the potential to reverse the emplacement process so that the waste containers can be retrieved.[64] Retrievability may be attractive to policymakers so that decisions can be informed by technological developments, new socio-political circumstances, as well as changes to regulation or national policy. Future fuel cycle options are kept open, should a decision be made to reuse the spent fuel. Public confidence could be secured by providing reassurance that unforeseen safety problems could be addressed... Much of the debate about retrievability concerns the conflict between two important principles: decisions for radioactive waste

(continued)

---

[62]President Donald Trump's administration has proposed reviving the long-stalled Yucca project, but the plan faces bipartisan opposition from the state's governor and congressional delegation. In May 2018, the U.S. House of Representatives approved a bill to revive the mothballed nuclear waste dump at Nevada's Yucca Mountain despite opposition from home-state lawmakers.

[63]The Royal Society, Fuel cycle stewardship in a nuclear renaissance, The Royal Society Science Policy Centre report 10/11, 2011.

[64]NEA, Reversibility and retrievability for the deep disposal of high level radioactive waste and spent fuel. Radioactive Waste Management Committee. OCED Nuclear Energy Agency: Paris. 2011. Oecd-nea.org/rmw/rr/documents/RR_report.pdf

TextBox 2.2 (continued)
management should be made now rather than being left for, and imposing undue burdens on, future generations; whilst options should be preserved for future generations to make their own waste management decisions. Retrievability should not be an excuse for indefinite delay of decision making about geological disposal. The safety case for a geological disposal should not depend on the potential retrieval.

2.7 Proliferation Issues

There has been a debate about the relationship between civil nuclear power and proliferation. Proliferation is the misuse of nuclear knowledge, materials and technology for weapons of mass destruction. The debate has reemerged with the nuclear renaissance and related developments of enhanced nuclear systems, including advanced closed fuel cycles. However, it should be stressed that it is control over the fuel cycle and not nuclear power per se that is important for making weapons grade material. Basically, one needs a source of uranium and yellow cake,[65] and a gas centrifuge.

High-caliber attempts have been made to develop a consistent approach to minimize the proliferation risk of innovative reactor concepts. Within the International Generation development program, a project was set up to develop a methodology to evaluate proliferation resistance and robustness of physical protection, coping with the multidimensionality of the problem. To give a flavor of the methodology (see Ref. [66]), it recognizes the importance of the isotopic compositions of uranium and plutonium mixtures and requires the definition of the threat space (the full spectrum of potential threats, considering all relevant actors, objectives, and so on). And because threats evolve over time, reasonable assumptions are necessary about the spectrum of threats that facilities and materials in the system could be subjected to over their full life cycles. Fissile Material Type is used as a measure to express the utility of the fissile material prior to fabrication of an explosive (see Table 2.5). This information can be used for a rank-ordered listing of specific material types, by their relative overall attractiveness (see Table 2.6 to follow). The full pathway by which a proliferant State obtains a nuclear explosive device is divided into three major stages: (1) Acquisition of nuclear material in any form, (2) processing to convert the acquired material and (3) fabrication starting from the

---

[65]Yellow cake is a type of uranium concentrate powder obtained in an intermediate step in the processing of uranium ores. It is a step in the processing of uranium after it has been mined but before fuel fabrication or uranium enrichment.

[66]GIF PRPPWG, Evaluation Methodology for Proliferation Resistance and Physical Protection of Generation IV Nuclear Energy Systems, Revision 6, September 15 2011.

Table 2.5 Intrinsic properties of fissionable materials, in metallic form, separated from spent fuel for typical operating conditions (burn up, etc.) and normalized to weapons grade plutonium

	Bare Metal Sphere Critical Mass	Heat Generation Rate	Spontaneous Neutron Emission Rate	Dose Rate of a 4 kg metal sphere at 1 m from center
Weapons-Grade Pu	1	1	1	1
GMR (Magnox) Pu	1.2	2	3.5	–
HWR (Candu-NU) Pu	1.2	2	6.7	–
LWR (LEU) Pu	1.4	6.5	7.3	3.5
^{233}U	1.6	Negl.	Negl.	–
LWR (MOX) Pu	1.7	16.6	3.7	–
ESFR Pu	1.75	50.9	1.67×10^3	–
MHTGR Pu	2.4	7.7	37.3	3.3
MSR Pu	4.7	23.5	94	7.5
HEU	5	Negl.	Negl.	Negl.
^{241}Am	5.9	278	0.1	375
^{237}Np	6.3	0.05	1.5×10^{-5}	Negl.

GMR Graphite Moderated Reactor, *HWR* Heavy-Water Reactor, *LWR* Light-Water Reactor, *MOX* Mixed Oxide Fuel, *ESFR* Example Sodium Fast Reactor, contains 2.3 wt% of ^{237}Np, 3.2 wt% of ^{243}Am, and 1.4 wt% of ^{244}Cm in plutonium, *MHTGR* Modular High Temperature Gas-Cooled Reactor; Plutonium Enrichment $= (^{239}$Pu $+ \ ^{241}$Pu$)/($Total Pu$)$, *HEU, MEU, and LEU* Highly, Medium and Low Enriched Uranium, resp. The 4 kg metal sphere is taken as a reference, assumed to be a criticality-safe mass for handling

Table 2.6 Ranking ordering of fissile materials by proliferation resistance (PR) (GIF/PRPPWG/2006/005, Technical Addendum, 01/2007)

Material	Ranking
Highly Enriched Uranium (HEU)	Very Low PR
Weapons Grade Plutonium (WG-Pu)	Low PR
Reactor Grade Plutonium (RG-Pu)	Medium PR
Deep-Burn Plutonium (DB-Pu)	High PR
Low Enriched Uranium (LEU)	Very High PR

processing stage or directly from the acquisition stage depending on the nuclear material.

In 2002, a joint project among the International Energy Agency (IEA), the OECD Nuclear Energy Agency (NEA), and the International Atomic Energy Agency (IAEA) spelled out eight criteria that should be satisfied by new systems to minimize proliferation risks:

- Simplifying the accommodation of IAEA safeguards equipment.
- Operating on a once-through fuel cycle: treat spent fuel as waste, eliminating the need for fuel reprocessing and associated safeguarding and monitoring requirements.
- Operating on a closed fuel cycle: use a reprocessing method that returns actinides to the reactor for consumption and thus avoid the need to store them.
- Operating a breeding ratio (near) 1: avoid need for uranium blanket around the core.
- Using features that make it difficult to extract fissile material from spent fuel.
- Using features that ensure a very low amount of actinides in the spent fuel.
- Reducing the amount of actinides produced.
- Reducing the amount of fuel stored at the site.

Proliferation issues are managed through the Non-Proliferation Treaty[67] and the verification of commitments by the IAEA. This safeguard mechanism and solid international governance have proven to be effective for "declared facilities". However, clandestine activities warrant permanent vigilance.

The good track record of non-proliferation safeguards suggests alternative pathways may be much more likely sources of proliferation than the diversion of nuclear material from civilian nuclear power programs.[68] Besides misuse of enrichment facilities and/or small research reactors to produce 'weapons grade' material,[69] potential proliferation threats are posed by the management of spent fuel of present reactors, in particular:

- Removal (theft) of spent fuel (containing fissile uranium and plutonium) from storage ponds and even replacement with dummy material, especially after long time storage so that the heat load and radioactivity is reduced.

---

[67]Under the NPT, all non-nuclear weapons states are required to conclude a comprehensive safeguard agreement with the IAEA. This involves declarations of the quantities and location of all nuclear materials and facilities; their correctness needs to be verified through "measures" by the IAEA (as currently illustrated in the Iran case).

[68]The Royal Society, Fuel cycle stewardship in a nuclear renaissance, The Royal Society Science Policy Centre report 10/11, 2011.

[69]Highly enriched uraniun-235 (HEU), while the maximum enrichment of fuel for civil/commercial power reactors is limited to 3–5%, or "pure" plutonium-239.

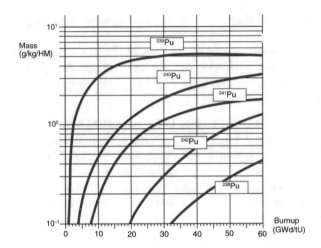

Fig. 2.20 Change of the fraction of plutonium isotopes in spent fuel as a nuclear reactor is operated at higher burn ups (source: The Royal Society Science Policy Centre report 10/2011). The build-up of isotopes other than Pu-239 makes civil reactor plutonium less attractive for nuclear weapon use. At high burnup, currently around 55 GWd/tU (gigawatt-days/metric ton of heavy metal), the total mass of plutonium in spent fuel is approximately 10 grams per kilogram of heavy metal

- Diverting separated plutonium at a licensed reprocessing facility, especially if stored and not further fabricated for re-use, e.g. as MOX fuel.
- Substitution of waste packages at a disposal facility by dummy canisters and following reprocessing/separation of weapons grade material, all at licensed facilities.
- Although difficult to conceal—running civil power reactors at very low burn up[70] to reduce buildup of isotopes other than plutonium-239,[71] see Fig. 2.20.

Safeguarding of nuclear material includes the disposal of radioactive waste. There is consensus that "deep geological" is the best practice; safety standards are provided by the IAEA for radioactive waste but do not explicitly consider disposal of plutonium. The mostly favored 'close and walk away' approach of the safety community requires disposal designs to be passively safe and environmentally sound, needing no active institutional oversight after closure. Yet, the safeguards community may require active monitoring and ongoing institutional control for spent fuel and other nuclear materials that cannot be considered to be practicably irrecoverable or otherwise suitable for the termination of safeguards. This is

---

[70]Burnup (also known as fuel utilization) quantifies how much energy is extracted from a primary nuclear fuel source. It can be measured as the fraction of fuel atoms that underwent fission in fissions per initial metal atom or as the actual energy released per mass of initial fuel in gigawatt-days/metric ton of heavy metal (GWd/tU).

[71]Pu-238 decays relatively rapidly, generating significant amounts of decay heat; Pu-240 could set off the chain reaction prematurely, substantially reducing explosive yield; Pu-241, although fissile, decays to AM-241, which absorbs neutrons and emits intense gamma radiation.

complicated by decisions (that have yet to be made in some countries) about retrievability of wasteforms[72] that can be designed into the disposal facility (see Textbox 2.2 for pros and cons).

There is international consensus that there is no proliferation proof fuel cycle, while the ranking order, based on the material type measure, places highly enriched uranium in the very low proliferation resistance category (see Table 2.6) and low enriched uranium (>20% U-235) in the very high proliferation resistance category; for plutonium, three rankings can be found, depending on the conditions of reactor operation and recycle of spent fuel, and the range of respective isotopic compositions.

Assessments have been made for thorium fuelled reactor systems: It turned out that those systems have the potential to fulfill all the eight criteria (above) spelled out by the three preeminent international organizations. This constitutes an important incentive to reconsider thorium from a proliferation standpoint, for which the starter fuel sequence deserves special attention. While the thorium fuel cycle would not fundamentally eliminate proliferation concerns (see end of Sect. 2.5), it presents a number of inherent obstacles to acquire weapons grade nuclear material and make explosives. To name a few, some important technical factors include: the presence of strong radioactive poisons, the radioisotope U-232 and its daughters, in spent thorium fuel; the possibility to "contaminate" fresh thorium with fresh uranium to accentuate the poisoning effect of U-232; and the technological challenges, even for advanced countries, to reprocess spent thorium fuel.

To maximize proliferation resistance, a balanced approach seems to be advisable including integrated nuclear governance and enhancement of intrinsic barriers by altering chemical, physical and radioactive properties as well as best practice approaches for reuse of sensitive material, notably plutonium, and spent fuel storage; all including the entire life cycle. There is no 'silver bullet solution': whether an open fuel cycle is more proliferation resistant than a closed fuel cycle with reprocessing/ separation of weapons grade material depends inter alia on the process characteristics and recycle its reuse strategy; other non-technical aspects have to be factored into the respective decision.

---

[72]The waste form is the key component of the immobilization process, as it determines both waste loading (concentration), which directly impacts cost (due to volume reduction), as well as the chemical durability, which determines environmental risk. See http://www.world-nuclear.org/infor mation-library/nuclear-fuel-cycle/nuclear-wastes/synroc.aspx

Annex

Table 2.7 Inventories of safety-relevant radionuclides in a reference canister containing 9 BWR UO2 fuel assemblies with a burn-up of 48 GWd/tHM, after 40 years decay (IAEA (International Atomic Energy Agency) (2009a) Geological Repositories. IAEA, Vienna)

Radio-nuclide	Fuel [Bq]	Structural materials [Bq]	Radio-nuclide	Fuel [Bq]	Structural materials [Bq]
^{3}H	7.0×10^{12}	5.4×10^{9}	^{228}Ra	3.2×10^{1}	–
^{10}Be	1.8×10^{7}	2.9×10^{3}	^{227}Ac	3.3×10^{5}	–
^{14}C$_{inorg}$	6.7×10^{10}	0	^{228}Th	1.9×10^{8}	6.8×10^{1}
^{14}C$_{org}$	0	5.7×10^{10}	^{229}Th	1.4×10^{4}	–
^{36}Cl	8.6×10^{8}	$1.4 \times 10^{*}$	^{230}Th	3.2×10^{7}	1.6×10^{1}
^{41}Ca	2.2×10^{8}	7.0×10^{7}	^{232}Th	4.1×10^{1}	–
^{59}Ni	8.9×10^{8}	1.3×10^{11}	^{231}Pa	7.6×10^{5}	–
^{63}Ni	9.2×10^{10}	1.3×10^{13}	^{232}U	1.8×10^{8}	6.5×10^{1}
^{79}Se	1.6×10^{9}	8.0×10^{3}	^{233}U	6.7×10^{6}	4.6×10^{0}
^{90}Sr	2.2×10^{15}	6.0×10^{9}	^{234}U	9.7×10^{10}	8.3×10^{4}
^{93}Zr	1.5×10^{11}	1.8×10^{10}	^{235}U	9.1×10^{8}	–
^{93m}Nb	1.2×10^{11}	1.4×10^{10}	^{236}U	2.1×10^{10}	6.0×10^{3}
^{94}Nb	2.1×10^{8}	1.3×10^{10}	^{238}U	1.9×10^{10}	2.2×10^{4}
^{93}Mo	8.3×10^{7}	3.3×10^{8}	^{237}Np	3.2×10^{10}	4.6×10^{4}
^{99}Tc	1.1×10^{12}	6.7×10^{7}	^{238}Pu	$2.6 \times; 10^{14}$	6.0×10^{8}
^{107}Pd	1.0×10^{10}	1.2×10^{5}	^{239}Pu	2.2×10^{13}	2.2×10^{8}
^{108m}Ag	1.4×10^{9}	4.3×10^{4}	^{240}Pu	4.0×10^{13}	3.0×10^{8}
^{126}Sn	3.0×10^{10}	3.2×10^{5}	^{241}Pu	1.5×10^{15}	1.9×10^{10}
^{129}I	2.7×10^{9}	1.9×10^{4}	^{242}Pu	1.9×10^{11}	2.9×10^{6}
^{135}Cs	3.7×10^{10}	1.8×10^{5}	^{241}Am	3.0×10^{14}	3.7×10^{9}
^{137}Cs	3.5×10^{15}	1.9×10^{10}	^{242m}Am	5.7×10^{11}	7.1×10^{6}
^{151}Sm	1.6×10^{13}	1.1×10^{8}	^{243}Am	2.2×10^{12}	3.8×10^{7}
^{166m}Ho	2.1×10^{9}	1.0×10^{4}	^{243}Cm	6.7×10^{11}	1.0×10^{7}
^{210}Pb	9.4×10^{4}	–	^{244}Cm	8.0×10^{13}	1.4×10^{9}
^{210}Po	9.1×10^{4}	–	^{245}Cm	5.6×10^{10}	1.0×10^{6}
^{226}Ra	2.9×10^{5}	–	^{246}Cm	1.2×10^{10}	2.6×10^{5}

Chapter 3
Governance: Organizations and Management Issues

Abstract Energy effectively constitutes a global public good, and nuclear energy belongs to its provision. The potential risks of high consequence accidents and the importance of strong safety culture, regulation, and governance, have been recognized early, and to an increasing extent over time. As a result, exceptional and leading efforts into risk and reliability analysis have and continue to take place within the nuclear community. However, the adequacy of existing governance and control mechanisms can be questioned. Indeed, authoritative organizations, notably the IAEA, provide standards, best practices, and platforms for exchange at the international level. And strong regulatory bodies provide assurances and verification of safety at the national level. However, ultimately minimum standards are not trivial to impose because they will only be adopted by states that consent. Further, case studies of major accidents show the pervasiveness of distorted information and sometimes risk information concealment—a common organizational deficit, highly problematic in this safety-sensitive domain.

Consulting governance principles, more specific improvements are put forth, relating to stronger and obligatory monitoring and standards, to help promote safe civilian operation of nuclear power. A better international collaboration and extensive peer-reviews are recommended; the role of the IAEA and related enforcement mechanisms should be strengthened.

3.1 Nuclear Energy Generation as a Public Good

As an extension of Samuelson's concept[1] from economics, global public goods (or public assets) are usually defined as non-rivaling and non-excludable, but their benefits (or, equivalently, reverberations) are also "quasi-universal in terms of

---

[1] Samuelson, P. A., The Pure Theory of Public Expenditure, Review of Economics and Statistics 36 (4), 387–389 (1954).

© Springer Nature Switzerland AG 2019
D. Sornette et al., *New Ways and Needs for Exploiting Nuclear Energy*,
https://doi.org/10.1007/978-3-319-97652-5_3

countries, people and generations".[2] While energy cannot strictly be defined in this sense as a public good, it is clearly a common necessity for everyone. And nuclear energy belongs to its provision as it contributes to the energy mix worldwide. However, unlike other energies in our energy mix, nuclear energy possesses a weakest-link technology character: a severe accident somewhere may impact the whole World both in terms of radioactive pollution, and much more in terms of emotional and political impacts.

The role of international organizations worldwide is in effect to provide assurances and verification for nuclear technology. This means that, for its efficient provision and governance, several actors must meet minimum standards. By actors, we include the owners and operators[3] of nuclear plants, relevant organizations, regulators/authorities but also other relevant stakeholders. Such minimum standards are not trivial to enforce because they will only be adopted by states that consent.

Considering the specific characteristics of nuclear energy—its weakest-link technology character, the several hazards and risks associated in case of bad management or the difficult waste management issues—effective control mechanisms for safety implementation are crucial. We address the question of whether current control mechanisms (i.e. local, national and international organizations and accepted standards) can—effectively and on a global scale—regulate nuclear energy to avoid it becoming a global "public bad" (or public liability). This includes developing commonly accepted safety and security measures, addressing proliferation and liability issues as well as securing a safety culture and putting in place long-term safety management processes. We look at how these current mechanisms are implemented, and from which hierarchical structure, with which tools and finally question whether this is adequate and sufficient.

We use Ostrom's framework of design principles as comparison criteria to assess the stable management of nuclear energy by governmental organizations, transnational actors and non-governmental organizations (NGO). Her polycentric governance approach is based on a "caring for the commons" idea, where governing and managing common-pool resources must be carefully thought about and collectively undertaken.[4] With respect to nuclear energy, not all of but most of the original design principles seem to be applicable or of relevance and in place as shown in Table 3.1.

---

[2]http://web.undp.org/globalpublicgoods/TheBook/globalpublicgoods.pdf (UN publication "Global Public Goods"); http://www.econ.yale.edu/~nordhaus/homepage/PASandGPG.pdf ("Paul Samuelson and Global Public Goods" by W. D. Nordhaus).

[3]It is worth mentioning that, in the nuclear field, nuclear law has very early assigned all liabilities to the plant owner exclusively, and none to the reactor vendor/supplier.

[4]Ostrom, E., *Beyond Markets and States: Polycentric Governance of Complex Economic Systems*. American Economic Review **100**(3), 641–672 (2010).

Table 3.1 Application of Ostrom's framework of design principles to the management of nuclear energy by governmental organizations, transnational actors and non-governmental organizations

Design principles and assessment of applicability	Explanation
1. Clear user and resource boundaries *Applicable*	Distinction between legitimate users and non-users (who is registered as user) Resource provision rules (who gets nuclear material)
2. Congruence with local conditions *Less applicable*	Can be considered but the overarching requirements accepted internationally should not be contingent on local conditions (e.g., safety should not be relaxed because a nuclear plant is sited in a non-populated area)
3. Collective choice arrangements *Applicable*	All the relevant stakeholders are authorized to participate in making/modifying rules about use of nuclear energy
4. Monitoring users *Applicable*	The users themselves (i.e. States) monitor the appropriation and provision levels that they declare to the IAEA
5. Graduated sanctions *Applicable*	Sanctions for rule violations start very low and become stronger with repeated violations
6. Conflict resolution mechanism *Applicable*	Rapid and cheap conflict resolution mechanisms are in place
7. Minimal recognition of rights *Less applicable*	Governments recognize the rights of the users to make their own rules
8. Nested enterprises *Applicable*	Governance activities are organized in multiple nested layers (i.e. the hierarchy around nuclear energy regulation)

3.2 Governmental Organizations and Transnational Actors

Such governmental organizations and transnational cooperation of the key actors worldwide, like the regulators and the operators, have the mission to guide and help countries and users of nuclear energy in its safe commercial use. We look below at three well-established organizations, i.e. the **IAEA** (International Atomic Energy Agency), the **OECD-NEA** (Nuclear Energy Agency of the Organization for Economic Cooperation and Development) and the **European Commission** (see Tables 3.2 and 3.7), and three transnational co-operations, i.e. **WENRA** (Western European Nuclear Regulators Association), **WANO** (World Association of Nuclear Operators) and **IRIS** (Integrated Review of Infrastructure for Safety). These organizations are based on non-binding guidance, as opposed to the non-proliferation treaty (NPT) of nuclear weapons, which is the only existing international signed, binding commitment related to nuclear power (see Table 3.6). The goals and tools of these organizations and co-operations are shown below; further information can be found in the annexed Tables 3.6 and 3.7.

The IAEA's portfolio of services and published standards can be found in Figs. 3.1 and 3.2.

International cooperation going beyond the responsibilities of national organisations are listed in Table 3.3.

Table 3.2 Comparison between international organizations

	IAEA	OECD-NEA	EU/EURATOM
Member states	168	31 (86% of world's installed nuclear capacity)	28 EU states + 1 associated state (Switzerland)
Goal(s)	1. Peaceful uses of nuclear technology 2. Safety and security 3. Non-proliferation	1. Uses of nuclear energy for peaceful purposes 2. Public protection 3. Promotion of liability and insurance	1. High level of nuclear safety (safety culture) and radiation protection 2. Effective safeguards in third-party countries 3. Safe management of radioactive waste
Tool(s)	Safety Standards series Technical Cooperation (TC) programs (including peer review and advisory services)	Committees responsible for action fields: CSNI (surveys, experts groups) CNRA (working and tasks groups)	Instrument for Nuclear Safety Cooperation (INSC) Nuclear Safety Directive

IAEA provides the safety standards series, which serve as reference for safety requirements worldwide. Both IAEA and NEA offer services based on collaboration and advice. The EURATOM provides "instruments" for supporting regulation and technical support with the EU

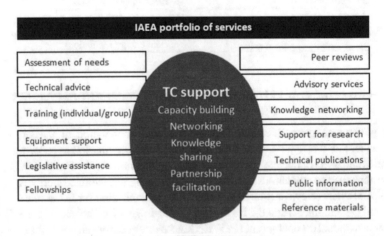

Fig. 3.1 IAEA Portfolio of Services ("What we do", IAEA Website, https://www.iaea.org/technicalcooperation/areas-of-work/index.html). The TC support programs are intended to increase sharing of competences and safety improvements through partnerships. Peer reviews and advisory services (see red arrows) are examples or services designed to share safety-knowledge and strengthen an adequate safety culture among members

Fig. 3.2 IAEA published standards (IAEA Safety Standards, Nuclear Safety and Security Program, Brochure. https://www-ns.iaea.org/downloads/standards/iaea-safety-standards-brochure.pdf) with a hierarchy based on increasing level of scope and detail: (**a**) Guides present best practices to help plant operators achieve high levels of safety; (**b**) Requirements are meant to be used by States for establishing national regulatory frameworks; (**c**) Fundamentals are accessible to the non-specialist reader and convey the basis and rationale for the safety standards

Table 3.3 Transnational co-operations compared. Both WENRA and WANO follow similar peer reviews and running groups than that of IAEA or NEA. IRIS is an IAEA methodology and is available to members for safety self-assessments

	WENRA	WANO	IRIS
Members	17 countries	130 members (operators)	168 (IAEA member states)
Goal(s)	1. Develop common approach to nuclear safety 2. Harmonize national regulatory approaches 3. Provide an independent capability to examine nuclear safety	1. Assurance of nuclear safety 2. Excellence in operational performance	Provide a methodology to implement IAEA requirements
Tool(s)	Running working groups that share experience and discuss issues among states	1. Peer-review system 2. Technical support and exchange 3. Professional and technical development	Framework for self-assessment of national infrastructure for safety

3.3 Non-Governmental Organizations (NGO)

The Greenpeace campaign states: "Greenpeace has always fought vigorously against nuclear power because it is an unacceptable risk to the environment and to humanity".[5] The organization communicates about the "real, inherent and long lasting" risks of nuclear energy and creates campaigns to spread its mission and make people aware of their view. The campaign against nuclear is sub-divided into topics, namely: Fukushima, Safety, Nuclear waste, Proliferation and "No more Chernobyls". Clearly, Greenpeace believes that there is no sufficient control over nuclear activities worldwide and that, where organizations or cooperation exist, the job done is insufficient and not transparent to the public. Greenpeace's reach is known to influence public opinion and hence the organization plays an important role in the public risk perception and debate around nuclear energy. The issue of "morally responsible risk communication" is discussed in Table 3.4 with respect to Greenpeace and governmental/international organizations.

Table 3.4 Framework for responsible risk communication

	Legitimate procedure	Ethically justified risk message	Outcome evaluation (assessing how the message is perceived)
Greenpeace	Arguable—who is involved in the decision-making? Are different stakeholders considered? Is the public actively participating in the debate, or only its own representatives?	Message is based on scientific data but also contains value judgments and strong feelings. How to keep the two apart? Is the extreme risk message justified with pertinent arguments?	Yes, encourages further campaigning, assesses how many people should become members, get involved in the movement, etc.
Governmental/International Organizations	Yes, to the extent that nuclear energy user self-determination is present.	Yes, risk message is present and justified.	Not completely, reverse feedback loop from actors and the public is not always efficient[a]

Morally responsible risk communication entails (1) a legitimate procedure, (2) an ethically justified risk message and (3) concern for and evaluation of the effects of the message and procedure (Fahlquist, J.N. and S. Roeser, *Nuclear energy, responsible risk communication and moral emotions: a three-level framework.* Journal of Risk Research **18**(3), 333–346 (2015).). This is the basis for trust-creation between the public and an organization's message

[a]One could argue that the public is not sufficiently well-informed to contribute to the decision making in any meaningful way. Some go as far as suggesting that the public is in fact brain-washed with inaccurate scare mongering circulated by NGOs

---

[5]http://www.greenpeace.org/international/en/campaigns/nuclear/ (Greenpeace campaign, "End the Nuclear Age").

It can be argued that an efficient safety culture for nuclear energy starts with responsible and accurate risk communication. For Greenpeace, the procedure and message are not always fully clear but always very strong. The outcome evaluation is invariably present and the engagement with other stakeholders makes Greenpeace a prominent actor. Governmental and international organizations focus their risk communication on the procedure and the message. For example, standards (i.e. IAEA standards) are developed from actual experience of use of nuclear energy. The outcome evaluation, however, is not complete because of the voluntary-basis nature of the safety standards and peer reviewing.

3.4   Risk Information Transmission Issues Related to Nuclear Disasters

In the presence of these diverse organizations, it is instructive to give a brief overview on how various coordination and management structures failed to fully function during the three most notorious nuclear accidents, which are described in further detail in Chap. 5. To put in context, managers are supposed to oversee other people by means of information.[6,7] They receive information from different sources, process it, make a decision, and translate this decision to subordinates and other audiences. The quality of the information being received about real conditions of the external and internal environment influences the quality of decisions, and later on the adequacy of an organization's response. However, the nuclear energy establishment has been severely tested in this respect during the Three Mile Island accident in the US (1979), the Chernobyl catastrophe in the former USSR (1986) and the Fukushima Daiichi disaster in Japan (2011); see also Table 5.1 for detailed descriptions. Detailed studies show that, before and after these accidents, executives and managers received distorted information from their staff and/or had developed year-long practice of risk concealment within their organizations. This is disturbing given that risk information deficit is a main cause for the inadequate actions of the organization and its staff during the normal practice of their duties. The magnitude of an accident is often influenced by inadequate decision-making just after the disaster caused by incomplete understanding of the severity of the crisis by involved parties, regulators, and victims. The following sections document the deficits of information transmissions that occurred during the three most famous nuclear accidents, complementing the summaries of Table 5.5.

---

[6]Hedberg, B., How organizations learn and unlearn, in: Nyström, P.C. & Starbuck, W.H., Handbook of Organizational Design, Oxford University Press, 1981.

[7]Mullins, L.J. and G. Christy, Management & Organisational Behavior, Financial Times Management, May 2010.

3.4.1   Three Mile Island (TMI)

On the 28th of March 1979, a loss-of-coolant accident occurred in reactor number 2 of Three Mile Island Nuclear Generating Station (TMI-2) in Harrisburg (Pennsylvania, USA). This happened following failures in the non-nuclear secondary system, followed by a stuck-open pilot-operated relief valve in the primary system, which allowed large amounts of nuclear reactor coolant to escape. No one was harmed, but the reputation of the nuclear industry was seriously damaged.

In the following, we present a brief account of the main human factors that led to this serious incident. On the road leading to this accident, one can identify two main sources of deficits of information on the risks and safety of this nuclear reactor.[8]

First, the post-WWII rapid expansion of the U.S. civil nuclear industry occurred at the expense of serious safety considerations. Indeed, the President's Commission report of the TMI-2 accident concluded that, because of the need to ensure national energy independence, *"the NRC* [Nuclear Regulatory Commission] *is so preoccupied with the licensing of plants that it has not given primary consideration to overall safety issues. . . . NRC has a history of leaving generic safety problems unresolved for periods of many years".*[9] The President's Commission report also stated: *"Two of the most important activities of NRC are its licensing function and its inspection and enforcement activities. We found serious inadequacies in both. In the licensing process, applications are only required to analyze 'single-failure' accidents. They are not required to analyze what happens when two systems fail independently of each other, such as the event that took place at TMI. . . . The accident at TMI-2 was a multiple-failure accident. . . .insufficient attention has been paid [by the NRC] to the ongoing process of assuring nuclear safety. . . . NRC is vulnerable to the charge that it is heavily equipment-oriented, rather than people-oriented. . . . [I]nspectors who investigate accidents concentrate on what went wrong with the equipment and not on what operators may have done incorrectly, in the lack of attention to the quality of procedures provided for operators, and in an almost total lack of attention to the interaction between human beings and machines."*

Second, there was a notable lack of communication about minor incidents within the civil nuclear industry until the TMI-2 nuclear accident. The President's Commission outlined this problem: *"The NRC accumulates vast amounts of information on the operating experience of plants. However, prior to the accident, there was no systematic method of evaluating these experiences, and no systematic attempt to look for patterns that could serve as a warning of a basic problem. . . The major offices within the NRC operate independently with little evidence of exchange of information or experience. For example, the fact that operators could be confused due to reliance on pressurizer level had been raised at various levels within the NRC organization.*

---

[8]Dmitry Chernov and Didier Sornette, Man-made catastrophes and risk information concealment (25 case studies of major disasters and human fallibility), Springer; 1st ed. (2016).

[9]Report of the President's Commission on the Accident at Three Mile Island: The Need for Change: The Legacy of TMI, October 1979.

Yet, the matter 'fell between the cracks' and never worked its way out of the system prior to the TMI-2 accident". Moreover, the President's Commission found out that the mistaken shutdown of the emergency cooling system was not unique to this incident, but a problem well known to representatives of the nuclear steam system suppliers. It had occurred at PWR plants on several occasions, but nobody had transmitted this information to other plants. In addition, the excessive complexity of control room design, which made it difficult for operators to quickly grasp the condition of a nuclear plant and so make decisions adequately, had been recognized at the design phase but ignored until the TMI case. There was a huge problem of risk information transmission between the different players during the development of the nuclear industry, and an inadequate response even to identified risks.[10] NRC's post-accident investigation confirmed the findings of the President's Commission: *"[Similar incidents to the TMI-2 accident] occurred in 1974 at a Westinghouse reactor in Beznau, Switzerland, and in 1977 at Toledo Edison's Davis Besse plant in Ohio, a Babcock & Wilcox (B&W) reactor similar in design to the one at Three Mile Island. Both involved the same failed open pressurizer relief valve, and the same misleading indications to operators that the reactor coolant system was full of water. In both cases, operators diagnosed and solved the problem in a matter of minutes before serious damage could be done. The NRC never learned about the incident at the Beznau reactor until after the TMI-2 accident, because Westinghouse was not required to report to the NRC such occurrences at foreign reactors."*[11]

In short, all the key organizations accountable for the safe operation and regulation of TMI-2 played their part in the accident, but none of them understood the whole picture of the risks involved in running a pressurized-water reactor at those times: no one fully grasped what could develop during a multi-failure hardware malfunction, under the control of staff who had not been trained for such failures.

After the accident, misreading of instruments and poor preparation of staff for interpreting abnormal measures led to mistakes by the operators at TMI-2. The resulting misjudgments of the status of TMI-2 led the operators and management to send misleading information to their supervisors at Met Ed and its parent company General Public Utility—who in their turn informed the NRC, the designers of the plant, federal, state and local government representatives and the general public about the unimportance of the accident.[9,15]

The misinformation about the real condition of TMI-2 led to inadequate crisis responses. The NRC concluded: *"In sum, . . .the evidence failed to establish that Met Ed management or other personnel willfully withheld information from the NRC. There is no question that plant information conveyed from the control room to offsite organizations throughout the day was incomplete, in some instances delayed, and often colored by individual interpretations of plant status. . . Lack of understanding*

---

[10]Dmitry Chernov and Didier Sornette, Man-made catastrophes and risk information concealment (25 case studies of major disasters and human fallibility), Springer; 1st ed. (2016).

[11]Three Mile Island: Report to the Commissioners and to the Public, M. Rogovin and G. Frampton, U.S. Nuclear Regulatory Commission, January 1980, Volume I.

also affected the public's perception of the accident because early reports indicated things were well in hand, but later reports indicated they were not. [Only on the third day after the accident started], when the continuing problems were generally recognized, the utility management and staff began effective action to obtain assistance, plan for contingencies, and direct daily plant operations to eliminate the hazards. The recovery effort was massive, involving hundreds of people and many organizations".[15] Thus, it was only on the third day that General Public Utility executives began to ask for scientific and operational assistance from other utilities, reactor manufacturers, firms of architects and engineers, and national nuclear laboratories.[15] In its turn, the President's commission declared that the NRC was not ready to conduct adequate response measures in such a situation: "*On the first day of the accident, there was an attempt by the utility to minimize its significance, in spite of substantial evidence that it was serious. Later that week, NRC was the source of exaggerated stories. Due to misinformation, and in one case (the hydrogen bubble) through the commission of scientific errors, official sources would make statements about radiation already released... The response to the emergency was dominated by an atmosphere of almost total confusion. There was lack of communication at all levels... The fact that too many individuals and organizations were not aware of the dimensions of serious accidents at nuclear power plants accounts for a great deal of the lack of preparedness and the poor quality of the response... Communications were so poor [more than 48 hours from the accident] that the senior management could not and did not develop a clear understanding of conditions at the site [due to lack of emergency measurements and instrumentation]. As a result, an evacuation was recommended to the state by the NRC senior staff on the basis of fragmentary and partially erroneous information... The President asked us to investigate whether the public's right to information during the emergency was well served. Our conclusion is again in the negative*".[9] The situation was aggravated by the fact that many decision makers were informed about the accident not by Met Ed managers or emergency agencies, but by media news representatives.*

The NRC concluded that: "*The TMI accident was a first of a kind for the nuclear power industry. Neither the utility nor the NRC was prepared to cope with the public's need for information. As a result, the residents around TMI were unduly confused and alarmed, and the level of anxiety nationwide about the safety of nuclear plants was unnecessarily raised. The information Met Ed and NRC provided to the news media during the course of the TMI accident was often inaccurate, incomplete, overly optimistic, or ultraconservative. Errors in judgment by Met Ed and NRC officials were major contributors to the inadequate public information effort at TMI... At the same time, the NRC failed to coordinate its internal flow of public information, resulting in speculative reports from Washington which conflicted with statements made by NRC to officials in Harrisburg. The NRC made the problem of conflicting reports even worse by refusing to participate in joint press conferences with the utility. The State's public information effort, which relied almost entirely on information from Met Ed and later the NRC, suffered accordingly. While both the public information performance of Met Ed and the NRC can be faulted in many instances, we found no evidence that officials from either the utility or the regulatory agency willfully provided false information to the press or public*.[15] According to a White House representative *"many conflicting statements about TMI-2 reported by the news media were increasing public anxiety"*.[9]

3.4.2   Chernobyl

On 26 April 1986, the Chernobyl disaster struck the No. 4 light water graphite moderated (RBMK) reactor at the site near the city of Pripyat of the Ukrainian Soviet Socialist Republic of the Soviet Union. The RBMK is an early Generation II reactor and the oldest commercial reactor design still in operation in Russia. It is rather unique in lacking containment structure. After the Chernobyl accident, some of the 11 RBMK reactors still in operation in Russia were retrofitted with a partial containment structure (and not a full containment building), in order to capture any radioactive particles released.

A detailed investigation of the Chernobyl case and of the causes of the disaster at unit 4 has revealed a deeply ingrained practice of concealment of the design mistakes made over previous decades on the class of RBMK reactors operated at Chernobyl.[12] The apparently careless actions of the plant staff, which were the proximate causes of the disaster, should be put in the broader context of personnel operating a highly dangerous object without actually understanding the whole picture of risks. Indeed, different actors withheld information on the design problems as well as on the existence of previous incidents, accidents and near-misses plaguing these RBMK reactors. The reconstruction of the history leading to Chernobyl's disaster leads one to conclude that the Chernobyl disaster was literally programmed to occur as a result of the Soviet civil nuclear energy organization planting the seeds of an inevitable disaster.[8] As a result, the inadequate actions of the Politburo during the Chernobyl disaster were deeply connected with the poor transmission of reliable information to subordinates about the condition of the system, not only after the disaster but also long before it happened.

In the decades preceding the Chernobyl accident, the Politburo created a "rush culture" in order to increase the speed of construction of nuclear power plants to meet urgent domestic energy needs, which was also fed by the nationalistic arrogance of Soviet nuclear executives, and their over-confidence in the infallibility of Soviet nuclear technology. In this climate, representatives of the Soviet civil nuclear industry were not prepared to consider even the remote possibility that a serious disaster could happen. As a result, the Politburo allowed the transfer of the control of nuclear power stations to the civil Ministry of Energy and Electrification, which was unprepared for such a compli-cated task. The Kurchatov institute, NIKIET, Minsredmash and the Ministry of Energy and Electrification focused only on their narrow departmental interests, which

---

[12]RBMK Reactors, Appendix to Nuclear Power Reactors, Updated June 2010, http://www.world-nuclear.org/info/Nuclear-Fuel-Cycle/Power-Reactors/Appendices/RBMK-Reactors/

prevented timely and adequate communication of risk information between different agencies. There was also a general environment of national security secrecy: before the accident, operators at the plant did not receive any information about the accidents that had occurred previously at other Soviet NPPs, or about international nuclear accidents. Finally, it was common practice among Soviet bureaucrats to present themselves to superiors in the best possible light, which created an organizational culture of "Success at Any Price" and "No Bad News" within the industry.

One of the most important causes of the aggravated severity of Chernobyl disaster was the major information distortion of the real severity of the accident at different levels of the Soviet hierarchy during the first days following the explosion of one of the nuclear power plant cores.[10] This led to an inadequate crisis response that magnified the severity and adverse consequences of the accident. The event became arguably one of the triggers of the collapse of the USSR,[13] as a result of the destruction of common people's faith in the ability of the Politburo to run the country adequately and fairly, since its behavior seemed to contradict the Glasnost initiative (literally "publicity"), a policy that called for increased openness and transparency in government institutions and activities in the Soviet Union.

3.4.3   Fukushima-Daiichi

On March 11 2011, a seaquake of magnitude 9.0 on the Richter scale occurred 70 km from the east coast of the Tōhoku region in Japan. The earthquake generated a large-scale tsunami, which flooded the Fukushima-Daiichi nuclear plant owned by Tokyo Electric Power Co. (TEPCO), leading to core melt accidents in three out of the six units of the site.

After interviewing 1167 people and organizing 900 hours of hearings, the Fukushima Nuclear Accident Independent Investigation Commission, established by the Japanese National Diet, concluded that *"the accident at the Fukushima Daiichi Nuclear Power Plant cannot be regarded as a natural disaster. It was a profoundly manmade disaster—that could and should have been foreseen and prevented. The accident was clearly "manmade". We believe that the root causes were the organizational and regulatory systems that supported faulty rationales for decisions and actions, rather than issues relating to the competency of any specific individual. We found an organization-driven mindset that prioritized benefits to the organization at the expense of the public".*[14] This conclusion rested above all on the fact that, prior to the disaster, several experts had warned that the Fukushima-Daiichi

---

[13]Mikhail Gorbachev, Turning Point at Chernobyl, Project Syndicate, April 14, 2006, http://www.project-syndicate.org/commentary/turning-point-at-chernobyl

[14]The official report of The Fukushima Nuclear Accident Independent Investigation Commission, The National Diet of Japan, Executive summary, July 5, 2012, p. 21.

plant was not adequately prepared for a high-wave tsunami, but the TEPCO hierarchy had failed to pass on these warnings.[10]

The organizational causes of the disaster can be found in the distinctive position of the nuclear industry within the Japanese economy and the misplaced loyalty of regulators concerning shortcomings in the design and operation of Japanese nuclear power plants, which allowed plant operators to neglect basic safety rules and conceal the occurrence of many safety violations from regulators and the public with impunity.[10] One should note the existence of a kind of national arrogance of both executives and regulators in the Japanese nuclear community, who refused to learn from the experience of other countries that had faced nuclear accidents, or to implement IAEA's recommendations and advanced safety requirements. Instead, they preferred to rely on their supposed technical superiority over the rest of the world. They assumed that falsifying data about minor equipment faults would never lead to catastrophic results and that the Japanese attitude toward work would always compensate for minor imperfections in reactor design during natural disasters. Similar to the previous cases, there was also the presence of a habituation and over-confidence among representatives of the Japanese nuclear industry concerning the extremely low probability of a severe nuclear accident caused by a tsunami. One should also mention TEPCO's focus on the short-term profitability of operations and on ongoing cost reduction, which provoked an unwillingness among executives to reveal the risks of its nuclear power plants—whether to IAEA specialists, representatives of local authorities or emergency services, investors or local residents—because this would entail additional expenses on advanced safety measures. Finally, the accident was managed particularly badly (the Prime Minister learned about the accident from TV channels), as it underscored the political struggle between the Democratic Party and the Liberal Democratic Party, which generated massive distortion of information about the real condition of the plant after the disaster. Both parties used the accident in their own political interests. Again, we refer to Ref. [8] for a detailed account.

3.5   Diagnostics and Recommendations

These three cases illustrate the severe mistakes that organizations have made in the past. As we will document in Chap. 4, the data indicates that the nuclear industry and its regulators have learned many lessons. But one should not be complacent and it is essential to affirm the role of international organizations worldwide to provide assurances and verification for the safe use of nuclear energy, as much as politically possible. This is very important because of the technical complexity and hazardous character of nuclear technology. International organizations and cooperation of transnational actors form a body of control, qualified to release information about safety standards, protocols and targets.

However, the reach of organizations and cooperation worldwide only goes as far as member countries are willing, as shown by the previous section, and the services

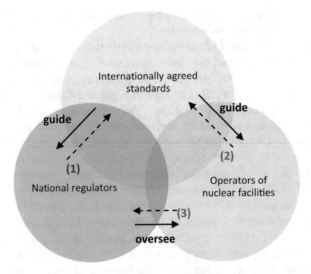

Fig. 3.3 Who controls the safe use of nuclear energy? Bold arrows: The operators of nuclear facilities are overseen by national regulators; both operators and national regulators are guided by the internationally agreed standards. Dashed arrows: Relationships that are non-binding and may be inadequately implemented: (1) National regulators to standards (e.g., member state requiring assistance or requesting peer review); (2) Operators to standards (e.g., newcomer state requiring help for safety implementation); and (3) Operators to national regulators (e.g., feedback loop so that national regulators are aware of all operators' activities, including events of safety and operational relevance)

offered (e.g. IAEA or WENRA peer reviews) are often not used due to their voluntary-basis. In the cases when services are used, it is often difficult to ascertain whether the operators or regulators implement the recommendations. Moreover, national regulators at the next position on the hierarchy are typically not aware of the inconsistencies between different implementation levels.

Figure 3.3 shows the actors involved around nuclear safety. Additionally, the figure above assumes that a strong national regulatory body is present in all countries using nuclear energy, which is not always the case (see later point of mandatory assistance to newcomer states).

Although the tools developed by international organizations seem to be the right ones, there are gaps in their adequate implementation, as became obvious in the case of Japan and TEPCO, the Fukushima plant operator. International safety cannot effectively rely on self-assessment because assessments are subject to interpretation and liberties can be taken. The non-binding nature of current control relationships does not seem sufficient to ensure a closed loop of information between all three bodies of control (standards, national regulation and operators). Additionally, neither an international nuclear liability framework nor sanctions, in case of non-compliance, exist today.[15] Operators are liable for the safety of their plants but, considering the potential damage associated with severe nuclear accidents, this is not enough. Further, the role and liability of the reactor suppliers should be rethought. Risk communication and perception appear not be adequately dealt with either. This impacts negatively the efforts targeting the establishment of a nuclear safety culture.

The following points[16] summarise where improvements to the critiques detailed above could be made:

- Binding agreement for adherence to rigorous safety standards and strengthened enforcement mechanisms;
- Operators and regulators being subject to international peer reviews, not just on request;
- Mandatory assistance to newcomer states (building on INIR (Integrated Nuclear Infrastructure Review) to assess national developments—e.g. UAE, Turkey, Belarus, and Poland);
- Fuel management: a more integrated network between countries for disposal facilities sharing and related transports.

Suggesting nuclear energy as a public good (and also as a potentially public global bad), the application of Ostrom's principles are presented in Table 3.5. Principles 1, 3 and 8 are present in all three international organizations. The IAEA is the overarching internationally agreed standard and principle 8 relates to the earlier Fig. 3.2. Principles 4 (monitoring users) and 6 (conflict resolution mechanisms) are rather unclear for the IAEA, NEA and EU Commission. This is in agreement with the previous conclusion that, although the tools seem to be adequate, there are gaps in their suitable implementation. Principle 4 recalls the idea of a closed loop of information—i.e. feedback from operators about nuclear reactors, provision of nuclear materials and safety issues. An inadequate supervision/feedback loop refers to an inadequate implementation of the control tool. In a nutshell, the IAEA plays a central role in the governance of nuclear energy. However, it provides revisions and reviews of best practice with however almost no enforcement mechanisms. It should thus be further strengthened.

---

[15]http://www.world-nuclear.org/information-library/safety-and-security/safety-of-plants/liability-for-nuclear-damage.aspx. The incredibly high level of performance in the aviation sector can be explained by the strict international safety regulations with immediate international sanction in case of non-compliance, namely through a denial of takeoff and landing of flights.

[16]Bunn, M. and M.B. Malin, *Enabling a Nuclear Revival—And Managing Its Risks*. Innovations: Technology, Governance, Globalization **4**(4), 173–191 (2009).

Table 3.5 Assessment of international organization against Ostrom's principles

	IAEA	OECD-NEA	EU Commission
1. User and resource boundaries	Legal framework for IAEA safeguards (no nuclear material used in disagreement with IAEA safeguards)	Regulatory framework differs per country	Safeguards: Reporting and verification on flows, processes and stocks of nuclear material; Inspection by commission officers
3. Collective choice arrangements	Yes, standards are developed on the basis of experience gained from the users	Nuclear law committee is in charge of updating documents based on current conventions	Consultations seeking stakeholders' and citizens' views on certain policies
4. Monitoring users	To some extent (e.g. peer reviews)	Assistance role, not monitoring	Euratom Supply Agency (see Annex)
5. Graduated sanctions	Yes	No, more of an assisting/helping role	Yes
6. Conflict resolution mechanisms	No, no competent court to resolve dispute between IAEA and a state	No, dependent on country's law	Yes, done by taking "appropriate steps", however no systematic description of the steps
8. Nested enterprises	Yes IAEA > member states > users in member states	Not applicable	Not applicable

Principles 2 and 7 are omitted because they are not directly applicable in the context of nuclear energy management, as presented in Table 3.1

Annexes

Annex 1: Non-Proliferation Treaty (NPT) of Nuclear Weapons

The only binding commitment on the use of nuclear energy is the NPT. With 190 state parties, it is the most widely adhered-to multilateral disarmament treaty.[17] The NPT came into force in 1970 and, since then, has the objective to "prevent the spread of nuclear weapons [...], to promote cooperation in the peaceful uses of nuclear energy and to further the goal of achieving disarmament [...].[18] States ratifying or acceding the NPT undertake not to transfer source material or equipment for nuclear fission to any other states. Safeguards to the NPT are implemented by the International Atomic Energy Agency.

The NPT was renewed in 2015 for an indefinite period. In addition, the 2015 Review Conference of the Parties decided on a measure for the review of the operation of the Treaty every 5 years. Not all nuclear-active countries have ratified

---

[17]2015 NPT Review Conference, http://www.un.org/en/conf/npt/2015/

[18]Treaty on the Non-Proliferation of Nuclear Weapons (NPT), United Nations Office for Disarmament Affairs, http://www.un.org/disarmament/WMD/Nuclear/NPT.shtml

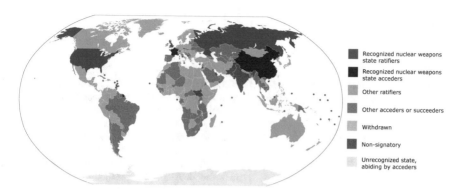

Fig. 3.4 List of parties to the Nuclear Non-Proliferation Treaty (NPT) (NPT Participation file, https://en.wikipedia.org/wiki/List_of_parties_to_the_Treaty_on_the_Non-Proliferation_of_Nuclear_Weapons)

the NPT. North Korea withdrew in 1993 but then suspended its withdrawal notice. It finally announced in 2003 that is was ending the suspension of its NPT withdrawal notification. The four UN member states that have never ratified the treaty are India, Israel, Pakistan and South Sudan. The world map below shows "ratifiers" and "acceders" to the NPT. The five recognized nuclear-weapon states (shown in blue) are the US, Russia, the UK, France and China (Fig. 3.4).

Annex 2: EURATOM (European Atomic Energy Community)

The EURATOM treaty ensures a regular and equitable supply of nuclear fuels to EU users. The Euratom Supply Agency (ESA) applies a supply policy based on the principle of equal access of all users to ores and nuclear fuel. The ESA has exclusive right to conclude contracts for the supply of ores, source materials and special fissile materials (Article 52) within the EU. The ESA has an advisory committee responsible for linking ESA and users/producers. The Committee members are appointed by the EU Member States on the basis of their degree of relevant experience and expertise in nuclear issues.[19]

---

[19]http://europa.eu/eu-law/decisionmaking/treaties/pdf/consolidated_version_of_the_treaty_establishing_the_european_atomic_energy_community/consolidated_version_of_the_treaty_establishing_the_european_atomic_energy_community_en.pdf (EURATOM consolidated version of treaty, 2010); http://eur-lex.europa.eu/legal-content/EN/ALL/?uri=celex%3A32012R0966 (2012)

Table 3.6 Governmental organizations

Factors	IAEA	OECD-NEA	European Commission Nuclear Safety Directive
Mission statement	"Help countries benefit from peaceful, safe and secure use of nuclear science and technology and prevent the spread of nuclear weapons" (1) Peaceful uses of nuclear technology (2) Safety and Security (3) Non-proliferation	OECD: "Promote policies that will improve the economic and social well-being of people around the world"; NEA: "Assist member countries in maintaining and further developing, through international cooperation, the scientific, technological and legal bases required for a safe, environmentally friendly and economical use of nuclear energy for peaceful purposes"	For "EU countries to give highest priority to nuclear safety at all stages of the lifecycle of a nuclear power plant" (1) Strengthen the role of national regulatory authorities by ensuring their independence from national governments (2) Create system of peer reviews (3) Safety re-evaluation of all NPP every 10 years (4) Increase transparency from NPP to public
Budget	EUR 344 million (of which 38% on nuclear verification)	EUR 11.1 million (2014)	EUR 9 million
Member states (MS)	168 as of Feb. 2016	31 (accounting for 86% of world's nuclear capacity)	28 EU states
Organization /Hierarchy	Primary blocks: 1. Nuclear energy technology, including fuel cycle 2. Nuclear safety and security 3. Nuclear sciences for human health, environment, and agriculture 4. Safeguard concepts and planning	Prioritized sectors: 1. Nuclear safety and regulation 2. Radioactive waste management 3. Radiological protection and public health 4. Nuclear science 5. Development and use of nuclear energy 6. Legal affairs	Part of ENSREG (European Nuclear Safety Regulators Group)
Means	– TC (Technical Cooperation) Program – Resolutions are finalized at IAEA General Conference – Secretariat is trusted to implement the actions in a prioritized manner within available resources – Secretariat encourages ongoing initiatives of MS and assists with	Committees: – CSNI (Safety of Nuclear Installations): forum for the exchange of technical information and for collaboration between organizations; working groups meet once or twice p.a. – CNRA (Nuclear Regulatory Activities): responsible for the regulation, licensing and	Not available

(continued)

Table 3.6 (continued)

Factors	IAEA	OECD-NEA	European Commission Nuclear Safety Directive
	guidance, training and assessment tools	inspection of nuclear installations with regard to safety; meets twice p. a., to exchange information on events with regulatory significance, evolution of regulatory requirements, and regulatory measures under consideration	
Legal demands	131/162 MS (33 least developed countries) are participating in the TC programs; MS must fulfill their obligations under UN Security Council resolution 1540 (i.e. proliferation of nuclear, chemical and biological weapons and means of delivery constitutes a threat to international peace and security)	None	Amended Directive from 2014 is to be transposed into Member States' legislation by 2017; (Directive = legislative act setting goal that EU countries must achieve; up to countries to devise their own laws to reach these goals)
Sanctions	IAEA first addresses States (not operators), then, failure by a State to take fully corrective action w.r.t non-compliance results in curtailment, suspension of assistance or membership		First warning Second withdrawal of assistance Third placing of the concerned member state under the ad-ministration of a board Last total/partial withdrawal of source materials

Table 3.7 Cooperations

	WENRA	WANO	IRIS (IAEA methodology)
Mission statement	"Become a network of chief nuclear safety regulators in Europe exchanging experience and discussing significant safety issues" (2003)	"Maximize the safety and reliability of nuclear power plants worldwide by working together to assess, benchmark and improve performance through mutual support, exchange of information and emulation of best practices"	"Support Member states in establishing their safety infrastructure"; "provide guidance on the establishment of a framework for safety in accordance with the IAEA safety standards"
Member states	17 countries represented	130 operators Three categories of membership: 1. Members having voting rights in general meetings 2. Members who are represented by another member within WANO 3. Members with ownership stake in a nuclear facility but don't officially represent the operating company	IAEA member states
Organization/ Hierarchy	By-lateral cooperation between countries represented in WENRA	Atlanta, Moscow, Paris, and Tokyo regional governing boards report to their respective regional centres, all being responsible to the main governing board and general assembly in London.	IRIS is a self-assessment methodology
Means	Three working groups to harmonize safety approaches between countries in Europe: – RHWG (Reactor Harmonization Working Group) – WGWD (Working Group on Waste and Decommissions) – WIG (WENRA Inspection Group) Working groups aim to compare individual national regulatory approaches with the IAEA Safety Standards	Five programs: 1. Peer reviews 2. Operating experience 3. Technical support and exchange 4. Professional and technical development 5. Communications (share WANO's mission and activities)	Framework for self-assessment of national infrastructure for safety

(continued)

Table 3.7 (continued)

	WENRA	WANO	IRIS (IAEA methodology)
Legal demands	European Nuclear Safety Directive requires (1) license holder of has prime responsibility for his nuclear installation, (2) Member states must establish and maintain a competent regulatory body in the field of nuclear safety	"All members accept their individual responsibility for the nuclear power plant they operate and accept their collective responsibility to assess, inform, help and emulate other nuclear operators" (WANO Charter)	Safety standards are used as "a reference for their national regulations in respect of facilities and activities". "Safety standards are binding on States in relation to IAEA assisted operations" "The way in which IAEA safety standards are applied to existing facilities is a decision for individual States"
Support to members	Two different regulator approaches in countries: (1) prescriptive, i.e. regulator establishes detailed requirements (2) non-prescriptive, goal setting, i.e. regulator establishes specific goals or outcomes for licensees to attain		(1) Assessing current situation and progress realized to build safety infrastructures (2) Identify gaps between current situation and expected status of national safety framework (3) Encouraging States to take appropriate actions to reach compliance with IAEA Safety Standards

Chapter 4
Risk in Nuclear Power Operation

Abstract Nuclear accidents are rare but costly and a common, but wrongly, held opinion is that, if the external cost of nuclear accidents were included in the price, then nuclear power would become economically unfeasible. Probabilistic Safety Analysis (PSA) is the standard method used to identify potential accident scenarios, estimate their probability, and their consequences with a focus on plant states as well as impact on health, the environmental, and certain direct costs. This methodology has developed substantially and is now used for regulatory purposes but is not without defects, notably when quantifying overall risk (level 3) and is best complemented and compared with statistical experience. More broadly, impacts of accidental releases are highly uncertain due to the difficulty of realistically estimating the effect of low doses of radiation to human health and mortality.

Here, the severity of nuclear accidents, quantified on an approximate integral cost basis and relying on the (conservative) Linear No Threshold (LNT) approach to estimate latent fatalities, is balanced with their rare probability, estimated based on accident precursor statistics. This, as well as more comprehensive sustainability assessments, indicate that the external cost of nuclear accidents is not worse than modern renewables, and far better than carbon-based energy. Further, a meaningful reduction in risk, and convergence of PSA estimates to match experience, is documented—especially in the US and North and Western Europe. However, the sheer potential for major accidents remains, leaving the pragmatic door open to seriously explore technologies with radically less severe potential consequences, as discussed in Chap. 6.

4.1 Notions of Risk

The consequences of a major nuclear accident can be severe and diverse, possibly causing high numbers of premature deaths due to cancer, and with "overall cost" projected to be on the order of hundreds of Billions of US dollars. Such losses are difficult to accept and should therefore be reduced to the lowest practicable level. This has long been argued at the highest levels of decision-making. However, decisions cannot be made solely on the basis of severe and worst-case imaginable

© Springer Nature Switzerland AG 2019
D. Sornette et al., *New Ways and Needs for Exploiting Nuclear Energy*,
https://doi.org/10.1007/978-3-319-97652-5_4

consequences alone. Probability must also be considered. More specifically, a proper measure of risk is required to provide information about potential health and environmental impacts, and to support a sufficiently complete accounting of costs, as well as comparison with other energy sources. In particular, risk concerns harmful events having two stochastic quantities:

1. *incidence (frequency):* the number of events per unit time, and
2. *severity:* the consequences of these events.

A full characterization of risk requires the probability distribution for these two quantities, as well as their uncertainty arising from limited experience, statistics or contradictory information. As practiced in PSA level 3, severity can be separated into different damage categories such as human health, environmental impacts, and standard economic costs. This is useful, allowing for a range of weightings of categories—capturing uncertainties and ideological differences—to provide an ultimate quantity to support a decision. See Sect. 5.3 for discussion of the cost of severe accidents.

A limited but objective and useful single quantity to summarize the risk is *the full expected cost per unit of time*, being the product of the event frequency and mean event cost. Further, when comparing with other energy sources, it can be converted to expected cost per unit of energy produced. This measure omits information about variation, but for a sufficiently large society and over a long-time period, it is the most relevant for overall cost minimization.[1] Moreover, to be fair and optimal, we must consistently balance the benefits with the negative consequences (both certain and stochastic) of all energy sources—and more broadly within the overall technological development and growth of our society.[2]

Box 4.1 Does an Economic Value for Human Life Really Exist?
Putting on ones' economist hat, price is defined at the level at which supply and demand equilibrate, or in the absence of a market, what level relevant agents are willing to pay (WTP) to have or willing to be compensated to accept (WTA). In a major technological (nuclear or other) accident, diverse consequences exist. Loss of capital, income and electricity produced have market values. Other consequences such as human suffering and death are more difficult to quantify in monetary terms. Many like to think of human life as

(continued)

---

[1]Because of the Law of Large Numbers, the average cost converges to the mathematical expected cost. Even when the distribution of costs is so heavy-tailed that the mean value would be mathematically not defined (or infinite), one can use extensions of the law of large numbers to deal with this extreme situation. See Sect. 4.6 below and, e.g., D. Sornette, Critical Phenomena in Natural Sciences, Chaos, Fractals, Self-organization and Disorder: Concepts and Tools, 2nd ed. (Springer Series in Synergetics, Heidelberg, 2006).

[2]E.g., incidents involving elevators and escalators kill about 30 and injure about 17,000 people each year in the United States. See also Sect. 4.4.3.

Box 4.1 (continued)

sacred and infinitely valuable. However, in the presence of finite resources to protect and save lives, the value for an individual life or suffering is finite, as we balance resources in a way to maximize the overall (quality/utility of the) situation, and also consider balancing the sacrifice of one life to save other lives[3,4] The large literature on WTP and WTA approaches to value such things confirms that values are finite (and sometimes surprisingly small). The approach is appealing insofar as it defines the value in a way that is consistent with individual decisions—e.g., taking the average as the value—and only violates/inconveniences individuals—forcing them to pay more, or receive less than they would prefer to the extent that there is variation within the population. Studies have emphasized the massive range in efficiency of cost to save lives, with a range of over 10 orders of magnitude in the cost per life saved.[5] Thus, the implied value is not always consistent. Aside from irrational decision-making, one may be willing to pay more for an individual—e.g., in a ransom negotiation—but extrapolating that quantity to a large population becomes unsustainable.

The value of a human life, even if highly uncertain, is a crucial part of rational decision-making and of both optimal and equitable distribution of limited resources (see Box 4.1). Taking hypothetical numbers, consider a per-reactor frequency of one large release of radioactive materials causing 10,000 latent fatalities (according to the LNT)[6] every hundred thousand years. For 400 reactors, such a large release on average happens every 250 years, and would thus shorten the lives of 40 people per year. A recent publication (see Fig. 4.3) has put this number at less than 2.5 premature deaths/year for the approximately 2500 TWh of annual output of the global nuclear fleet (i.e. less than 10^{-6} deaths/GWh) for Generation II units, and

---

[3]Everett, J.A., Pizarro D.A. and Crockett M.J., Inference of trustworthiness from intuitive moral judgments, J. Exp. Psychol. Gen. 145 (6). 772–787 (2016).

[4]That is, placing too much value on one individual would limit resources available to others. In human terms, infinite care for one individual would exclude care for all others.

[5]Tengs, T. O., et al., Five-hundred life-saving interventions and their cost-effectiveness, Risk analysis 15.3 (1995): 369–390.

[6]The LNT model was introduced by the International Commission on Radiological Protection (ICRP) (1977, 1991, 2007) as a practical system for radiological protection, especially in planning situations. In order to assess risks, one should not use collective dose in the "whole", but build up classes of collective dose with different dose ranges, consider organ doses and not "effective dose" and so on (Prof. Christian Streffer, private communication). Indeed, the ICRP (2007, para 66) writes explicitly: "Because of this uncertainty on health effects at low doses, the Commission judges that it is not appropriate, for the purposes of public health planning, to calculate the hypothetical number of cases of cancer or heritable disease that might be associated with very small radiation doses received by large numbers of people over very long periods of time". For long term prognosis of 100,000 years, see Streffer et al. Radioactive Waste—Technical and Normative Aspects of its Disposal. Springer-Verlag Berlin-Heidelberg-New York, 2011.

more than 100 times less for the Generation III EPR. Based on a rigorous and comprehensive analysis (see Sect. 4.2), these numbers are tiny when compared with deaths due to burning coal, and on a similar level to "new renewables", where low fatality events for wind power are well documented, while solar-panel fabrication involves numerous explosive and toxic chemicals, and rare high-fatality failures of hydro-electric facilities have been experienced. Note that fatalities associated with new renewables are real—i.e., persons have died—while the fatality statistics associated to nuclear power accidents are dominantly based on a hypothetical small shortening of life across a large group of people.

Despite this, the risk of nuclear accidents occurring is viewed by the public with a unique stigma. The risk perception is a complicated issue that we do not attempt to review here, but partly discuss in Sect. 5.5. However, clear and simple factors are the presence of aversion to risk and ambiguity (see Box 4.1), as well as the special dread of ionizing radiation (see Sect. 5.5). This skews the comparison with energy sources—with the preference for many "silent deaths" by burning coal, for instance, rather than few caused by radiation. In particular, a special dread of radiation needs to be invoked when one considers that the risk of carbon-emissions-driven climate change is potentially riskier and more ambiguous than nuclear accidents.

A variety of risk types exist within nuclear energy,[7] but the dominant one, which we will focus on here, is the risk of an accident at a commercial nuclear power facility, i.e., a severe core damage (CD) event with the potential of a large early release (LER) of radioactive material, and the associated probabilities. For this, we first introduce the standard methods to quantify the core damage probability (CDP) and large releases and then compare these predictions with real statistics.

Box 4.2 Risk, Uncertainty and Ambiguity Aversions

Societies' planning horizons, and lifetimes are finite; and subjective preferences can be justified. These can lead to the consideration of other summary risk measures, beyond the simple expected cost, which "penalize" variation. In particular, stochasticity, the pure probabilistic variation strongly present for rare events, and ambiguity, due to limited or conflicting information, are important as they relate to three well-documented tendencies of human decision-making:

1. Risk aversion is the preference for outcomes that are certain compared to alternatives that may occur even with a high probability (but are not fully

(continued)

---

[7]Other lower-order risks include: non-core-related incidents resulting in injury and irradiation of employees, small releases not related to a core damage (CD) state, accidents in (small) experimental and demo plants, and accidents in the back-end of the fuel cycle. Quite uncertain, but probably relatively minor are the loss and misuse of radioactive materials, and failure in the long-term storage of nuclear waste. It is also important to consider non-safety-relevant risks, such as cost overruns in construction, and business disruption/downtime.

Box 4.2 (continued)

certain). Maurice Allais famously documented this preference in what is now known as the "Allais paradox".[8] Individuals often prefer and will choose a smaller gain that has a higher probability to materialize over a larger gain that has a lower probability, even if the later choice has a significantly large expected value.

2. Uncertainty aversion is the dislike of options of known possible amplitudes but unknown or unquantifiable probabilities. In general, one prefers to play gambles with known outcomes and known odds and avoid gambles for which the odds are not known.

3. Ambiguity aversion is the dislike for situations in which even the amplitudes of the events are poorly constrained or even unknown: not only the probabilities are hard or impossible to estimate but one does not know all the events that could occur and their amplitudes.[9]

Taking a long-term societal view, we should not be too averse to these things, as e.g., any risk aversion will, in the long run, result in a higher total cost of risky events. This is because we will take alternatives with known (or estimated) lower and more certain costs, but with higher frequency that could more than compensate the reduction in cost.

4.2   The Standard Analytical Framework: Probabilistic Safety Assessment (PSA)

4.2.1   Key Methods and Results

The framework of Probabilistic Safety Assessment (PSA) has been developed to systematically analyze the behavior of the reactor and its safety features as a complete system and to provide plant-specific risk information at three sequential levels of end states (Fig. 4.1). It is the primary framework for studying and regulating safety within the nuclear community.

PSA Level 1 corresponds to the assessment of the risk of core damage accidents, being quantified by the core damage frequency (CDF), typically per reactor-year. PSA Level 2 characterizes the release of radioactive substances from the reactor

---

[8]Allais, M., Le comportement de l'homme rationnel devant le risque: critique des postulats et axiomes de l'école Américaine, Econometrica 21 (4), 503–546 (1953).

[9]Example: Say there exists a debate on the health effects from exposure to small amounts of radiation. One party claims that the risk is low—or at least acceptable—and the other party claims that the risk is alarmingly high. Assume that both parties are perceived as having some authority, and that the general public is unwilling or unable to resolve these claims on its own. The result of this will be two sources of public pessimism towards nuclear power: risk aversion due to the existence of a claim of high risk, and ambiguity aversion due to the existence of conflicting claims.

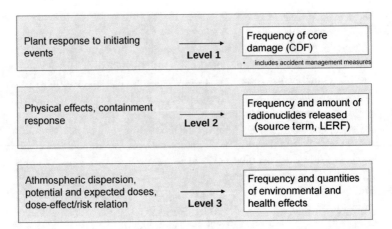

Fig. 4.1 Structure and sequential levels of Probabilistic Safety Assessment (PSA) for Nuclear Power Plants

building to the environment in the event of an accident (full "source term" or at least large early release frequency, LERF), in order to develop accident management strategies and identify potential design weaknesses in the containment design and performance. CDF and LERF are regarded as representative surrogates to steer the design and operations of systems towards a high safety level and achieving quantitative and qualitative safety goals. PSA Level 3 evaluates the impact of such releases on public health and the environment, as well as direct costs, and is used mainly for emergency planning.[10]

The methods to determine the CDF and LERF are theoretical: One must identify the multitude of initiating triggers, and event sequences, that can lead to the core damage or large release end-state, and assign probabilities to all of these chains in the absence of statistical experience at the system level, but with statistical information at the components level. In principle, all relevant internal and external triggers, as well as human and physical effects, should be built-in. A classical set of initiators and contributions to core damage probability are exemplified in Table 4.12.

In the nuclear domain, PSA levels 1 and 2 are required to support regulatory activities in most countries, e.g., in the USA since 1995 to complement the deterministic approach within the framework for risk informed regulation. For instance, in Switzerland, PSA is integrated into licensing and inspection processes aimed at assessing overall safety, requiring demonstration of compliance with target values,[11]

---

[10]W. Kröger, "Risk analyses and protection strategies for the operation of nuclear power plants," Chap. 2 in Alkan et al., Landolt-Börnstein: Numerical data and functional relationships in science and technology—new series, advanced materials and technologies, *Nuclear Energy*, pp.186–235: Springer (2005).

[11]Target Values, set by the Swiss Nuclear Ordinance of 10.12.2004, SR732.11: Core damage frequency due to internal and external initiators less than 10^{-5}/year for new plants, also for existing plants, if reasonably achievable; Hazards due to natural events such as earthquakes should be

and whether risk contributions and related safety measures and specifications are balanced[12] (see Ref. [8] for overview). PSA Level 3, providing a basis for the estimation of at least radiological consequences of an accident, is not required, and such studies have been done rarely and mainly by academics. This is seemingly in contradiction with the legal requirement that plants must be made as safe as reasonably achievable—which is regulated via cost-benefit analysis of safety investments (see Ref. [13] for regulatory analysis in the US).

Given the regulatory need to use PSA for risk assessment, the requirements regarding quality and validity of PSA are very high, calling for a continuous process building on past experience, use of huge data, and of advanced knowledge gained by safety research. PSAs produced by utilities need to be peer reviewed by the regulatory body for which guidance has been provided.[14]

The PSA methodology is complex and has evolved to have national and international guidelines[15,16,17]; over 200 PSAs have been conducted around the world. PSA studies are subject to periodic review, mostly as part of the Periodic Safety Review process,[18] or are even "alive", taking into account the most appropriate plant-specific data, new insights gained by experience and safety research, modified design and operational rules, and so on. As a general rule, a full-scope PSA needs to take into account internal and external events as well as normal operation and non-full power (shutdown) states. The system analysis part of PSA is mainly based on linked fault and event tree analysis (FTA/ETA), human reliability analysis (HRA) and dependent or common cause failure (CCF) analysis.

System codes, such as SAPHIRE 8 (System Analysis Programs for Hands-on Integrated Reliability Evaluations) and versions of MACCS (MELCOR Accident Consequence Code System) are widely established and used.

Current full-scope PSA comprehend a large set of initiating events and a huge number of logic trees[19] and are difficult for non-specialists to understand. Therefore,

---

analyzed site-specifically; and safety hazards with a frequency $\geq 10^{-4}$/reactor-year need to be considered. Note: PSA level 1 comprises accident scenarios leading to core damage within 24 h; PSA level 2 considers accident scenarios leading to loss of containment function within (at least) 48 h.

[12]ENSI Richtlinie ENSI-A06, PSA: Anwendungen.

[13]US NRC, Regulatory Analysis Guidelines of the US Nuclear Regulatory Commission, NUREG/BR-0058, Rev. 4 (2004)

[14]Richtlinie ENSI-A05, PSA: Qualität und Umfang, January 2009.

[15]IAEA, "Development and application of level 1 probabilistic safety assessment for nuclear power plants," *IAEA safety standards no. SSG-3, Specific Safety Guide* (2010).

[16]IAEA, "Development and application of level 2 probabilistic safety assessment for nuclear power plants," *IAEA safety standards no. SSG-4, Specific Safety Guide* (2010).

[17]IAEA, "Procedures for conducting probabilistic safety assessments of nuclear power plants (level 3), offsite consequences and estimation of risks to the public", *SSG-25, Specific Safety Guide* (2011).

[18]IAEA, "Periodic Safety Review for Nuclear Power Plants", *Safety Series* No 50-P-12 (1996).

[19]As an example, the PSA for the Swiss Leibstadt NPP comprises about 202 initiators, 2000 fault trees and 300 event trees.

presentation of results should be made as "transparent" as possible and include *inter alia* presentation of uncertainties, scope and coverage and omitted phenomena.[20]

Quantitative uncertainty analysis is restricted to data variation that assumes traditional (Gaussian) distributions. Fault Tree Analysis follows the principle of reductionism and assumes that system failure can be estimated by the sum of failures of its parts. The axiom system of Kolmogorov probability theory serves as the basis for quantification; the numerical effort is limited by rare event approximation and minimal cut-sets with cut-off values. Event Tree Analysis follows the principles of causality, arranges binary events induced by an initiating event in a chronological order and deals with conditional probabilities. Logic trees are semi-static and linear. Cut-offs and binning techniques are applied to cope with a large number of sequences. Classical Human Reliability Analysis incorporates unintentional errors of omission (not doing something you should have done) but not errors of commission (doing something that you should not have), and not the strong, highly dynamic interaction between the crew and the plant, especially under severe accident conditions. Special attention must be paid to the analysis of potential dependent failures as they can dominate the failure probability of highly redundant systems (see annexed Table 4.7 for more details).

PSA uses in general the assumption that a nuclear power plant is essentially a "closed" system, even under catastrophic accident conditions. PSA is mostly limited to single units, embedded in an ideal environment, in which sufficient safety culture is ensured and infrastructure is available for accident mitigation and recovery. Severe accident management guidelines (SAMG) are taken into account in the PSA framework as long as they are planned and trained. Regarding the consequence assessment (PSA level 3), atmospheric transport and dispersion is considered to dominate (thus neglecting other pathways like water and soil), weather conditions are assumed to be stable over critical times and emergency countermeasures are taken into account deterministically, as planned beforehand (without feedback on the "source-term").

Plant and site specific core damage frequencies (CDF) for light water reactors obtained by PSA vary between 10^{-4}/reactor-year and 10^{-5}/reactor-year, and go down close to 10^{-6}/reactor-year for new builds (CDF of 2×10^{-6}/reactor-year for EPR in Finland) or plants well retro-fitted (3.9×10^{-6}/reactor-year for the Leibstadt plant in Switzerland). The uncertainty of these estimates is often quoted to be about a factor of ten, although the uncertainty quantification is limited and seldom represented as a formal statistical confidence interval. Aside from design, the different levels of exposure to external hazards are dominant causes of the persistent variation in high quality full scope PSA. These PSA values are accepted by licensing authorities after thorough review. Compared with these general CDF values, plant-specific large early release frequencies (LERF), resulting from containment capabilities, are typically one order of magnitude lower, i.e. in the range 10^{-5} to 10^{-6}/reactor-year and lower. The CDF and LERF values have decreased remarkably over time and design generations

---

[20]IAEA Safety Series No.75-INSAG-6, PSA 1992

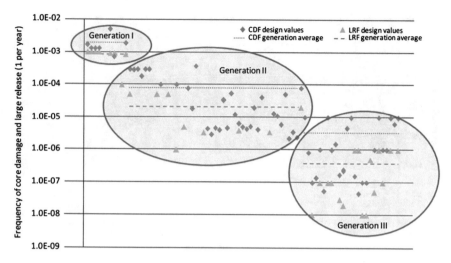

Fig. 4.2 Evolution of core damage frequency and large release frequency for reactor generations I, II and III. This data include both early generation reactors and new builds (TECDOC, IAEA, "Status of Advanced Light Water Reactor Designs", IAEA, Vienna, Austria (2004))

(see Fig. 4.2) due to safety improvements, of which PSA were an integral component; and for which some argue that intensified use of PSA was a driving force.

PSA level 3 studies make deterministic assumptions about population behavior, after threshold values have been reached, for taking actions such as alarming/sheltering, evacuation, relocation and ban on foodstuff. They take deterministic (early) and stochastic (late cancer effects) into account for submersion, inhalation and ingestion exposure pathways. They are site-specific but extend calculations of consequences to large distances (800 km) to capture low dose latent effects. Figure 4.3 depicts the results of a relatively new study[21] using generic, simplified techniques to model/estimate risks for advanced nuclear power plants operated at sites in five countries. While the number of early fatalities is small or even zero, the numbers shown for latent fatalities are extraordinarily high due to an assumed linear dose-risk relationship (see discussion in Sect. 4.3), albeit at an extremely low frequency level, i.e., 10,000 or even more latent fatalities at a cumulated frequency of 10^{-10} per GWe-year or even lower after a steep decrease.[22] These results, based

---

[21]C. Lecointe, D. Lecarpentier, V. Maupu, D. Le Boulch, and R. Richard, "Final report on technical data, costs and life cycle inventories of nuclear power plants," NEEDS deliverable D (2007).

[22]In the older "generation II" LWR, despite major methodological improvements since, similarly high fatalities have been predicted to be possible, but with uniformly higher frequency: In the original 1975 WASH-1400 study, about 2000–8000 fatalities would be expected from an accident with frequency 10^{-6}/reactor-year, and up to 10 times worse for an 10^{-9}/reactor-year "worst case" event. The controversial 1982 follow-up, NUREG/CR-2239, stated that, in the event of a major release, the mean fatalities would be on the order of 100–1000 (site depending), and there would be a 1/100 chance (i.e., with frequency 10^{-7}/reactor-year) of more than 1000–30,000 fatalities (site depending). The later 1990 NUREG-1150 study provided similar results for these figures, claiming

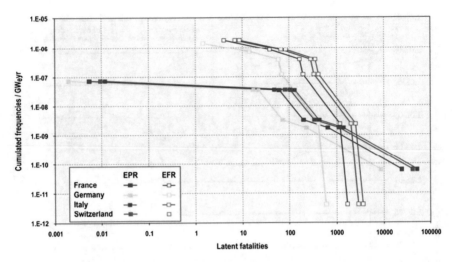

Fig. 4.3 Cumulated Frequency-Consequence Curve for latent fatalities for European Pressurized Reactor (EPR) and European Fast Breeder Reactor (EFR) (Burgherr, Peter, Stefan Hirschberg and Erik Cazzoli. Deliverable n D7. 1–RS 2b "Final report on quantification of risk indicators for sustainability assessment of future electricity supply options (2008), Project no: 502687 NEEDS New Energy Externalities Developments for Sustainability.) Note that the high numbers of latent fatalities (>1000) are associated with extremely low probabilities $<10^{-10}$ per GWeyr. Smaller numbers of latent fatalities (1–10) have higher probability but it is still a very low probability of ~10^{-6} per GWeyr. A 1.2 GW reactor will produce ~1 GWe/yr

on a simplified PSA level 3, should be viewed with some caution[23] and not interpreted as entirely realistic.

4.2.2   Criticism and Limitations

PSA—in particular levels 1 and 2—has been successful in identifying vulnerabilities (weaknesses) and possible improvements in plant safety through design and operation, as well as in evaluating the completeness and balance of the design for safety. Thus, PSA can be thought of as a plant-specific platform for technical exchanges on safety matters between regulators and the industry, among peers, and between designers and operators. However, PSA information is often used for decision making within the public and corporate sector, which need sound assessments of

---

on average thousands of fatalities for the 10^{-6}/reactor-year event, and tens of thousands for the 10^{-7} /reactor-year event.

[23]To put in perspective, the German strategy to stop their nuclear reactors following Fukushima led to a massive growth of their coal-based production. Nowadays, it is estimated that coal plants in Germany cause 2500 premature deaths each year in Europe due to particle pollution [62].

the full risk of nuclear power, often acknowledged to be broader in scope than PSA level 3. The bottom-up theoretical approach is not necessarily well suited for these needs, and will be contrasted with statistics in Sect. 4.5.

However, the use of PSA is subject to limitations, related to the quality of the methods, process and people involved in performing such investigations, introducing an element of subjectivity. Sufficient completeness is an issue often raised by critics, substantiated by operating experience: A review of precursors that occurred in US plants during 2001 to 2009 revealed that 30% of the identified incidents involved initiators or failure causes that were not explicitly modeled in the associated PSA plant model of the regulator.[24]

Based on theoretical considerations and lessons learned from past experience, notably "Fukushima", the following concerns must be kept in mind[25]:

- Uncertainties are an intrinsic part of PSA results despite all attempts to reduce them; completeness cannot be conceptually proven in a mathematical sense.
- Assuming single initiating events and developing linear causal event chains by standard application of Event Tree Analysis (in combination with Fault Tree Analysis) at single plant level cannot sufficiently account for the indirect, non-linear, and feedback relationships that characterize accident sequences in complicated and, at times, complex systems like a nuclear power plant.
- Restricting accident management measures to "planned, trained actions" may lead to unrealistic CDF (core damage frequency) and LERF (large early release frequency) estimates, as severe accident conditions may differ from what was assumed in the PSA; this calls for flexibility and computer-based "scenario generators".
- Unscrutinized application of simplifications such as rare event approximation, cut-sets with rigorous cut-off values, the normal distribution for parameter variability and so on, may lead to analytically incorrect quantification of fault trees; this includes common-cause or common mode failures (CDF/CMF)[26] that are particularly difficult to capture.
- Basic modelling of nuclear units as "closed systems", while they turn into "open systems" under severe accident conditions, intensively interacting with and depending on their potentially hostile environment, may lead to an unrealistic "big picture" and unrealistic overall PSA results.
- Isolated analysis of multiple units at one site may lead to questionable results as they may share common features/equipment and depend on shared resources and personnel.

---

[24]US Office of Regulatory Research, Status of the Accident Sequence Precursor Program and the Standardized Plant Analysis Risk Models, SECY-10-0125, 2010.

[25]W. Kröger and D. Sornette, Reflections on Limitations of Current PSA Methodology, Proc. PSA 2013, International Topical Meeting on Probabilistic Safety Assessment and Analysis, Columbia, South Carolina, USA, September 22–26, 2013.

[26]An example of a common-cause failure is what occurred due to the tsunami generated by the Tohoku earthquake in March 11, 2011, which was a single "attack" to all defensive elements of the Fukushima Daiichi plant.

– PSA levels 1 to 3 are carried out in a decoupled manner and by different groups, although feedback loops are imaginable and may influence the course of an accident. The usual consequence (damage) indicators of PSA level 3, which is focused on radiation-induced health and environmental impacts and on related financial losses, may be too limited. Psychological effects due to "fear", worsening diseases and so on, turned out to be significant, as illustrated by the Fukushima Daiichi accident.[27]

As a general remark, state-of-the-art PSA application for nuclear power plants, often combining fault and event trees, results in huge and potentially non-transparent logic trees, therefore hard to be modified under abnormal conditions, e.g., to take re-arranged or damaged systems into account under accident conditions. Furthermore, the basic approach that a system behavior can be modeled by the "sum of the behavior of its parts" becomes highly questionable under severe accident conditions when system interdependencies cause cascades,[28] regime shifts occur, and so on. These complex phenomena become increasingly important, and make the plant behavior harder to understand, model, and tackle.

Finally, we should be aware that the operating environment is assumed to be politically and economically stable, free of major concessions to economic pressure, and that safety standards and safety culture are adhered to.

To avoid misinterpretation, we do not argue against PSA as a systematic approach to determining whether safety systems are adequate, the plant design is balanced, the defense in depth requirement has been realized and the risk being as low as reasonably achievable. We rather recommend further developments of and careful attention to the concerns identified above.

4.2.3   Need for Peer Review

Probabilistic safety assessments (PSA), generally carried out by the plant owner and operator, provide essential information, which is legally required in many countries. This risk-related information complements deterministic analysis and is often used within public debates on the safety of nuclear power plants. Therefore, assurance of high quality is important and careful peer review, conducted by a team of independent experts with technical competence and experience in the areas of evaluation, is mandatory.

Historically, the first multi-site review of methods and results of PSA, conducted for five nuclear power plants in the USA, was given in the late 1980s in the report

---

[27]S. Kondo, "Lessons learned from Fukushima PSAM Community," PSAM Topical Conference, Tokyo, Japan (2012).

[28]D. Sornette, T. Maillart and W. Kröger, Exploring the limits of safety analysis in complex technological systems, International Journal of Disaster Risk Reduction 6, 59–66 (2013), for a simple illustrative cascade model applied to nuclear risk modeling.

NUREG-1150.[29] However, none of the many subsequent PSA were peer-reviewed internationally in a similar way. Nowadays, peer reviews are conducted at a national level by the regulatory body and contracted experts or technical support organizations; as PSA are complex and hard to audit, peer reviewers often use their own models and data.

The review should aim at the following (according to Ref. [[30]]):

- Confirmation that the modeling reflects the current design and operation features, takes into account all relevant operating experience, includes all modes of operation and has a scope agreed with the regulatory body.
- Check for completeness against an appropriate set of postulated initiating events and hazards.
- Verification that omissions of hazards are based on site-specific justifications and that they do not weaken the overall risk assessment of the plant.
- Verification that the analytical methods and computer codes as well as adopted verification standards are appropriate.
- Securing that the potential for unidentified cross-links and effects of common cause events are taken into account in the models and that human reliability analysis is carried out in the PSA on a plant-specific and scenario-dependent basis and that current methods are applied.

Although the structure of PSA and presentation of results follow certain standards, they are hard to understand and check by non-experts and remain non-transparent. Furthermore, they include detailed confidential information about the plant design and operational procedures. Thus, the responsibility for peer review of required PSA lies with the regulator and, finally, the public has to have confidence in the independence and competence of that body.

4.3 Toxicology of Ionizing Radiation

A key element of nuclear risk is the effect of exposure to ionizing radiation on human health. As is true for toxicology in general, the effect depends on the dose: "Sola dosis facit venenum".[31] As mentioned in Sect. 4.2, of primary relevance is the response (effect on human health) due to relatively low doses of radiation, which tend not to cause deterministic observable effects in the short-term,[32] but in the long-

---

[29]US Nuclear Regulatory Commission. "Reactor risk reference document", NUREG-1150, (1987).

[30]IAEA, Periodic Safety Review for Nuclear Power Plants, IAEA SAFETY STANDARDS SERIES No. SSG-25, (2013).

[31]"The dose makes the poison"—Paracelsus, Swiss physician and chemist, 1530.

[32]E.g., doses larger than 3 Gray are considered to be lethal, with probability 1, leading to death shortly after (Health risks from exposure to low levels of ionising radiations (BEIR VII Report). National Academy of Sciences. National Academy Press, Washington, DC (2006)).

term may contribute to the risk of developing cancer—hence the term latent sto-chastic effects. The most basic dose measure is the Gray, being the energy absorbed in Joules per Kilogram. However, the response of tissues and organs varies for different types of radiation, and have different sensitivities. The Sievert (Sv) is then a derived measure, combining these different exposures, to obtain an overall whole-body measure of the health effect of ionizing radiations—also called equivalent dose.

Due to the crucial importance for radiation protection, the dose-response rela-tionship—typically in terms of cancer incidence and fatality caused by radiation exposure—is its own field of study, with guidelines published by the International Commission on Radiological Protection (ICRP), the United Nations Scientific Committee on Effects of Atomic Radiation (UNSCEAR), the World Health Orga-nization (WHO), and various national regulatory bodies.

To give some points of reference, Fig. 4.4 shows the different contributions of public radiation exposure in the US, Germany and globally. The average person receives 3–4 mSv (millisieverts) per year from naturally occurring so-called "back-ground radiation". The majority (about two thirds) of this comes from radon gas, which is emitted from the decay of uranium in soil, rock, and water. Radon exposure thus varies geographically, and e.g., Denver, Colorado sees concentrations five times higher than average. Further, at high altitudes one receives about 1.5 mSv more cosmic radiation per year than if living near sea level. Medical imaging is the highest non-natural contributor to radiation exposure, with typical doses for different pro-cedures being: X-ray of extremity 0.001 mSv; mammography 0.4 mSv; and com-puted tomography of the chest 7 mSv.

Concerning the dose-response relationship, according to the US NRC, a "lethal dose" is an intense dose of 4–5 Sv, expected to give a 50% probability of fatality within 30 days. The dose rate is also important, as exemplified by Albert Stevens who, due to a Plutonium injection, accumulated a colossal 64 Sv of exposure

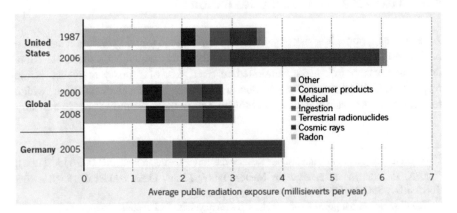

Fig. 4.4 Contributions of public radiation exposure in the US, Germany and globally. Taken from Abbott, A., Researchers pin down risks of low-dose radiation, Nature 523, 17–18 (2015)

continually over 20 years, before passing away at 79 years of age (cardiorespiratory failure).[33]

Concerning smaller doses, the so-called Linear No Threshold (LNT) model predicts a roughly 5% increase in the frequency of *eventual* death from cancer, with each Sv dose of accumulated exposure.[34] For radiological protection purposes, U.S. guidelines recommend extrapolating dose-response risk linearly, all the way down to dose zero. This is attested to be conservative (pessimistic) for low doses (<100 mSv), and strongly criticized by the French Academy of Sciences,[35] given the biological fact that we are adapted to living in a naturally radioactive world, and endowed with DNA repair mechanisms.[36] Indeed, the International Commission on Radiological Protection (ICRP) has always realized and pointed out that LNT in the low dose range (<100 mSv) has a high uncertainty and should not be used for calculating cancer fatalities especially in the very low dose range of 1 mSv. Many argue therefore to introduce a lower threshold dose, as shown in Fig. 4.5, along with other proposed response curves.

Quoting the ICRP, "the epistemological limitations of the sciences of radiobiology and radio-epidemiology, and their influence on the attribution of health effects to low-dose exposure situations, are often ignored".[37] To exemplify a rough test of the Linear No Threshold (LNT), with up to 10 mSv/year higher exposure in Denver, Colorado from background radiation than in Florida and California, one would expect 11% higher cancer incidence in Colorado.[38] In contrast, there is only about 1% difference in age-adjusted cancer incidence between the three states.[39] However, this prediction and comparison are relying on heavy over-simplifications: there are different types of radiation, internal and external exposure pathways, and other confounding factors that cause cancer, which need to be taken into account—for instance, lower air pollution in Colorado. In fact, the death-rate for cancer is very different in various US states and has no measurable correlation with background radiation. For the years (2011–2015), the death rate (per 100,000 people) was for the

---

[33]Welsome, E. "The plutonium files: America's secret medical experiments in the Cold War." (1999).

[34]"Health risks from exposure to low levels of ionising radiations (BEIR VII Report)." National Academy of Sciences. National Academy Press, Washington, DC (2006).

[35]Tubiana M., Dose-effect relationship and estimation of the carcinogenic effects of low doses of ionizing radiation: the joint report of the Académie des Sciences (Paris) and of the Académie Nationale de Médecine, Int. J. Radiat. Oncol. Biol. Phys. 63(2), 317–319 (2005).

[36]Russo, G.L., I. Tedesco, M. Russo, A. Cioppa, M.G. Andreassi and E. Picano, Cellular adaptive response to chronic radiation exposure in interventional cardiologists, European Heart Journal 33 (3), 292–295 (2012); Gori. T. and T. Münzel, Biological effects of low-dose radiation: of harm and hormesis, European Heart Journal 33 (3), 292–295 (2012).

[37]ICRP. "Report of ICRP Task Group 84 on Initial Lessons Learned from the Nuclear Power Plant Accident in Japan vis-à-vis the ICRP System of Radiological Protection", ICRP ref. 4832-8604-9553, 2012 November 22.

[38]Assuming LNT: 0.05 deaths/Sv, 40% cancer lifetime incidence, 20% lifetime fatality rate, and 80-year average lifetime.

[39]https://statecancerprofiles.cancer.gov/incidencerates/

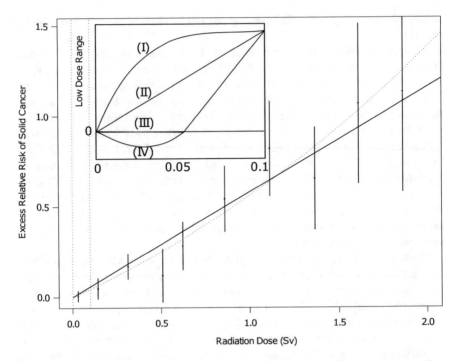

Fig. 4.5 Excess relative risk is the percentage increase of frequency of solid cancer above baseline, based on Japanese atom bomb survivors, adapted from Fig. ES-1 of Ref. (National Research Council. "Committee to Assess Health Risks from Exposure to Low Levels of Ionizing Radiation, Health Risks from Exposure to Low Levels of Ionizing Radiation: BEIR VII Phase 2." (2006)). E. g., doses around 2 Sv are expected to double the typical frequency, but this conclusion is highly uncertain. The data are summarized by 95% confidence intervals and a linear regression. Note that some Japanese subjects were exposed to around five hundred times the typical annual dose in a short time, not to mention the thermal radiation. The inset shows, for low doses, (I) supralinear, (II) linear, (III) threshold, and (IV) hormetic dose-response relationships. The hormetic zone (Mattson, M.P. and E.J. Calabrese, eds., Hormesis (A Revolution in Biology, Toxicology and Medicine), Springer, New York (2010)), if it exists, is characterized by favorable biological responses to low exposures to toxins and other stressors, here radiations

two highest states: Kentucky (199) and Mississippi (195) and for the two lowest states: Hawaii (136) and Utah (127) (from the Surveillance, Epidemiology, and End Results (SEER) Program, an authoritative source for cancer statistics in the United States). It is also trivial to illustrate the uncertainty of the LNT at very low doses: Consider 1 mSv, which is about one quarter of the average annual dose, for which the LNT predicts an increased probability of eventual death by cancer of 0.05/1000. This is four thousand times smaller than the average lifetime cancer fatality rate[33]— impossible to detect in epidemiological studies, given dominant spatial and temporal variation in lifestyle, environment, genetics, and other risk factors. As a result, on a statistical epidemiological basis, for low doses (<100 mSv), the null hypothesis of zero risk cannot be rejected, and neither can the alternative LNT hypothesis and moderate deviations around it.

No matter how unrealistic, the sheer possibility that an added dose as low as 1 mSv can lead to an increased risk necessitates consideration of the consequences of potential exposures of large populations in the event of a large release. For instance, in a population of 100 million persons exposed to this additional dose of 1 mSv, the LNT hypothesis would predict $5 \ 10^{-5} \times 10^8 = 5000$ additional deaths by cancer. This illustrates that the combination of low dose and high population can completely dominate the quantification of expected fatalities from additional nuclear radiation exposure. Therefore, in this LNT framework, the dominant potential risk of nuclear accidents becomes that of the long-term exposure of large populations to low doses.[40] In relation to these possible additional 5000 cancer deaths, it is necessary to quote the total number of cancer deaths in such a population. On the basis of the US cancer register (SEER) for the years 1975–2015, the yearly average number of cancer deaths for 100 Million persons during this period was 193,840. The highest number appeared in 1991 with 215,060 deaths and the lowest in 2015 with 158,950 deaths. In fact, the number of cancer deaths changes from year to year and in different states. With such variations, it is not possible to measure these possible 5000 deaths given the huge uncertainties for the very low dose ranges, making such figures meaningless. In response, the ICRP has stated that "the computation of cancer deaths based on collective effective doses involving trivial exposures to large populations is not reasonable and should be avoided".[41] However, this remains the standard approach to estimating long term stochastic fatalities (see 4.2.1).

To examine radiological exposure impacts in the real world, three major accidents analysed within the LNT framework are discussed below. More information on these major accidents and their consequences are given in Chap. 5, including radioactive releases in Table 5.5.

Three Mile Island, 1979 There was a small release from the plant's auxiliary building of a size such that less than one eventual death from cancer related to the accident was expected.[42] While there was a full-scale core meltdown, the primary containment did its job and prevented a substantial release.

Chernobyl, 1986 There was a massive release and the Chernobyl Forum was created by representatives of the IAEA, UN, WHO and others, to scientifically clarify the radiological, environmental, and health consequences of the Chernobyl accident. From these studies, it is estimated that there were 56 immediate radiological fatalities, and that eventual fatalities from premature death due to exposure will

---

[40]In an accident, the number of people exposed to high doses is relatively low. Further risk to residents surrounding the plant to emissions below regulatory limits during normal operation are insignificant.

[41]ICRP (2007). Assessing dose of the representative person for the purpose of radiation protection of the public and the optimisation of radiological protection: Broadening the process. ICRP Publication 101. Ann ICRP, 36 (3). ICRP stands for International Commission on Radiological Protection.

[42]Hatch, M. "Cancer near the Three Mile Island nuclear plant: radiation emissions." *American Journal of Epidemiology* 132 (3), 397–412 (1990).

approach 4000. However, these studies restricted the scope of exposures to Belarus, Ukraine and the Russian Federation. For those same regions, another LNT study estimated twice the fatality number and, for the full population of Europe[43] (near 600 million in 1986), it predicts about 16,000 cancer fatalities by 2065, i.e., 79 years after the accident. Highlighting the sensitivity to assumptions about no threshold, 90% of the European population considered in that study, and half of the predicted deaths, are from estimated exposures of less than 1 mSv. It is also important to note that, by 2006, only 10% of the predicted fatalities are expected to have occurred, as the majority of the eventual fatalities are projected to take place at an old age. In this regard, years of life lost would be a more appropriate measure, as discussed in 4.4.

Fukushima-Daiichi, 2011 No immediate radiological fatalities were reported, but a large total release, about an order of magnitude smaller than that of Chernobyl, occurred over populated areas. A major international committee of experts at the UNSCEAR continue to refine dose estimates,[44] where early dose estimates were published by the WHO.[45] The long-run predicted cumulative dose is 48 thousand Sv across the entire Japanese population (128 million at the time). This is on average 0.38 mSv per person across many years—an insignificant amount. However, while doses beyond the neighboring prefectures of Fukushima are indeed estimated to be negligible (<1 mSv per person), higher doses are estimated closer to the site, where lifetime dose can be roughly three time that of the dose in the first year following the accident. The highest doses were to emergency responders and site workers, with 25 thousand workers exposed to on average 10 mSv (250 Sv total, LNT giving 12.5 eventual deaths), but with few (only 175 workers) exposed to doses moderately above 100 mSv (about 20 Sv total, LNT giving 1 eventual deaths). Estimates provide confidence that most of the two million inhabitants of the Fukushima prefecture have been exposed to less than 10 mSv in the first year, with the exception of about 25 thousand inhabitants receiving between 10 and 50 mSv (middle value implying 750 Sv total, LNT giving 37.5 eventual deaths). Applying LNT for doses only above 10 mSv indicates 51 eventual fatalities. Allowing for doses in the range of 5 mSv for the residents of Fukushima would predict 100 additional eventual fatal

[43]Cardis, E., et al. "Estimates of the cancer burden in Europe from radioactive fallout from the Chernobyl accident." International Journal of Cancer 119 (6), 224–1235 (2006).

[44]UNSCEAR. "Developments since the 2013 UNSCEAR Report on the levels and effects of radiation exposure due to the nuclear accident following the Great East-Japan Earthquake and Tsunami. A 2015 white paper to guide the Scientific Committee's future programme of work." *United Nations Scientific Committee on the Effects of Atomic Radiation (UNSCEAR): New York, NY, USA* (2015).

[45]World Health Organization. "Preliminary dose estimation from the nuclear accident after the 2011 Great East Japan Earthquake and Tsunami." http://www.who.int/ionizing_radiation/pub_meet/fukushima_dose_assessment/en/index.html (2013).

cancers, which is on the same order as the results of an early academic study using LNT.[46]

4.4 Nuclear Energy Compared to Other Energy Sources: Statistical Life Losses

The World Health Organization (WHO) projects that, in 2012, nearly 400,000 deaths were caused by ambient air pollution in high income countries alone, to which energy generation is a major contributor[47,48] The WHO also projects more than four million deaths in 2012 due to indoor air pollution related to burning coal, biomass, etc. for heating and lighting, mostly in low-income countries, which could be dramatically reduced by greater availability of electricity. It is worth noting that these unintended costs pales in comparison to the enormous benefits from development enabled by increased energy consumption—e.g., considering covering health (e.g., premature deaths avoided due to modern healthcare), education and standard of living[49,50]

It is recognized that burning coal produces significantly more CO_2 emissions than burning oil or natural gas to produce the same amount of electricity[51,52] More important for health than CO_2 emissions is that the particulates produced from burning coal (2.5 μm in size) have detrimental impacts on human health that extend beyond the borders of the country burning it. It is important to note that these health impacts vary by country and technology. In the OECD, coal fired power stations are normally in the countryside, away from large concentrations of population and may be fitted with equipment that removes particulates and sulphur dioxide. In China, old fashioned unabated coal fired power plants have been located near densely populated urban centres (like they were in the OECD during the 1950s) and these do result in smog and have chronic impact on the health of the inhabitants.[56]

---

[46]Ten Hoeve, J.E., and Jacobson, M.Z. "Worldwide health effects of the Fukushima Daiichi nuclear accident." Energy & Environmental Science 5 (9), 8743–8757 (2012).

[47]WHO, Burden of disease from Ambient Air Pollution for 2012 (2016), Retrieved 08 24, 2016, from http://www.who.int/phe/health_topics/outdoorair/databases/AAP_BoD_results_March2014.pdf

[48]Wolf, R., Why wealthy countries must not drop nuclear energy: coal power, climate change and the fate of the global poor. International Affairs, 91(2), 287–301 (2015).

[49]Mazur, A. "Does increasing energy or electricity consumption improve quality of life in industrial nations?." Energy Policy 39.5 (2011): 2568–2572.

[50]Pasternak, A.D. "Global energy futures and human development: a framework for analysis." US Department of Energy, Oak Ridge (2000).

[51]Epstein, P. R., & et al. (12 authors), Full cost accounting for the life cycle of coal, Issue: Ecological Economic Reviews, Annals of the New York Academy of Sciences, 1219, 73–98 (2011).

[52]There are two reasons for this. In natural gas, it is C–H bonds that are broken while in coal it is C–C bonds. Of equal importance, modern combined cycle gas turbines (CCGT) have higher energy efficiency than coal plants.

Effects on human health result from both normal operation and accidents—with normal operation dominating in terms of magnitude of health effects. Sustainability assessments of different energy-producing technologies are used to evaluate their effects on human health. As seen earlier in Chap. 1, technology-specific evaluation criteria and indicators have been established by the EU Needs project[53] to assess the environmental, economic, and social dimensions of sustainability.

4.4.1 Methodological Approach

In this section, we look at statistical life losses stemming from normal operation and accidents occurring in energy production plants. This includes risk to workers, to the public and to emergency workers in the case of an accident. We focus on the risk category of the evaluation criteria of the NEEDS project, as shown in Table 4.1 below; while the normal risk and severe accidents have cardinal units (a score), the perceived risk category has ordinal units (a ranking).

There are two frameworks from which one can approach the statistical life losses assessment. The first one looks solely at the operation of power plants. The second one examines the full-life cycle of an energy production technology. We consider

Table 4.1 Risk categories developed within the NEEDS Project

Normal risk	Normal operation risk (Source: NEEDS Research Stream 2b for life cycle risk data)	
Mortality	Years of life lost (YOLL) by the entire population due to normal operation compared to without the technology.	YOLL/ kWh
Morbidity	Disability adjusted life years (DALY) suffered by the entire population due to normal operation compared to without the technology.	DALY/ kWh
Severe accidents	Risk from severe Accidents (Source: NEEDS Research Stream 2b for severe accident data)	
Accident mortality	Number of fatalities expected for each kWh of electricity that occurs in severe accidents with 5 or more deaths per accident.	Fatalities/ kWh
Maximum fatalities	Based on the reasonably credible maximum number of fatalities for a single accident for an electricity generation technology chain.	Fatal./ accident
Perceived risk	Perceived risk	
Normal operation	Citizens' fear of negative health effects due to normal operation of the electricity generation technology.	Ordinal scale
Perceived acc.	Citizens' perception of risk characteristics, personal control over it, scale of potential damage, and their familiarity with the risk.	Ordinal scale

---

[53]NEEDS, Final report on the monetary valuation of mortality and morbidity risks from air pollution (2006); NEEDS, Final report on sustainability assessment of advanced electricity supply options (2009).

Table 4.2 Methods and units matrix

Method	Unit	Explanation
The Impact Pathway Approach (IPA)	YOLL[a]	Years of life lost (years/unit of energy produced)
Life Cycle Analysis (LCA)	DALY[a]	Disability-adjusted life years = YOLL plus years lived with disability (years/unit of energy produced)
Monetary valuation	VOLY	Value of a life year lost by air pollution mortality (monetary)
Value of statistical life[b]	VSL	Estimate of the monetary value that society places on increased risk of death

[a]YOLL = Years of life lost by the entire population due to normal operation compared with/without the technology; DALY = Disability adjusted life years suffered by the entire population due to normal operation compared with/without the technology
[b]The value of statistical life corresponds to the rate at which people are willing to trade increased risk of death for other goods and services

here the full-life cycle approach because it allows for a better representation of the true effects from normal operation and/or accidents.

The Life Cycle Analysis (LCA) calculates "cradle to grave" energy, environment, material and economic resources used by the most relevant power supply options. The Impact Pathway Approach (IPA) (see annexed Fig. 4.10 for the steps involved) calculates the monetary values of the external costs associated to the supply of electricity and heat, based on different—current or future—technological options. The results of IPA and LCA are not directly comparable: IPA allows for site-specific effects[54] while LCA does not.[55] Table 4.2 relates methods to units and provides pertinent explanations.

4.4.2 Assessment of Energy Technologies

Hirschberg et al.[53] and Treyer et al.[56] have compiled estimated statistical life losses for current technologies as well as projected technologies for 2030. Life Cycle Analysis (LCA) data comes from the ecoinvent LCA database, presented in Table 4.3. While carbon capture and sequestration (CCS) may reduce CO_2 emissions, its impacts on human health are ambiguous. We thus report non-CCS values.

---

[54]ReCiPe and IMPACT 2002+ LCA do not allow for site-specific impact assessments (i.e. population density, meteorological conditions). IPA, on the other hand, allows for site-specific assessments (Treyer, K., Bauer, C., & Simons, A., Human health impacts in the life cycle of future European electricity generation, *Energy Policy* 74, 31–44 (2014)).

[55]Hirschberg, S., Bauer, C., Burgherr, P., Cazzoli, E., Heck, T., Spada, M., Treyer, K., Health effects of technologies for power generation: Contributions from normal operation, severe accidents and terrorist threat, *Reliability Engineering and System Safety* 145, 373–387 (2014).

[56]Treyer K., Bauer C. and Simons A., Human health impacts in the life cycle of future European electricity generation, *Energy Policy* 74, S31–S44 (2014).

Table 4.3 Life Cycle Analysis (LCA) study for current technologies (see annexed Table 4.10), measured in disability-adjusted life years (DALY/TWh)

	Current technologies[a]		Technologies for 2030
	LCA	IPA	Hierarchist LCA
Fossil			
Hard coal	319	59	320
Lignite	1000	–	1000
NGCC	30	27	29
Renewables			
Hydro reservoir	5	1	5
Hydro run-of-river	6	2	6
Wind offshore	30	6	30
Crystal Photovoltaic	–	19	–
Amorphous Photovoltaic	–	16	–
Solar thermal	8	–	18
Geothermal	30	28	30
Nuclear			
EPR	49	5	49

The results of the Impact Pathway Approach (IPA) study for current technologies are given in years of life lost (YOLL/TWh). Hierarchist LCA is a study for 2030 technologies (see annexed Tables 4.8 and 4.9), with its results measured in disability-adjusted life years (DALY/TWh). For comparison, a 1 GW electric nuclear reactor, working at 80% of its capacity, produces about 7 TWh per year. NGCC is Natural Gas Combined Cycle (CCGT is also often used as combined cycle gas turbine). No statistics are given for onshore wind, the most common form of renewable energy in Europe
[a]The selection of technologies was made by estimating an average European situation (see Treyer et al. (2014) for more details)

The 2030 numbers are following the "hierarchist" view, representing current policy and consensus,[57] and without taking into account the anthropogenic component of climate change.

Table 4.3 shows the mortality (IPA) and the Mortality + disability (LCA) for different energy sources Impacts on human health are the lowest for hydro, in the middle range for nuclear, natural gas and other renewables, and highest for coal. These observations are similar for projected European technologies used in 2030. Another report on electricity generation and health by Markandya and Wilkinson, summarized in Table 4.4, arrive at similar conclusions regarding air pollution-related effects for workers and the public.[58] The units are different from Table 4.3 (deaths instead of YOLL and DALY), however, ranking per energy source is the same, with coal and oil largely outstripping nuclear and gas in terms of air pollution related

---

[57]The life cycle analysis (LCA) ReCiPe method has three cultural perspectives: Hierarchist (most common policy principles, 100-year timeframe), Individualist (short-term interest, 20-year timeframe) and Egalitarian (strongest precautionary perspective, 500-year timeframe).

[58]Markandya, A., Wilkinson, P., Electricity generation and health, *The Lancet 370* (9591), 979–90 (2007).

Table 4.4 Health effects of electricity generation in Europe by primary energy sources (quantities are given per Terra Watt hour, TWh $= 10^{12}$ Watt hours.), as summarized by Markandya et al. (2007) based on data from impact pathway analysis (IPA) of the European Commission

	Deaths from accidents		Air pollution-related effects		
	Among the public	Occupational	Deaths	Seriousillness	Minorillness
Lignite	0.02 (0.005–0.08)	0.10 (0.025–0.4)	32.6 (8.2–130)	298 (74.6–1193)	17,676 (4419–70,704)
Coal	0.02 (0.005–0.08)	0.10 (0.025–0.4)	24.5 (6.1–98.0)	225 (56.2–899)	13,288 (3322–53,150)
Gas	0.02 (0.005–0.08)	0.001 (0.0003–0.004)	2.8 (0.70–11.2)	30 (7.48–120)	703 (176–2813)
Oil	0.03 (0.008–0.12)	..	18.4 (4.6–73.6)	161 (40.4–645.6)	9551 (2388–38,204)
Biomass	..	..	4.63 (1.16–18.5)	43 (10.8–172.6)	2276 (569–9104)
Nuclear	0.003	0.019	0.052	0.22	..

95% CI means '95% confidence interval'. Deaths includes acute and chronic effects. Serious illness includes respiratory and cerebrovascular hospital admissions, congestive heart failure, and chronic bronchitis. For nuclear power, they include all non-fatal cancers and hereditary effects. Minor illness Includes restricted activity days, bronchodilator use cases, cough, and lower-respiratory symptom days in patients with asthma, and chronic cough episodes

effects and health consequences. Overall, we can see very roughly that LCA is approximately 5 times larger than IPA, which seems reasonable for most energy technologies, except for NGCC and geothermal where mostly fatal accidents occur.

Figure 4.6 links two aspects: (1) impact on human health (quantified either by deaths or cases of serious illnesses) and (2) CO_2 emissions. It is interesting that the same trend is observed for the two dimensions of human health.

For a simple comparison of figures, we consider health consequences related to burning coal, omitting those related to carbon emissions and ecological damage: Coal and lignite, producing around 900 TWh per year in Europe, with coal fatalities from Table 4.4, gives 22,500 annual early fatalities in Europe per year; and the IPA from Table 4.3 gives 53,100 YOLL, and substantially higher DALY. The European Environment Agency estimates that, in 2014, air pollution in Europe led to 428,000 premature deaths due to particulate matter, and 78,000 due to NO_2—for which energy production contributes 5% and 19%, respectively.[59] This indicates 36,200 deaths in that year, mainly attributable to fossil fuels. The Europe Dark Cloud report (Table 4.11), an NGO publication, provides similar numbers, finally indicating resultant health costs of €62.3 billion per year[60]—one third of a percent of European Gross Domestic Product.

---

[59]European Environment Agency, Air quality in Europe 2017, EEA Report No 13/2017 (2017).

[60]*Europe's Dark Cloud: how coal-burning countries are making their neighbors sick*, published by CAN Europe, HEAL, WWF European Policy Office, Sandbag, Brussels, Belgium (2016).

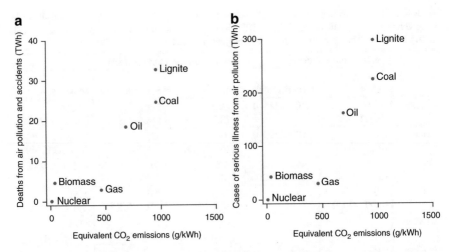

Fig. 4.6 Impacts of human health versus CO_2 emissions: the worst performers (coal and oil) are in the right-hand corner while the better performers (nuclear, biomass and gas) are in the lower left corner. Figure taken from Markandya et al. (2007). The biomass technologies refer to state of the art plants that meet EU environmental standards, in contrast to residential burning of wood, which is a main contributor to air pollution

To end with a cautionary note, there are significant differences in interpretation between actual mortality statistics—e.g., the difference between a person falling off a roof installing a solar panel and a hypothetical shortening of life based on rather uncertain assumptions.

4.4.3   Conclusions

The research sampled here arrive at the same conclusions in terms of which primary energy source has the greatest human health impact: coal and oil, while renewables and nuclear tend to have more moderate effects overall. Nuclear has the problem of high-perceived risk and the public has difficulty in accepting that it does in fact have very low human health impacts. Having nuclear replaced by coal should therefore be unthinkable in terms of both added emissions and impact on human health[61,62] But, since the decision in 2011 to shut all 17 of Germany's nuclear power stations by 2022, in the wake of the Fukushima Daiichi disaster, despite a rising share of

---

[61]WHO. (2014). *Burden of disease from Ambient Air Pollution for 2012*. Retrieved 08 24, 2016, from   http://www.who.int/phe/health_topics/outdoorair/databases/AAP_BoD_results_March2014.pdf

[62]Wolf, R., Why wealthy countries must not drop nuclear energy: coal power, climate change and the fate of the global poor, International Affairs 91 (2), 287–301 (2015).

renewable power, reliance on coal has remained high. A similar trend is observed in the USA,[63] where nuclear power plants have been closing as a response to political pressure and distorted markets, even though they provide the majority of their region's clean power, and are often significant contributors to the economic strength of their local economies.

Mortality data related to other man-made activities such as those of the transport sector allow us to put these findings into perspective. There are many transport modes, for instance: air, road, rail, maritime and inland waterways. The EU Mobility and Transport Road Safety[64] reported over 25,000 road casualties in Europe in 2013, and four times as many permanently disabling injuries. These accident numbers alone (ignoring substantial air pollution emitted by road traffic) surpass the numbers for human health impacts of primary energy sources by a factor of almost 100,000 for public deaths from accidents, and by a factor of 61 for pollution-related deaths.[65]

4.5 Testing PSA Estimates with Experience

4.5.1 Core Damage Events and Precursor-Based Assessment

A PSA core damage frequency of 10^{-5} per reactor-year, whose value may also be taken as the probability of core damage (CDP) in one reactor-year, predicts that an average of 0.15 events should occur during 15,000 reactor-years of experience—which is the actual amount of reactor-years of operation within the human nuclear age. To express this another way, there should be a 86% probability of zero core melting events during that time.[66] This seems at odds with multiple core damage events having been experienced. Here, we perform an empirical test of PSA CDF figures by comparing with experienced events relevant to core damage states.[67] Due to the rare nature of such

---

[63]Conca, J., Natural Gas – Not Renewables – Is Replacing Nuclear Power, Forbes, May 16, 2016. (http://www.forbes.com/sites/jamesconca/2016/05/16/natural-gas-is-replacing-nuclear-power-not-renewables/#659ef1404abb)

[64]EU Commission, *Mobility and Transport, Road Safety, Statistics - accidents data*. (2016). Retrieved 10 10, 2016, from http://ec.europa.eu/transport/road_safety/specialist/statistics/index_en.htm

[65]For the factor calculations, we take the highest numbers in health effects, which are obtained for lignite.

[66]The CDF for a single unit is small, so the sum of core damages for multiple units is approximated by a Poisson distribution.

[67]A common definition is a melt of 1% core inventory, for a typical commercial unit, beyond which fall-out and acute health effects become a possibility. See discussion in IAEA. Low Level Event and Near Miss Process for Nuclear Power Plants: Best Practices. IAEA, Safety Reports Series No. 73 (2012).

events, an empirical study thus requires an appropriate pooling of experience to obtain a sufficient sample size.[68]

Some pooling of experience has taken place in the nuclear community,[69] however, the majority of studies is site and unit specific. There is also an aversion to pooling based on a belief that each unit is incomparably unique to others, and to itself over time. However, the majority of existing units belong to the same generation, with a limited number of design classes, from a limited number of vendors, and so on. Further, as evidenced by PSA, the dominant risks are common ones (external hazards, transients including loss of off-site power, and so on). The commonality is further demonstrated when "history repeats" in the case of forerunner events,[70] which exist for all major accidents, including nine forerunners for Three Mile Island.[71] For Chernobyl, the "positive SCRAM effect", which led to the catastrophe, first came to light during operation of the first RBMK reactor near Leningrad, in an incident in November 1975. At Fukushima, it was known that Tsunami's of a size that would breach the defenses had occurred before. Moreover, the IAEA has indicated the potential for greater pooling of and learning from lower level events (see Fig. 4.7).

With this in mind, bona fide core damage events are Fukushima-Daiichi, 2011; Chernobyl, 1986; and TMI, 1979. While a number of events have happened at experimental, demo/prototype facilities,[72] and weapons facilities, it is improper and misleading to include them.[73] Finally, there have been a number of minor (<1% of core) melt events at commercial units,[74] which although being of high safety significance, are clearly of a lower order than the core damage accidents. For statistical estimation of the CDF, one can count Fukushima Daiichi as 3, TMI as 1, and Chernobyl as 1, out of roughly 15,000 reactor-years; or omit Chernobyl and use 13,500 reactor-years of non-former-USSR operational experience. Both

---

[68]E.g., with zero events in a fleet with 250 reactor years' experience, the CDF (per reactor-year) point estimate is zero, but the 95 percent confidence interval is (0,0.02)—too wide to rule out even a very high CDF, and is thus meaningless.

[69]E.g., on the frequency of initiating events, see NUREG CR-3862, CR-6143, and CR-689; and at WANO in particular.

[70]Experienced safety-relevant initiators and sequences that later on occurred, often elsewhere, and with worse consequences. The existence of a forerunner implies a failure to learn from the past.

[71]Kemeny, John George. "Report of the President's commission on the accident at Three Mile Island." The need for change: The Legacy of TMI (1979).

[72]Chalk River,1952; EBR-I, 1955; SRE, 1959; SL-1, 1961; Fermi, 1966; Lucens, 1969; and Ågesta, 1968.

[73]Experimental and demo facilities can be left out due to their small size. Weapons facilities were designed and operated under radically different conditions and oversight, and with a different purpose—the 1957 Windscale fire being a good example.

[74]E.g., Greifswald-5, 1989: 10–30 fuel elements—being a small fraction of a single assembly, out of more than 100 assemblies in the core—overheated, but did not melt. St. Laurent A-2, 1980: melting of one channel of fuel (out of >1000). St. Laurent A-1, 1969: 50 kg of fuel melted (out of >100 tons, similar fuel damaged to above event).

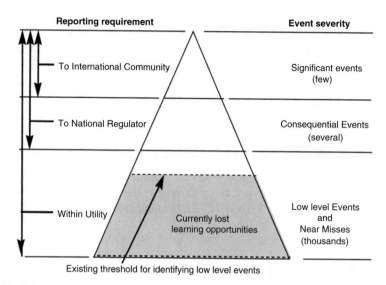

Fig. 4.7 Schematic of the reporting of events at different levels, taken from the IAEA (IAEA. Low Level Event and Near Miss Process for Nuclear Power Plants: Best Practices. IAEA, Safety Reports Series No. 73 (2012)), who stated "[to improve safety] the reporting threshold should be lowered from incidents to anomalies with minor or no impact on safety. This will provide an insight on precursors, which are near misses or low level events that provide information for determining advance warnings"

estimates give about the same historical average CDF of 3.3×10^{-4}, with 95% confidence interval $(10^{-4}, 10^{-3})$.

To enable more specific analysis, we look to accident precursors. For instance, consider Fukushima-Daini, a four-unit site and neighbor to Daiichi, where the 2011 earthquake and tsunami also caused major flooding. Unlike at Daiichi, one of the four external power lines remained, allowing the plant operators to maintain control, and due to outstanding accident management, a core damage accident was avoided. This was a near-miss, and the essential question is: with what probability would such an event result in a core damage? PSA methods provide a way to estimate this probability.

In detail, a core damage accident precursor[75] is an event in which core damage would have resulted if additional failures and/or initiators had occurred. With precursor analysis, a conditional application of PSA, these events are quantified by a conditional core damage probability (CCDP). There are two types of events that, in some combination, form a precursor: (1) an initiating event itself and (2) degradation of equipment or systems without an immediate impact on plant operation (a vulnerability). Both can be mapped onto a PSA model. Taking an example from Table 4.12, an observed initiating event would have a probability P_1 set to 1, and the

---

[75]Sheron, B. (2010). Status of the accident sequence precursor program and the standardized plant analysis risk models. NRC SECY-10-0125

probability of the failure of required safety functions, P_2, would be the CCDP. In the case of the precursor event involving a vulnerability, all relevant initiators and sequences need to be identified and counted over the duration of the vulnerability.

According to Ref. [76], precursor analysis has many uses, including:

- identification of safety issues that might have been overlooked/underestimated within PSA;
- consideration of quite complex events or combinations of events, including effects of potential improvements;
- confirmation or monitoring a level of safety, including trending over time.

As a hybrid approach, precursor based assessment combines the best features of statistical analysis and PSA approach by expanding the sample size of core damage relevant events, and reduces the scope of PSA to more narrowly defined specific events.

4.5.2   Safety Improvement and Comparative Risk

Precursor analysis takes place within many countries. In particular, concerning precursors at US plants, the US NRC Accident Sequence Precursor Program (ASP) provides open information: e.g., in 1998, precursor causes were 50% equipment failure, 40% human error, 20% component-out-of-service, and 10% weather-related.[77] Annexed Table 4.13 provides the top precursors from the ASP, and Table 4.5 the number of ASP events that we could find. Having precursor events dramatically enlarges the sample relative to only having bona fide core damage events—although the distribution of precursor CCDP values is very extreme, such that statistical analysis will be relatively uncertain.[78]

The sum of all US precursor CCDP gives the expected total number of US core damage events,

Table 4.5 Count of US NRC precursors by conditional core damage probability (CCDP) order, for all time. There are 141 events with CCDP on the order of 10^{-6}

CCDP magnitude	10^{-6}	10^{-5}	10^{-4}	10^{-3}	10^{-2}	10^{-1}	1
Count	141	103	52	37	10	3	1

---

[76]Proceedings of the Workshop on Precursor Analysis, AVN, Brussels. OECD NEA/CSNI/R (2003)11 (2001).

[77]NRC NUREG/CR-4674 Vol. 27, 1998; The sum is larger than 100% because of multiple simultaneous causes.

[78]TMI, 1979 contributes more than half of the sum of all US precursors, and the top 5 events about 85%.

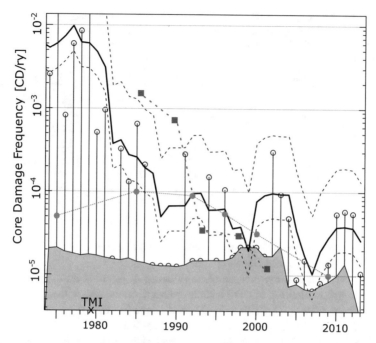

Fig. 4.8 Reduction in estimated CDF based on precursor assessment, combined with quoted PSA CDF values. The black line gives the CDF estimate based on US precursors, with 90% confidence bands; and the hollow dots give the same estimates done on annual windows. The gray area gives the contribution to the precursor assessment from long-term vulnerability precursors. The red square points give the PSA CDF results for the two Swiss Beznau units and the orange dots give the mean PSA CDF results for the US fleet (Gaertner, John, Ken Canavan, and Doug True. "Safety and operational benefits of risk-informed initiatives." An EPRI White Paper, Electric Power Research Institute (February 2008))

1 (TMI) + 0.5 (other events in Table 4.13) + 0.4 (not in the table) = 1.9,

which gives CDF = $1.9/4300$ reactor-years $= 4.4 \times 10^{-4}$, with 90% confidence interval $[1.7–8.0] \times 10^{-4}$, which is not far from the historical international average CDF per reactor year.

Importantly, this broader sample enables the study of the evolution of the CDF over time, as summarized in Fig. 4.8 for the US fleet. A high CDF prior to TMI is assessed, after which a drop of two orders of magnitude took place, attributed to retrofits and learning arising from the NRC TMI Action Plan.[79] This roughly agrees

---

[79]US NRC. *NRC action plan developed as a result of the TMI-2 accident.* Vol. 1. The Commission, 1980.

Table 4.6 For the four regions [(1) the US, (2) North and Western Europe, plus Canada (NWE), (3) Japan, Korea and India (JKI), and (4) Eastern Europe (EE)], the estimated historical core damage frequency (and for US* from 1980 onwards) are given by the estimated mean and 0.1 and 0.9 quantiles, scaled by 10^4

Region	Reactor years	INES 0 + 1	INES 2	INES$\geq$3	CDF $\times$ 10^4
US	4300	245	87	15	2.3 + {1.4, 2.5, 3.6}
US*	3440	245	75	7	{0.4, 0.6, 0.8}
NWE	5500	86	78	7	{0.7, 1.1, 1.5}
JKI	2800	57	13	12	12.1 + {0.7, 1.4, 2.1}
EE	2300	12	27	15	4.3 + {3.0, 4.8, 6.1}

For the US, the contribution to the estimate for TMI (2.3) is given separately, as well the Fukushima, 2011 related events (4 INES = 3 at Daini and 3 INES = 7 at Daiichi) for JKI, and Chernobyl for the EE category. The INES columns give the counts of INES scores corresponding to the precursor events (see Table 4.14)

with the analysis by the NRC, who estimated a level of 2.3×10^{-3} for 1969–1979,[80] and similarly for the Swiss Beznau plant.[81]

Insights from this study include clear evidence of learning and improvement: a very meaningful reduction of CDF is demonstrated, coinciding with capacity factors growing from 60% in the 1980s to 80–90% since 2000.[82] Also, the assessment supports current PSA CDF estimates for US plants below 10^{-4}, and perhaps as low as 10^{-5} given low external hazards and superior safety systems. However, relevant limitations include that this assessment is not plant specific, relies on the PSA based precursor analysis, and may not fully capture the risk of rare beyond design basis accidents.

On a comparative note, a rough precursor based assessment of CDF is done for four different regions[83]: (1) the US, (2) North and Western Europe, plus Canada (NWE), (3) Japan, Korea and India (JKI), and (4) Eastern Europe (EE). Aside from the US, precursor CCDP values are very imprecise, being based on a mapping from INES (International Nuclear Event Scale) scores (see Table 4.14 for the mapping and Sect. 6 for more on the INES). The results are summarized in Table 4.6, indicating:

---

[80]Sattison, Martin B. "Nuclear accident precursor assessment. *"Accident precursor analysis and management: reducing technological risk through diligence* (2004): 89.

[81]For instance, the big drop from pre- to post- 1990 was due to the 1/2 Billion USD invested in bunkered safety systems.

[82]e.g., https://www.nei.org/Knowledge-Center/Nuclear-Statistics/US-Nuclear-Power-Plants/US-Nuclear-Capacity-Factors

[83]NWE is North and Western Europe, plus Canada. JKI is Japan, Korea and India. Unfortunately, we do not have data for China. EE is "Eastern Europe", including Russia, former USSR, Armenia, Bulgaria, Hungary, Romania, Slovakia, Slovenia, and Ukraine. This is a rather conventional grouping, which has some justification based on relatively common technology and governance/regulation. However, this "clustering" of countries has not been tested here, and within the clusters not all national fleets and units are the same.

- For the full operating history, NW Europe has had a significantly lower CDF than the US. However, post-TMI, the estimated levels are similar, around 10^{-4} per reactor year.
- For Japan, Korea, and India, Japanese external BDBA-triggered events (incl. Fukushima, 2011 and Kashiwazaki, 2007) drive up the CDF, which would otherwise be similar to the US & NWE.
- For Eastern Europe, even without Chernobyl, the estimated CDF is around 5×10^{-4}. However, it is likely that improvements have taken place, not captured by this historical figure.

4.6 Accident Externalities

The LCOE (levelized cost of electricity) is the average cost over the full lifecycle of an energy source, per unit of energy produced. It is intended to capture the full cost of production, but in practice tends to omit a range of externalities[84]—notably, among other things, the cost of accidents for nuclear energy, the system costs for intermittent renewables,[85] and carbon emissions for fossil fuels. For a point of reference, the OECD NEA puts the LCOE of nuclear power generation at 6–9 USD cents/kWh.[86]

The accident externality can also be computed on a per kWh basis. The cost of historical accidents (see Sect. 5.3) is estimated to amount to 1–2 USD cents/kWh of historical nuclear electricity production.[87] For current US LWR, our conservative estimate is around 0.1–0.2 USD cents/kWh.[88] Other accident externality estimates are compared in Fig. 4.9: Estimates from the nuclear community prior to Fukushima, 2011, give a level about 100 times lower than our estimate, due to perhaps overly

---

[84] An externality, or external cost, is a cost of an activity incurred by the public rather than the producer.

[85] Costs imposed on the overall energy system, relating to dispatch-ability, intermittency, etc.

[86] Capital cost is from 4–6 USD cents/kWh, operation and maintenance about 1 cent/kWh, and fuel less than 1 USD cents/kWh, for a total of 6–8 cents/kWh, using discount rates between 3 and 6 percent. Setting the discount rate for the longer-term back end of the fuel cycle to zero, one can add about 1 cent to that. Source: Nuclear Energy Agency, "The Economics of the Back End of the Nuclear Fuel Cycle", NEA/OECD Report, No.7061, Paris, 2013.

[87] Perhaps up to USD 1 Trillion cost of accidents and about 50,000 TWh of historical production.

[88] We construct a rough estimate of the accident externality for the post-TMI, 1979, time period: Take a pessimistic CDF for a typical Western European or US reactor to be 10^{-4}. For severity, consider a case where, given a core damage, there is an average cost of USD 100 Bil. The risk is then 1×10^{-4}/reactor-year $\times$ USD 100 Billion = USD 10 Million/reactor-year, therefore giving an externality of about 0.1 USD cents/kWh, for a unit with annual generation of 10 TWh/reactor-year.

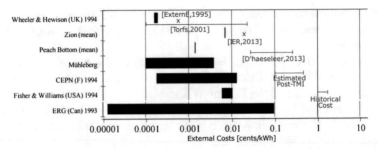

Fig. 4.9 Nuclear accident externality (2017 USD cents per kWh) with figure adapted from NEA (OECD/NEA, Methodologies for Assessing the Economic Consequences of Nuclear Reactor Accidents, 2000). The figure is modified by including more modern estimates, indicated by × marks for point values, and intervals when available: [ExternE, 1995] being 0.0005 euro-cents/kWh (CEPN, "ExternE – Externalities of Energy – Vol. 5: Nuclear", European Commission, Brussels/Luxemburg, 1995. http://www.externe.info/externe_d7/sites/default/files/vol5.pdf), [Torfs 2011] applied specifically to Belgium (R. Torfs, "Externe kosten van elektriviteitsproductie – Fase 3 van het CO2-project", Final report for Electrabel and SPE, Regulatory Committee for Gas and Electricity, Belgium, June 2001) yielding a range of 0.0001–0.035 euro-cents/kWh, [IER, 2013] giving 0.023 €/kWh (Die Risiken der Kernenergie in Deutschland im Vergleich mit Risiken anderer Stromerzeugungstechnologien", P. Preiss, S. Wissel, U. Fahl, R. Friedrich, A. Voss, authors, Universität Stuttgart, IER, February 2013.), and a review giving a range 0.03–0.3 euro-cents/kWh. The rightmost two intervals are our estimates

optimistic values for both frequency and severity of accidents. In a comprehensive review, D'haeseleer[89] put the current accident externality in a range of 0.04–0.4 USD cents/kWh, as part of an overall current externality of nuclear power of less than 1 cent/kWh. The accident externality is therefore significant, but likely to be less than a few percent of the LCOE.

Further, the total externality for nuclear, according to Fig. 1.15, is similar to wind and PV, and many times smaller than fossil fuels—on the basis of health impacts from emissions alone, and more so when the potential of anthropogenic climate change is considered. Indeed, on this basis, nuclear power seems like a more rational option then fossil fuels. However, the risk is *heavy-tailed*—not evenly incurred over time and across the fleet and population, but instead being extremely rare with severe consequences primarily on the country where the accident takes place. Given the natural aversion to such risks, it is reasonable and desirable to make efforts for further reductions, in particular lowering the potential severity of accidents to the greatest extent possible.

---

[89]D'haeseleer, W. D., "Synthesis on the economics of nuclear energy." Study for the European Commission, DG Energy, Final Report, European Commission, Brussels (2013).

Annexes

Table 4.7 Basic approach, scope and simplifications of key probabilistic safety analysis (PSA) Methods[a,b]

Key PSA Method	
Fault Tree Analysis (FTA)	Top-down approach for failure analysis, starting with a failed state ("top event") and determining deductively all the ways it can be caused by lower level ("basic") events, connected through logic gates, this following the concept of reductionism. The approach is static, includes just one (regarded representative) failure mode of basic events, applies the axiom system of Kolmogorov probability theory and minimal cut-sets for quantification.
Event Tree Analysis (ETA)	An inductive procedure that starts with an initiating event and "propagates" it by considering all possible ways in which it can affect the behavior of the system in a chronological order. The usual binary modes of an event tree represent inter alia the possible functioning or malfunctioning of a (sub-) system. The approach is a (semi-) static and linear, based on causality and conditional probabilities.
Human Reliability Analysis (HRA)	Usually restricted to unintentional errors of omission, i.e. single operator or a team fails to fulfill a clearly defined task within a given time, although striving for success. First generation methods (e.g., THERP, SLIM) offer human error probabilities (HEPs) while 2nd generation methods (e.g., ATHEANA) focus on improved understanding of error-forcing contexts. – THERP (Technique for Human Error Rate Prediction): Breakdown of actions into sub-actions until estimators are available (like FTA), which consider particularly the influence of time and training with consideration of interdependencies between actors and actions afterwards. – SLIM (Success Likelihood Index Methodology): Questioning of experts in order to assess "performance shaping factors" influencing HEPs. The identification of numerical values is then based on a calibration of the expert opinions by means of experience. – ATHEANA (A Technique for Human Event Analysis): Designed to support the understanding and quantification of Human Failure Events (HFEs), assuming that HFEs occur when the operators are placed in an unfamiliar situation where their training and procedures are inadequate or do not apply.
Dependent Failure (DF) Analysis	Distinction between common cause initiating events, common cause/common mode failures (CCF/CMF) and causal/cascading failures. Inclusion of DF in a model (e.g. fault tree) and their consideration in an explicit (e.g., structural or functional dependency) and/or implicit way (unspecified cause by, e.g., β-factor or MGL-factor technique).

[a]W. Kröger and D. Sornette, Reflections on Limitations of Current PSA Methodology, Proc. PSA 2013, International Topical Meeting on Probabilistic Safety Assessment and Analysis, Columbia, South Carolina, USA, September 22–26, 2013

[b]W. Kröger, "Risk analyses and protection strategies for the operation of nuclear power plants," Chap. 2 in Alkan et al., Landolt-Börnstein: Numerical data and functional relationships in science and technology – new series, advanced materials and technologies, *Nuclear Energy*, pp. 186–235: Springer (2005)

Fig. 4.10 The principal
steps of an impact pathway
analysis (IPA), for the
example of air pollution

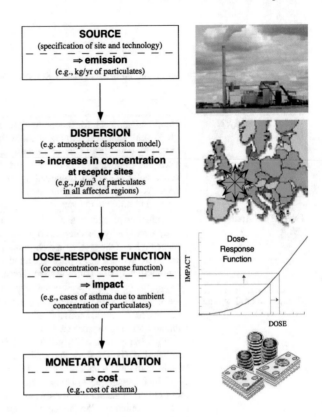

Table 4.8 Technologies for year 2030: Comparison of the ReCiPe results of Life Cycle Analysis (LCA) in DALY per TWh (DALY stands for Disability-adjusted Life Years), excluding climate change, for three cultural perspectives: Hierarchist (most common policy principles, 100-year timeframe), Individualist (short-term interest, 20-year timeframe) and Egalitarian (strongest precautionary perspective, 500-year timeframe)

	Hierarchist (H) [nDALY/kWh]	Egalitarian (E) [nDALY/kWh]	Individualist (I) [nDALY/kWh]
Fossil			
Hard coal	320	11,118	60
Hard coal, CCS	390	13,434	77
Lignite	1000	37,497	99
Lignite, CCS	1193	45,636	91
NGCC	29	142	27
NGCC, CCS	32	182	28
Nuclear			
EPR	49	1024	25
Renewables			
Hydro, reservoir	5	23	4
Hydro, run-of-river	6	24	5
Wind, offshore	30	824	10
Solar thermal	18	179	14
Geothermal	30	208	26

The terms Hierarchist, Individualist and Egalitarian refer to three sets of subjective choices on time horizon and assumed manageability, as described by Thompson et al. (1990).[a] Table from Treyer et al. (2014)[b]

[a]Thompson M, Ellis R., Wildavsky A.; Cultural Theory, Westview Print Boulder (1990)

[b]Treyer K., Bauer C. and Simons A., Human health impacts in the life cycle of future European electricity generation, *Energy Policy* 74, S31–S44 (2014)

Table 4.9 Technologies for year 2030: List of technologies for LCA analysis, from Treyer et al. (2014)

Table 1 List of technologies chosen for the comparative LCA in this paper. All datasets are valid for estimated average European conditions in 2030

| Category | Technology | Description | Source | Technology specifications | | | |
				Net electric capacity [MWel]	Full load hours [h/a] (Availability)	Life time [years]	Net electric efficiency [%]
Fossil	Natural gas combined cycle (NGCC), without CCS	500 MW combined cycle power plant with installations for the reduction of NO$_x$ (85%), SO$_2$ (0%), PM (0%).	Volkart et al. (2013)	720	7500 (86%)	25	62%
	Natural gas combined cycle (NGCC), with post-combustion CCS, pipeline 200 km, storage depth 1000 m	See above. Chemical CCS technology used (MEA[a] 30 wt%) with a CO$_2$ capture rate of 90%.	Volkart et al. (2013)	366	7500 (86%)	25	54%
	Pulverised hard coal, without CCS	500 MW power plant with installations for the reduction of NO$_x$ (94%), SO$_2$ (99%), PM (96%), Mix of 25% flow-through cooling and 75% wet cooling	Volkart et al. (2013)	736	7500 (86%)	40	46%
	Pulverised hard coal, with post-combustion CCS, pipe-line 200 km, storage depth 1000 m	See above. Chemical CCS technology used (MEA[a] 30 wt%) with a CO$_2$ capture rate of 90%.	(Volkart et al., 2013)	576	7500 (86%)	40	36%
	Pulverised lignite, without CCS	Mix of 100 MW (30%) and 400 MW (70%) power plants with installations for the reduction of NO$_x$ (74%), SO$_2$ (95%), PM (99.9%). Mix of 25% flow-through cooling and 75% wet cooling	Volkart et al. (2013)	989	7500 (86%)	40	43%
	Pulverised lignite, with post-combustion CCS, pipeline 200 km, storage depth 1000 m	See above; Chemical CCS technology used (MEA[a] 30 wt%) with a CO$_2$ capture rate of 90S.	Volkart et al. (2013)	759	7500 (86%)	40	33%

Nuclear	European Pressurised reactor (EPR), Gen III	Includes uranium extraction via 50% in-situ leaching, 30% underground mining and 20% open pit mining. Only centrifugal enrichment is taken into account.	Simons and Bauer (2012)	1530	8000 (91%)	60	34%
Hydro	Hydropower, reservoir plant	The dataset describes the average operation of all major dams in Switzerland, including emissions from reservoir lakes.	ecoinvent (2013)	n.s.	n.s.	150/80[b]	78%
	Hydropower, run-of-river plant	The dataset describes a representative mix of flow-through plants in Switzerland and Austria.	ecoinvent (2013)	n.s.	n.s.	80	82%
Wind	Wind, offshore, 20 MW turbine	Wind turbine with the tower and monopole made from steel. 10 m of depth. 30 km distance to the shore. The single wind turbines will function within a wind-park consisting of many turbines.	Roth et al. (2009)	20	4000 (46%)	20	n.s.
Solar thermal	Solar trough, PCM-storage	Phase Change Material (PCM) is a material having a high heat storage capacity per volume/mass and is used for storage of the heat.	Roth et al. (2009)	200	6400 (73%)	30	19%
Geothermal	Enhanced Geothermal System (Hot-Dry-Rock) plant.	The plant is modelled based on data from a single Hot-Dry-Rock binary cycle geothermal plant.	Roth et al. (2009)	36	7000 (80%)	20	11.3%

[a]Monoethanolamine is used for chemical absorption
[b]Structural part/turbines

Table 4.10 Total mean health impacts of current technologies

Total man health impacts of different electricity producing technologies in mDALY/GWh for results estimated using ReCiPe and in mYOLL/GWh for results assessed with IPA. Some values are missing in the table; though they are available for all types of technologies presented in the table the performance characteristics for some cases are such that consistent comparison is not possible.

Human health impact	Social perspective	Fossil						Nuclear	Renewables						
		Hard coal, no CCS	Hard coal, post CCS	Lignite, no CCS	Lignite, post CCS	NGCC, no CCS	NGCC, post CCS	Nuclear, EPR	Hydro, reservoir	Hydro, run-of-river	Wind, offshore	Crystal, PV,CH	Amorph, PV,CH	Solar thermal	Geothermal
LCA-based total human health, with CC [mDALY/GWh]	R(H)	1449	660	2300	1399	598	213	56	11	10	42	–	–	49	64
	R(E)	13,819	13,956	40,738	46,122	1527	570	1039	38	35	851	–	–	245	288
	R(I)	1121	429	1220	284	553	230	31	11	9	21	–	–	42	60
LCA-based total human health, w/o CC [mDALY/GWh]	R(H)	319	390	1000	1192	30	32	49	5	6	30	–	–	18	30
	R(E)	11,117	13,434	37,497	45,636	146	177	1024	23	24	824	–	–	179	208
	R(I)	60	77	99	91	27	28	25	4	5	10	–	–	14	27
IPA-based [mYOLL/GWh]		59	–	–	–	27	–	5	1	2	6	19	16	–	28

For the Life Cycle Analysis (LCA), units are mDALY/GWh; for the Impact Pathway Approach (IPA), units are mYOLL/GWh. NGCC means Natural Gas Combined Cycle. From Ref. [a]

Abbreviations: ReCiPe, R(H) = Hierarchist, R (E) = Egalitarian, R (I) = Individualist; CC = Climate Change; IPA = Impact Pathway Approach; CCS = Carbon Capture and Storage; EPR = European Pressurized Reactor.Abbreviations: ReCiPe, R(H) = Hierarchist, R (E) = Egalitarian, R (I) = Individualist; CC = Climate Change; IPA = Impact Pathway Approach; CCS = Carbon Capture and Storage; EPR = European Pressurized Reactor

aReCiPe and IMPACT 2002+ LCA do not allow for site-specific impact assessments (i.e. population density, meteorological conditions). IPA, on the other hand, allows for site-specific assessments (Treyer, K., Bauer, C., & Simons, A., Human health impacts in the life cycle of future European electricity generation, *Energy Policy* 74, 31–44 (2014))

Table 4.11 Breakdown of health costs from coal plant emissions, taken from EEA (2017)[a]

Health impacts from emissions of EU coal powered plants		Associated health costs in million Euros (VSL, median/ high value, 2013 prices)
Premature deaths from $PM_{2.5}$	19,000	23,900/48,500
Premature deaths from ground-level ozone	200	200/500
Premature deaths from NO_2 (scaled to 2/3rd's of model results to avoid double-counting with $PM_{2.5}$)	3800	4700/9600
Infant mortality	40	70/100
Hospital admissions (respiratory or cardiovascular)	21,000	50
Cases of chronic bronchitis (adults)	11,800	700
Workdays lost	6,575,800	1000
Additional restricted activity days	23,502,800	2500
Minor restricted activity days	1,166,700	60
Asthma symptom days in children	538,300	30
Bronchitis in children	51,700	40
Total health costs		32,400/62,300

[a]European Environment Agency, Air quality in Europe 2017, EEA Report No 13/2017 (2017)

Table 4.12 Classic basic event tree analysis from the 1979 German Risk Study[a]

Table 1: Summary of the results of event tree analysis

Accident Initiating Event	Probability of Occurrence of the Initiating Event per Reactor Year (P_1)	Failure Probability of Required Safety Functions (P_2)	Probability of Occurrence of Core Melt per Reactor Year ($P_3 = P_1 \times P_2$)
Large LOCA	2.7×10^{-4}	1.7×10^{-3}	5×10^{-7}
Medium LOCA	8×10^{-4}	2.3×10^{-3}	2×10^{-6}
Small LOCA	2.7×10^{-3}	2.1×10^{-2}	5.7×10^{-5}
Loss of Off-Site Power	1×10^{-1}	1.3×10^{-4}	1.3×10^{-5}
Loss of Main Feedwater Supply	8×10^{-1}	4×10^{-6}	3×10^{-6}
Emergency Power Case with Small Leak at Pressurizer	2.7×10^{-4}	2.6×10^{-2}	7×10^{-6}
Other Transients with Small Leak at Pressurizer	1×10^{-3}	2×10^{-3}	2×10^{-6}
ATWS-Events	3×10^{-5}	3×10^{-2}	1×10^{-6}

[a]Birkhofer, Adolf. "The German risk study for nuclear power plants." IAEA Bulletin 22.5/6 (1980): 23–33

Table 4.13 Incomplete list of the major American precursors, evaluated by the US NRC.[a] When the date is given in the format exemplified by '01(>30), this indicates a vulnerability event discovered in 2001 with duration of at least 30 years

Date	Site and unit	CCDP	Brief description
28/3/79	Three Mile Island-2	1	Loss of feedwater; pressure operated relief valve failed open; operator errors led to major core damage.
22/3/75	Browns Ferry-1	0.2	Cable fire disabling systems related to the control of unit 1 (and 2).
20/3/78	Rancho Seco	0.1	Failure of non-nuclear instrumentation and steam generator dry-out.
24.09.77	Davis Besse	0.07	Stuck open pressure operated relief valve caused reduced water level in steam generator.
'01 (>30)	Point Beach-1-2	0.042[b]	Design deficiency in air-operated minimum-flow recirculation valves of the EFW pumps (not modeled in PSA).
'11 (>40)	Oconee-1-3	0.036	Potential failure of Jocassee Dam would likely cause accident.
8/5/74	Turkey Point-3	0.02	Failure of three emergency feedwater pumps to start during test.
20/7/76	Millstone-2	0.01	Loss of offsite power from grid disturbance; errors in emergency diesel generators loading, and failure of the emergency core cooling systems.
29/4/75	Brunswick-2	0.009	Multiple valve failures; reactor core isolation cooling inoperable as a result of stuck-open safety valve.
07/04/74	Point Beach-1	0.005	Inoperable auxiliary feed water pumps during shutdown.
5/11/75	Kewaunee	0.005	Inoperable EFW pumps during startup as a result of leaks from demineralizer into the condensate storage tank.
'05(30)	Kewaunee	0.0015	Design deficiency could cause unavailability of safety-related equipment during postulated internal flooding.
'08(30)	St. Lucie-1	0.0015	Air intrusion into component cooling water system causes pump cavitation.
'04(11)	Palo Verde-1-3	0.0013	Containment sump recirculation potentially inoperable.
12/1/71	Point Beach-1	0.001	Failure of containment sump valves.
27/2/02	Davis-Besse	0.006	Cracking of CRDM nozzles, RPV head degradation, potential clogging of the emergency sump, and potential degradation of the HPI pumps.
9/12/86	**Surry-2	INES 3	Catastrophic rupture of feedwater line due to corrosion, resulted in 8 workers injured, four died later; the event cascaded from the non-nuclear-part across safety-grade systems causing accident management problems.[c] Appears to be omitted from 1986 NRC analysis.[d]

(continued)

Table 4.13 (continued)

Date	Site and unit	CCDP	Brief description
26/12/85	**Rancho Seco	INES 3	Following loss of the integrated control system (ICS), the unit tripped and an overcooling transient occurred due excessive feedwater flow. Finally, the operator attempted to close the manual isolation valve but it could not be moved because no maintenance on that valve during life of the plant. Restoration of power within 26 min

Asterisks on the unit name indicate two potential precursors not found in the ASP documentation
[a]Sheron, B. (2010). Status of the accident sequence precursor program and the standardized plant analysis risk models. NRC SECY-10-0125
[b]CCDP = 2 units $\times$ 30 reactor-years $\times$ 7 $\times$ 10^{-4}/reactor-year
[c]US NRC, Erosion/Corrosion-Induced Pipe Wall Thinning in US Nuclear Power Plants, NUREG–1344 (1989)
[d]Forester, J. A., et al. Precursors to potential severe core damage accidents: 1982–1983, A status report. Vol. 24. NUREG/CR-4674

Table 4.14 Conversion from CCDP (above called ICCDP for "incremental conditional core damage probability") to the defense in depth based scores from the INES (see Chap. 6)[a]

$ICCDP_{Event}$	INES
$1 > ICCDP_{Event} \geq 1 \cdot 10^{-2}$	3
$1 \cdot 10^{-2} > ICCDP_{Event} \geq 1 \cdot 10^{-4}$	2
$1 \cdot 10^{-4} > ICCDP_{Event} \geq 1 \cdot 10^{-6}$	1
$1 \cdot 10^{-6} > ICCDP_{Event} \geq 1 \cdot 10^{-8}$	0

Note that the full INES scale runs from 0 to 7
[a]Guideline ENSI-A06/e, Probabilistic Safety Analysis (PSA): Applications, November 2015

Chapter 5
Severe Accidents: Singularity of Nuclear Disasters?

Abstract The risk of nuclear accidents has proven to be low in absolute and relative terms. Nevertheless, the high-energy density fission process and current technology make today's reactors vulnerable to very severe albeit very rare accident scenarios. The three disasters experienced demonstrated the importance of site conditions, containment systems and severe accident management measures. Not being triggered by a single or combined technical failure in a classical sense, disasters experienced are partly explained as a product of five hierarchical levels of individual and societal human factors. The potentially severe consequences, including costs, of nuclear accidents have played a decisive role in the development of the nuclear power sector, and dominate nuclear risk analysis. However, in terms of cost or loss of life, these accidents are not singular—as other industrial and energy generation sectors have comparable severe accidents.

Despite this, there is a widespread and exceptional human dread of low level radiation exposure that prevents us from facing the real issue of how to improve world prosperity while burning less fossil fuel. To deal with this pragmatically, and operating on the principle that "nuclear power plant safety requires a continuing quest for gain in excellence", we identify enhanced requirements to take the dread out of nuclear, and to rely less on social stability and long term husbandry of wastes.

5.1 The International Nuclear and Radiological Event Scale (INES)

The INES, represented in Fig. 5.1, was developed in 1990 by the IAEA and OECD/NEA to classify events at facilities associated with civilian nuclear power. It is the primary official statistic used to communicate with the public about event severity. Orders exist requiring that licensees report safety-relevant events to the IAEA (International Atomic Energy Agency), which are given INES scores. INES has discrete escalating levels from 0 to 7. An event between 4 and 7 is termed an accident, between 1 and 3 an incident, and level 0 below scale with no safety significance. With the INES, an event is scored along three tracks:

© Springer Nature Switzerland AG 2019
D. Sornette et al., *New Ways and Needs for Exploiting Nuclear Energy*,
https://doi.org/10.1007/978-3-319-97652-5_5

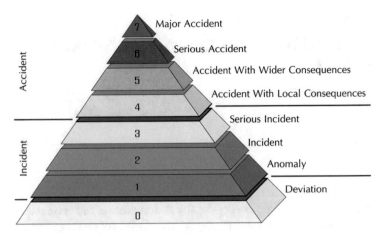

Fig. 5.1 International Nuclear Event Scale, standard pyramid representation of the 8 levels (INES, IAEA. "The international nuclear and radiological event scale user's manual 2008 edition." *IAEA and OECD/NEA* (2008))

1. People and Environment: on and off-site human radiological exposure, with scores 3–7.
2. Radiological Barriers and Controls: contamination and fuel damage, with scores 2–5.
3. Degradation of Defense in Depth: near-miss/precursor risk, with scores from 1 to 3.

Along each track, each point increment corresponds to an order of magnitude increase in the relevant quantity, and the final INES score is the highest score across the three tracks. See Annex A.1 for a more complete overview of the INES nomenclature and experienced events. Among other limitations,[1] which make study of INES scores problematic, until recently, there has been no complete public database of official scores, and little retro-active scoring of events prior to 1990, when the scale was introduced. However, based on our newly constructed database[2], filtering on events with core-safety relevance at commercial nuclear power stations, plotted in Fig. 5.2, shows a relatively large number of precursor events, as described in Sect. 4.5.

---

[1] It mixes frequency and severity; it limits the score of alarming near-misses to three, a level that can also be attained by e.g. a small release of radioactive gas without a connection to a core damage accident sequence; its discrete limited set of values and lack of measurement of integral consequences make it of limited use for measuring event severity; and different styles of severity assessment allow for subjectivity in scoring of incidents.

[2] Wheatley, S., Kröger, W., and Sornette, D. "Comprehensive Nuclear Events Database: Safety & Cost Perspectives". ESREL 2017. CRC Press (2017).

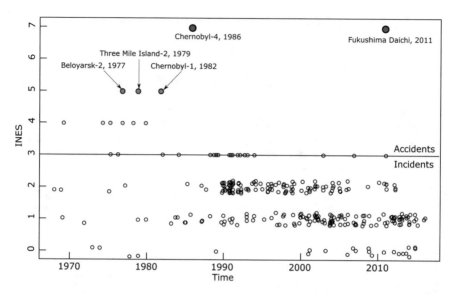

Fig. 5.2 INES scores for events at commercial nuclear power stations, deemed to have core-safety relevance (e.g., excluding Tokai-mura 1999), from the database of Wheatley, Kröger, and Sornette (2017). The incident points are spread around their INES value for visibility. The INES 4 events include: 1969 St Laurent-1; 1974 and 1975, Leningrad-1; 1977 Bohnice-1; 1978, Beloyarsk, which has INES 3–4; and 1980, St Laurent-2. Note the low number of incidents prior to 1990, when INES was introduced, and a tendency for fewer large INES scores over time

5.2   Brief History and Lessons Learned from Nuclear Accidents

The use of nuclear energy has come along with the following high INES-scaled accidents[3]:

- Two level 7 events: Chernobyl (1986, former Soviet Union), Fukushima Daiichi (2011, Japan);
- One level 6 event: Kyshtym MCCC (1957, former Soviet Union);
- Three-to-four level 5 events: The Windscale fire[4] (1957, United Kingdom), Three Mile Island (TMI; 1979, USA), Chalk River (1952, Canada), and perhaps Lucens (1969, Switzerland) rated at a level of 4–5;
- Numerous lower level accidents including the level 4 accident at Tokai-mura (1999, Japan).

The most severe accidents are described in detail in Tables 5.1 and 5.2 (see Table 5.5 for comparison of releases), for which broad lessons learned are

---

[3]See appendix A.1 for accident descriptions and further INES details.

[4]Omitted from the count for being a military weapons related facility.

Table 5.1 Summary and evaluation of three selected major accidents in commercial civilian nuclear power facilities

	TMI-2, USA; 28.3.1979	Chernobyl-4, former USSR, 26.4.1986	Fukushima 1, 1-4, Japan, 11.13.2011
Type; thermal power; start operation	PWR; 2575 MW; 19781	RBMK; 3200 MW; 12/1978	BWR; 1380–2380 MW; 3/1971-10/1978
Location; units at site	5 km Harrisburg, 140 km Washington, DC; two single units (unit 2 shut not in operation)	10 km Pripyat, 140 km Kiev; two twin units	250 kms NE of Tokyo; two twin units (1–4) plus. Single units (5–6) nearby, units 1–3 in operation.
Type of accident	Partial core melt under pressure after temporary loss of core cooling	Power excursion (explosion); graphite ignition and burning	Core meltdown units 1–3, hydrogen explosions and structural damage
INES scale (1–7)	5 (wider consequences)	7 (major)	Single unit: 5 (wider consequences), whole event: 7 (major)
Triggering event [time]	[4:00 am] Inadvertent closure of condensate check valve and shut-off of feed water pumps—emergency reactor shutdown—auxiliary system failure due to closed valves—primary loop pressure increase triggered PORV to open but it stuck open	[1:23 am] Simulated emergency shutdown to test residual steam capability to generate electric power in the event of failure of on-site diesel generators back up power	[2:46/3:45 pm] Beyond design 9.0–9.2 magnitude earthquake/ 14–14.5 m tsunami; total loss of AC/DC power due to flooding; loss of core cooling
Course of events	No indication of stuck open PORV; loss of coolant and pressure, misleading readings, turning off the emergency core cooling pumps; overfilling of pressurizer relief tank; shut down of reactor coolant pump—loss of water circulation; steam formation—nuclear fuel melting with hydrogen production and troublesome bubbles	Drop of core power below 700 MWth (min allowable level); withdrawal of rods to overcome reactor poisoning; unstable core temperature, coolant flow and neutron flux; with the beginning of the test water flow decreased while steam formation (voids) and (due to positive void coefficient) power increased; insertion of emergency rods added reactivity; power excursion(s)/steam explosion(s) destroyed reactor (no containment dome); uncontrollable graphite fire	Unit 1—as worst: No seismic induced equipment failure; emergency generators provided power after loss of offsite power; emergency and hillside backup generators (switch station) failed due to flooding; passive core cooling system stopped manually, valves could not be re-opened; no cooling water injected for 1 day; core meltdown began 3¼ h after tsunami struck, erosion of concrete base (0.7 m); hydrogen explosion (March 12)

(continued)

Table 5.1 (continued)

	TMI-2, USA; 28.3.1979	Chernobyl-4, former USSR, 26.4.1986	Fukushima 1, 1-4, Japan, 11.13.2011
Key failures (flaws)	Reactor operation with closed auxiliary feed pumps (violation of regulatory rule); mechanical failure of PORV; flawed design of PORV indicator, lack of dedicated instruments created confusion of operators; lack of situational awareness and knowledge/training of beyond design basis accidents (misperception of the real situation, i.e. loss of core cooling, hydrogen formation and behavior inside the reactor primary circuit)	Boost of reactor power inconsistent with approved procedure; less than 28 rods inserted to prevent prompt criticality (contradictory to regulations); peculiarities of physics and in reactor construction (dangerously large positive coefficients of reactivity), flawed control rod design, no adequate core simulation model; noncompliance with accepted standards and requirements of nuclear safety (transfer of automated functions to human operator, lack of safety culture)	Combination of extreme natural events systematically underestimated; no separation of promoting and regulatory bodies; lack of safety awareness, knowledge, skills and means regarding beyond design severe accidents; no thorough plant-specific full-scope probabilistic safety assessment (PSA)
Radiation exposure/ radioactive releases Off-site impact	First line of containment breached by operators (pumping of radioactive coolant to an auxiliary building for short time); small release of noble gases (93 PBq) and radioiodines (560 GBq) resulting to an average dose of 14 μSv to 2 million people nearby	Open vault to atmosphere with release of 4% (7.7 tons) uranium oxide fuel, 5200 PBq radioactive substances (mainly Xe-133, I-131, Cs-137, Sr-90); more than 116,000 people evacuated from 30 km zone; 100,000 km^2 contaminated in Belarus, Ukraine, Russia; worldwide traceable effects; 28 workers/firemen died shortly after the accident, 19 in the following 15 years, 9 children due to thyroid cancer; 4000 people could prematurely die from radiation in the surrounding countries.	Release from containment vessel due to deliberate venting, discharge of coolant water into the sea, uncontrolled events (explosions), no fuel ejection, release of 900–940 PBq radioactive substances (mainly Xe-133, I-131, Cs-137, atmospheric), 20 km exclusion zone, 150,000 people continuously displaced (as of 2013), health effects primarily psychological rather than physical (note: 18,500 people died due to quake/tsunami).
Findings, lessons learned	Confirmation of physical/chemical susceptibility of reactor to insufficient core coolingImportance of safety culture and	Significant design deficits—not well communicated to the plant operators and hidden from the Western nuclear community	Nuclear disaster was *"man-made as all direct causes were foreseeable"* (NAIIC) and no new phenomena were revealedImportance of

(continued)

Table 5.1 (continued)

	TMI-2, USA; 28.3.1979	Chernobyl-4, former USSR, 26.4.1986	Fukushima 1, 1-4, Japan, 11.13.2011
	training beyond routine work and design base accidents at plant levelKey role of situational awareness and of clear, unambiguous instrumentation in the control room.	(i.e. institutional failure) Lack of sophisticated and safe technology as well as economic and political pressure (cold war) may lead to dangerous conditions, a lack of safety culture, and the need to keep whole spectrum of major potential accidents in focus	socio-cultural attitudes and institutional deficits Dangerous overestimation of own capabilities as well as lack of imagination that accidents can happen and international conventions need to be followed, the latter calling for international controlImportance of safety margins and passive/inherent systems

See Sect. 4.3 for details on projected fatalities due to cancer caused by exposure to ionizing radiation. See Box 5.1 for definitions of terms used here

documented below, with the exceptions of Kyshtym (still not fully understood) and Chalk River (very early days of nuclear reactors).

1. Out of a total of about 440 nuclear power plants being commercially operated (roughly 15,000 reactor-years of experience), three reactors (Fukushima Daiichi units 1–3 counted as one) of different types (PWR, RBMK, BWR) faced a nuclear disaster—all with multiple units at the site, located in countries with extensive nuclear programs (North Western European countries, with near 100 units have not yet had a serious event).

2. Besides strong human factor elements, serious accidents in other industrial sectors, e.g. Challenger 1986 and Deep Water Horizon, Gulf Oil Spill 2010 were caused by single or combined technical failures in a classical sense. In contrast, the nuclear disasters have had more complex underlying multiple causes (e.g., pumps fail to run, valves fail to open); technical components failed when operated beyond design limits (e.g., stuck open pilot relief valve when exposed to water, condensate check valve closed and feedwater pumps consequently stopped due to false maintenance (TMI), emergency diesels failed after flooding (Fukushima)) or due to peculiar design (e.g. emergency shut down rods (Chernobyl)).

3. Lack of knowledge and safety awareness (Windscale, Lucens), experience and training (Tokai-mura) as well as communications skills (TMI) contributed as triggers, notably in the earlier days of nuclear power.

4. Clear underestimation of the possibilities of severe beyond design base accidents, and a lack of awareness, preparedness and training.

5. Individual and societal human factors at five overlapping hierarchical levels played a decisive role before, during and after the disasters:

Table 5.2 Summary and evaluation of three selected major accidents in non-commercial civilian nuclear power facilities

	Early phase; experimental-pilot reactor		Other nuclear facilities
	Windscale , UK; 10.10.1957 (military purpose facility)	Lucens, CH; 21.1.1969	Tokai-mura, Japan, 30.9.1999
Type; thermal power; start operation	Graphite moderated, air-cooled pile for Pu-production; 180 MW; 1950/51	Heavy-water moderated, CO_2-cooled; 30 MW; 1966	Reprocessing plant (fuel fabrication for experimental fast reactor); 1/1981
Location; units at site	NW coast; two pile military facility	50 km from Geneva/ Lausanne; single rock cavern plant	120 km NE of Tokyo
Type of accident	Graphite heatup and burning (1300 °C)	Loss-of-cooling, partial core melt during start up	Workers induced criticality when preparing a small batch of fuel
INES scale (1–7)	5 (wider consequences)	4–5	4 (local consequences)
Triggering event	Trial to release buildup energy in graphite by annealing, unexpected temperature rise in one channel due to fire	Collection of water in Uranium fuel elements with Magnesium cladding, accumulation of corrosion products, blockage of CO_2-coolant flow	Precipitation tank reached criticality after 7th bucket of aqueous uranyl nitrate (18.8% enriched U-235) added
[time]	[5:40 am]	[...]	[10:35 am]
Course of events	Speeding up of fans— whole reactor on fire for more than one day; shutting off all cooling and ventilation air	Overheating of fuel channel(s); split of surrounding pressure tube; loss of coolant; release of molten Uranium and heavy water into the rock cavern	Self-sustaining chain reaction with intense gamma and neutron radiation, fission products released into the building; stop of reaction by draining water (moderator) from the tank's cooling jacket (20 h later)
Key failures (flaws)	Inadequate handling of buildup (Wigner) energy; misdiagnosis of temperature increase	Penetration of water into the reactor circuit and core due to problems with the sealing water gasket of the recirculation fan	Lack of proper qualification and training; breach of safety principles; tank was not designed/configured to prevent criticality
Radiation exposure/ radioactive releases.Off-site impact	Reactor tank remained sealed but the fire released about 740 TBq I-131, 22 TBq Cs-137; spread all over Western Europe	Massive contamination of the rock cavern which was then sealed; no off-site impact	2 workers died, many workers exposed to elevated levels of radiation; basically no off-site release/impact; 161 people within 350 m radius evacuated for 2 days (high radiation level)
Findings, lessons learned	Importance of sufficient knowledge and		Importance of experience, training and safety

(continued)

Table 5.2 (continued)

	Early phase; experimental-pilot reactor		Other nuclear facilities
	Windscale , UK; 10.10.1957 (military purpose facility)	Lucens, CH; 21.1.1969	Tokai-mura, Japan, 30.9.1999
	experience (here with graphite exposed to neutrons) as well as of technical and organizational deficits	Lack of knowledge/ experience Early days technical arrogance	awareness at workers level

1. At the cultural/ideological—socio/political meta level:

   - In Japan, reluctance to question authority, nuclear accidents were regarded unthinkable;
   - in the former Soviet Union, people were assumed reliable and not susceptible to make any mistake;
   - "insularity thinking" (Japan and the former Soviet Union did not follow international standards and conventions; there was a feeling of superiority in the USA).

2. At the higher institutional level, there were clear institutional deficits:

   - in Japan, e.g., no separation of operational from regulatory bodies,
   - in the former Soviet Union, the "design institute" had a dominating role and there was a culture of central control—"Moscow",
   - in the USA, there were deficits within the regulatory body,
   - in the UK, there was a lack of sufficient oversight and control in plutonium-production for bombs in times of the Cold War,
   - in Switzerland, there was over-ambition aimed at developing its own reactor design.

3. At the company—plant level: Lack of safety awareness aggravated by inadequate design:

   - in Japan, TEPCO (the operating company) tended to ignore new insights from research and experience gained elsewhere and to delay updates of safety requirements and establishment of accident management measures at Fukushima;
   - in the former Soviet Union, the "design institute" did not adequately communicate design flaws and sensitivity parameters to the Chernobyl plant operators;
   - in the USA, inadequate indicators and poorly designed instrumentation created confusion in the TMI-control room.

4. At the lower plant level: Lack of situational awareness and notably of skills to manage accidents beyond design base and in adverse circumstances:

- at TMI, inadequate communication between maintenance and operating crew, non-safety oriented human actions;
- at Chernobyl, ignorance of guidelines and unsafe working practices;
- at Fukushima Daiichi, shutting off the passive core cooling system and no water injection for the first day (unit 1). Workers were totally overwhelmed and lacked training to manage the situation and hostile environment.

5. Accidents in commercial nuclear power plants did not reveal unforeseeable causes or unknown phenomena. A lack of imagination worsened by missing full-scope probabilistic analyses and sufficiently detailed plant simulation models (the latter particularly for large RBMK) are implicated in each accident.

6. Accidents demonstrated the importance of site conditions and of properly designed and operated containment systems as well as the potential of large radioactive releases with huge (partially over stated) radiological off-site impacts. At Fukushima, the significance of non-radiological, psychological health impacts associated with the evacuation and fear of radiation[5], as well as broader economic and political consequences, were underestimated.

Box 5.1 Useful Definitions

LWR: Reactor using light-water as coolant and moderator of neutrons (Light Water Reactor), either under high pressure (around 155 bar) to avoid boiling (Pressurized Water Reactor with secondary feedwater-steam cycle including steam generator(s), **PWR**) or under moderate pressure (around 75 bar) to allow boiling and formation of steam (Boiling Water Reactor with steam separator and direct steam cycle, **BWR**); both equipped with a steel reactor pressure vessel, RPV); small core design due to high power density (80 MWth/m^3); power range 600–800 MWe for demo/early second generation plants and 1000–1600 MWe for late second generation and new third generation plants; typical thermal efficiencies about 33% or slightly higher depending on technology and heat sink; uranium dioxide fuel with low U-235 enrichment (3–4%), sometimes mixed uranium-plutonium oxide fuel (MOX). LWRs currently dominate commercial nuclear power parks, notably in OECD countries.

PHWR: Reactor using pressurized heavy water as coolant and moderator to allow for natural uranium fuel, such as the Canadian CANDU type. Normal water is H_2O while heavy water is D_2O where D = deuterium the heavy isotope of hydrogen ^2H where "normal" hydrogen comprises a single proton ^1H.

(continued)

---

[5]I. Fairlie, Summing the health effects of the Fukushima nuclear disaster, Chain Reaction 125, 24–25 (Nov 2015). (http://search.informit.com.au/documentSummary;dn=794793518968591;res=IELAPA).

Box 5.1 (continued)

RBMK: Reactor using light water as coolant and graphite as structural material and moderator; huge number (around 1500) of pressurized (65 bar) channels; large cores due to moderate power density (4.2 MWt/m^3); power ranges from 1000 to 1500 MWe; low U-235 enrichment (2 to 2.5%). RBMKs were only built and operated in the former Soviet Union; last built was in 19xx?. The RBMK had not concrete secondary containment, and once the reactor at Chernobyl caught fire, it vented straight to atmosphere. There are xx RBMK units still in operation in yyy.

PORV: Pilot operated relief valve, e.g., to stabilize the pressure of the primary reactor circuit (it failed stuck open in TMI-2 and played an important role in the early phase of the accident).

Bq: Becquerel, measure of radioactivity, defined as radioactive decay per second ; G (10^9)/T (10^{12})/P(10^{15}) Bq (Ci = Curie, a precursor unit to the Bq; 1 Ci = 37 GBq i.e. 37 billion Bq).

Reactivity coefficient: Change of neutron balance associated with a given change, in physical parameters, e.g., in fuel temperature, void fraction, power. A negative fuel temperature coefficient—for example—means that rising fuel temperatures has a negative effect on the neutron balance and makes the reactor subcritical (i.e. more safe) while a positive coefficient (like in Chernobyl) causes the opposite.

PSA: Probabilistic Safety Analysis is a systematic and comprehensive technique used to evaluate the risks associated with large-scale engineered systems such as nuclear power plants (see Sect. 4.2).

5.3   Cost as a Measure of Full Consequences of Nuclear Accidents

As mentioned in Sect. 4.2, estimation of the consequences of hypothetical accidents in nuclear power is not a regulatory requirement, but is sometimes done with PSA level 3. The study of the full consequences of experienced accidents, best measured by cost, is less common still. However, a realistic and full accounting of actual and potential costs should be a key input into assessment of safety improvements on a cost-benefit basis (see e.g., regulatory analysis guidelines from the NRC[6]), external-ity figures (Sect. 4.6), and comparative assessments.[7]

---

[6]US NRC, Regulatory Analysis Guidelines of the US Nuclear Regulatory Commission, NUREG/BR-0058, Rev. 4 (2004).

[7]Baxter, M. (2001). Health and Environmental Impacts of Electricity Generation Systems: Procedures for Comparative Assessment: International Atomic Energy Agency, Technical Reports Series No. 394, IAEA, Vienna, December 1999 (STI/DOC/010/394 TRS 394), 193 pp., price 630 Austrian Schillings (45.78 EUR) paperback, ISBN 92-0-102999-3.

In 2010, an OECD report[8] summarized the wide range of estimates for the cost of major hypothetical accidents. For instance, about USD 10 Billion was given as the estimated hypothetical cost, excluding radiological health impacts, for a major accident at Mühleberg—a small single unit near Bern, Switzerland. After "Fukushima", the estimates tend to be higher: The French IRSN estimated the cost of a hypothetical major (Chernobyl-sized release) accident in France, to be USD 500 Billion.[9,10] In a conservative German study[11], the rare worst-case accident ("Unfallkategorie 1") was assigned a cost of 14,000 Billion Euros, being four times the German GDP. This is a massive range, not only depending on the site and specific accident scenario considered, but heavily on the assumptions and methods used.

To provide a point of reference, focusing on the cost of eventual fatalities due to ionizing radiation, in Sects. 4.2 and 4.3, it was shown that rare severe accidents may lead to thousands of eventual cancer fatalities. According to UNSCEAR, for those that do die from the resultant cancer, the average years of life lost (YOLL), also called loss of life expectancy (LLE), is 10–20 years.[12] Taking a conventional value of a full human life of 6 million USD[13] with life expectancy of 80 years, gives a loss of 0.6–1.2 million USD for the average victim. Thus ten-thousand eventual fatalities would be given a cost of 6–12 billion USD. NRC guidelines are similar, but ignore YOLL, treating an eventual fatality as a full life lost.[14]

Concerning experienced accidents, there are difficulties around agreement on what types of costs to include. A 2013 OECD/NEA workshop,[15] and the NRC Regulatory Analysis[8] provide useful guidelines. Following these and compiling various sources, rough estimated costs of the historical three major accidents, and a French hypothetical one,[16] are summarized in Table 5.3, with details in Annex A.2.

---

[8]OECD/NEA, Methodologies for Assessing the Economic Consequences of Nuclear Reactor Accidents, 2000.

[9]Pascucci-Cahen, L. and Momal, P., The cost of nuclear accidents. OECD/NEA International Workshop on the Full Costs of Electricity Provision. 2013.

[10]IRSN. Méthodologie appliquée par l'IRSN pour l'estimation des coûts d'accidents nucléaires en France. PRP-CRI/SESUC/2013-00261 (2013).

[11]Die Risiken der Kernenergie in Deutschland im Vergleich mit Risiken anderer Stromerzeugungstechnologien", P. Preiss, S. Wissel, U. Fahl, R. Friedrich, A. Voss, Universität Stuttgart, IER, February 2013.

[12]See Tables 29 and 30 of "Sources and effects of ionizing radiation." UNSCEAR (Vienna, Austria) (1994).

[13]E.g., See EPA Mortality Risk Valuation: https://www.epa.gov/environmental-economics/mortality-risk-valuation.

[14]US NRC "Reassessment of NRC's dollar per person-rem conversion factor policy." NUREG-1530, November (1995).

[15]The workshop was called OECD/NEA Workshop on Approaches to Estimation of the Costs of Nuclear Accidents, with materials at https://www.oecd-nea.org/ndd/workshops/aecna/presentations/

[16]With a 10^{18} Bq massive release, similar to Chernobyl.

Table 5.3 Estimated full cost, in Billions of 2017 USD, of major nuclear accidents, including the French hypothetical major accident[a, b]

Cost (USD Bil.)	ISRN 2011	TMI 1979	Chernobyl 1986	Fukushima Dai-ichi 2011
On-site	10	5-10	**25–35**	**20–30**
Life and Health + Public-Economic	<60 + 110	0.1	26–33 + **150–250**	14–15 + 50–100
Replacement + Retrofits	110	5–15 + **100–200**	10–30 + 2–8	**100–150 +** 60–120
Beyond…	200 "reputation"	**Sector inflection point**	**Political instability?**	**German nuclear exit, etc.**
Approx. Total	< 500	110–225	213–356	244–415

The interval in the approximate total is the sum of the upper and lower bounds of the individual costs. Comparatively large costs given in bold. Life and health impacts include deterministic and projected radiological fatalities, and evacuation related trauma. Replacement is the incremental cost of replacement power. The "beyond" category identifies potential costs that are outside of the scope of the estimation. Arguably, the TMI retrofit costs could be excluded on the basis that they were beneficial. Deeply uncertain costs and potential benefits relating to impacts on energy policy are not quantified. See details in Annex A.2
[a]IRSN. Méthodologie appliquée par l'IRSN pour l'estimation des coûts d'accidents nucléaires en France. PRP-CRI/SESUC/2013-00261 (2013)
[b]"Die Risiken der Kernenergie in Deutschland im Vergleich mit Risiken anderer Stromerzeugungstechnologien", P. Preiss, S. Wissel, U. Fahl, R. Friedrich, A. Voss, authors, Universität Stuttgart, IER, February 2013

This rough accounting indicates that a major accident can cost hundreds of billions of USD to society. Perhaps most notably:

- In terms of public and on-site impacts, TMI cost on the order of USD 10 billion. It had a major measurable influence, costing ten times more, through the increased safety requirements worldwide (see Sect. 5.6)—and is hence has substantial benefit associated with it.
- Chernobyl showed the potential of massive public economic costs, claimed to have exceeded even the major toll on life and health due to exposure to ionizing radiation.
- Fukushima gave insight into the non-radiological trauma caused by evacuation and continued displacement. Perhaps most costly is the disruption of nuclear power, which continues to cost Japan dearly as natural gas is imported to replace the lost nuclear electricity production.
- The IRSN hypothetical estimate is similar to the estimated costs of Chernobyl and Fukushima.

Although the toll on life and health of a major accident are severe, they form a minority of the full costs: For both the IRSN study, and Fukushima, it is less than 10% of the total; and for Chernobyl—a unit with design weaknesses that do not exist in modern units—it is likely to be less than 25% of the total, especially as this relies on the Linear-No-Threshold model for long-term low doses.

According to our cost estimates, based on a comprehensive database of about one thousand events in nuclear energy facilities,[2] the cost of these three major events dominate the total, implying that the most severe accidents effectively define the accident externality.

5.4   Disasters in Other Industrial Sectors

For the purpose of comparison, four major accidents in industries other than the nuclear sector are briefly summarized and characterized (See Chernov and Sornette[17] for a detailed analysis).

- Vajont Dam disaster (Italy, 1963): A huge landslide filled the dam reservoir, caused a mega-tsunami (max. wave height 250 m) and which went over the dam, wiping out several villages, killing 1917 people and destroying the valley below.

   - *Natural/hazardous site conditions—politically ignored evidence (level 2 human factor as introduced before).*

- Bhopal pesticide plant gas leak (India, 1984): Release of toxic methyl isocyanate (MCI) nearby shanty towns, killing 3787 people directly and causing *inter alia* 3900 severely and permanently disabling injuries (other estimates show much higher numbers). The cause remains unclear; however, a backflow of water, either due to slack management and deferred maintenance or sabotage, entered into a MCI tank triggering a fierce chemical reaction and finally the disaster.

   - *Slack management/maintenance (level 3 (company-plant), level 4 (lower plant) human factor, outlined before).*

- Deepwater Horizon oil spill (Gulf of Mexico/USA, 2010): Explosion and sinking of the oil rig, discharge of about 780,000 m^3 of oil, largest accidental spill (180,000 km^2 directly impacted) caused by oil at high pressure and temperature entering the annulus of the well following failure of a concrete seal and failure of the "blowout preventer" (BoP) on demand to stop the flow; 11 workers missing/ declared dead.

   - *Inadequately assured quality of the concrete and insufficient designed/ maintained vital component (BoP)—human factors at level 2 (institutional), level 3 (company-plant).*

- Challenger Space Shuttle disaster (USA, 1986): Orbiter broke apart 73 s into its flight, leading to death of 7 crew members and grounding of the space shuttle

---

[17]Chernov, D. and Sornette, D. Man-made catastrophes and risk information concealment (25 case studies of major disasters and human fallibility), Springer (2016).

program for almost 3 years, caused by an O-ring in the right rocket booster that failed at liftoff, finally allowing gas to reach the outside and to impinge the external fuel tank.

– *Technical failure—NASA, organizational culture (level 2,3 human factor).*

It may be appropriate to also compare with what could be termed "slow disasters", which are less or almost invisible but nevertheless of comparable severity when properly accounted for. For instance, as discussed in Sect. 4.4.3, extraction, transport, processing, and combustion of the life cycle of coal induce high external health and environmental costs. In the US, these costs to the public have been estimated at a third to over one-half of a trillion dollars annually.[18]

As shown before, the direct health impact of nuclear disasters is comparably small[19] and the dimension of environmental impact (land losses and contamination) is in part shared with other industrial disasters. However, the kind of exposure (i.e. radiation) as well as potentially large, long lasting effects (area contamination, permanent displacement) and, in particular, potential latent health effects (cancer) make nuclear accidents somehow unique in the psyche of the general public who have become fearful of nuclear power.

It is important to emphasize that severe nuclear accidents are rare in absolute and relative terms due to disproportionate, far-reaching design and operational measures. Nevertheless, the physical process (surplus of neutrons, decay heat, etc.) and current technology (high power density and size, meltable fuel cladding and structural material, etc.) make todays reactors vulnerable to perturbations and deficits of the operational environment; although of substantially low frequency, the potential of large radioactive releases and associated frightening consequences for present generation reactors cannot be ignored.

5.5 The Dread Factor

Nuclear (fissile) energy is plagued by general misgivings from the public associated with the fear of extreme nuclear catastrophes (probably in part linked with the association with the horrors of nuclear bombs), the dread of invisible radiations, a public relation issue on its safety track record and the problem of long-lived nuclear wastes.[20]

---

[18]Epstein, P.R. et al. "Full cost accounting of the life cycle of coal." Annals of the New York Academy of Sciences 1219: 73–98 (2011).

[19]Almost 50 (Chernobyl, 28 workers/firemen died within first three months, 19 patients in the years 1987–2004) compared to almost 2000 (Vajont) to 4000 or more (Bhopal).

[20]Gómez Cadenas J.J., Nuclear Power, No Thanks? In: The Nuclear Environmentalist, pp 103-124, Springer, Milano (2012).

The reader should be well aware by now that nuclear fission splits up the exceptionally heavy elements uranium and thorium into medium-sized nuclei such as xenon and strontium, releasing energy. This fission "chemistry" at the level of the atoms' nuclei implements routinely in civil nuclear plants what was for many centuries the most sought-after goal in alchemy and the pinnacle of enlightenment—the philosopher's stone. The philosophers stone was a legendary alchemical substance that was supposed to be capable of turning base metals such as mercury (atomic number 82) into gold (atomic number 79).[21] We hypothesize that one factor contributing to the dread associated with nuclear energy has its roots in the image of the "mad" or irresponsible alchemist playing with the mysterious forces of Nature (see Box 5.2). In other words, nuclear energy is identified in the collective consciousness with playing with the occult dimensions of the Universe, which should remain untouched as they lay beyond our full ability for understanding and control.

Box 5.2 Transmutation and the Philosopher's Stone

The philosopher's stone was the central symbol of alchemy, symbolizing perfection, enlightenment, and heavenly bliss. For instance, Isaac Newton is known to consider his scientific achievements as relatively secondary to his quest for alchemy and mysticism. For the scientists and philosophers of those centuries, the problem was that, until the discovery of radioactivity by Henri Becquerel in 1896, the known chemistry could only exploit processes involving reconfigurations of the clouds of electrons orbiting around the atoms' nuclei, which involve energy changes of the order of 1 eV (electron-volt). In contrast, the energy associated with just one fission of one nucleus or one fusion event between two nuclei is of the order of several MeV (Mega or million eV). These energies and reactions could only become controllable once completely different technologies were progressively developed in the first half of the twentieth century. Nowadays, it is possible to transmute lead into gold,[22] or in principle any element into a different one, using nuclear reactors or specifically designed particle accelerators, albeit the cost of transmutation to create gold is many times larger than the market price of gold.

The fear and dread that many people feel towards anything related to nuclear has a long history, loaded with epic historical events loaded with emotions, from the dropping of The Bomb up to the recent crisis at the Fukushima reactor in Japan. We refer to Weart,[23,24] which dissect the interplay between emotions and reason in

---

[21]The atomic number of an atom is defined by the number of protons its nucleus possesses.

[22]John Matson, Fact or Fiction?: Lead Can Be Turned into Gold (Particle accelerators make possible the ancient alchemist's dream—but at a steep cost), Scientific American, January 31, 2014.

[23]S.R. Weart, Nuclear fear, a history of images, Harvard University Press, first edition 1988, 550 pages.

[24]S.R. Weart, The Rise of Nuclear Fear, Harvard University Press, Reprint edition (March 19, 2012), 381 pages.

the atomic debate of the past 100 years. He convincingly demonstrates that the potent images associated with all things nuclear prevent us from facing the real issue of how we are to *"improve world prosperity while burning less fossil fuel?"*

Box 5.3 The Power of Image and the Rise of Fear of Nuclear Systems in Society

Weart shows how imagery has dominated the nuclear debate. From initial scientists' dreams of nuclear-powered clean cities, to anti-nuclear fears of radioactive mutated monsters, he uncovers how the public atomic dreams and nightmares form *"one of the most powerful complexes of images ever created outside of religions."* After the Bomb, the collective public consciousness believed that, somehow, exposure to radiations could lead to utter catastrophes or to impossible feats, such as mutations with new powers, by the way, feeding a large science fiction literature on superheroes with superpowers derived from exposure to radiation.[25] The emotional relationship not just with atomic weapons but with nuclear radiation in general is well-illustrated by the case of MRI (magnetic resonance imaging) used in medicine. In fact, the original physics literature of the 1940s that first described the method called it "nuclear induction" and later "nuclear paramagnetic resonance", which finally evolved in the late 1950s to "nuclear magnetic resonance" (NMR). This term accurately refers to the precessions of the spins of the nuclei at the center of atoms in reaction to applied magnetic fields (a static one and an oscillating one, the later creating a resonance in the presence of the former, hence the first historical name). When imaging methods using NMR signals were first developed, the term NMR imaging was applied to them. In reaction to the patients' concerns over the dangers of nuclear radioactivity, by the mid-1980s, the word "nuclear" had been largely dropped and replaced by the more neutral and appeasing "magnetic resonance imaging" (MRI).

Another protagonist who has been deeply involved in the nuclear debate is Mr. Patrick Moore (not to confuse with the well-known British media astronomer), cofounder of Greenpeace and former director for 15 years. In an interview given in 2008, he explains one of the reasons why he left Greenpeace[26]: *"...we were so focused on the destructive aspect of nuclear technology and nuclear war, we made the mistake of lumping nuclear energy in with nuclear weapons, as if all things nuclear were evil. And indeed today, Greenpeace still uses the word "evil" to describe nuclear energy. I think that's*

(continued)

---

[25]For instance, the fictional Marvel character called Spider-Man owes his super strength, agility and sixth sense to having been bitten by a radioactive spider. Another fictional Marvel (green super strong) Hulk character is a scientist who was exposed accidentally to gamma radiations. See https://en.wikipedia.org/wiki/X-Men

[26]Fareed Zakaria, A renegade against Greenpeace, Newsweek 4/12/08 (2008) (http://www.newsweek.com/renegade-against-greenpeace-85579 accessed 19 Sept 2019).

> **Box 5.3** (continued)
>
> *as big a mistake as if you lumped nuclear medicine in with nuclear weapons.*
> *Nuclear medicine uses radioactive isotopes to successfully treat millions of*
> *people every year, and those isotopes are all produced in nuclear reactors."*
> He continues with: *"How many Americans know that 50% of the nuclear*
> *energy being produced in the U.S. is now coming from dismantled Russian*
> *nuclear warheads?"*

On August 6th and 9th 1945, the United States dropped two nuclear bombs on the Japanese cities of Hiroshima and Nagasaki. With the help from the national government through the Hiroshima Peace Memorial City Construction Law passed in 1949, Hiroshima was rebuilt after the war, and is today a pleasant and prosperous city of 1.2 million people. It was initially thought that it would take half to a full century before it would be inhabitable again.

On 26 April 1986, the Chernobyl disaster struck the No. 4 light water graphite moderated reactor at the site near the city of Pripyat of the Ukrainian Soviet Socialist Republic of the Soviet Union. During the disaster, four hundred times more radioactive material was released than at the atomic bombing of Hiroshima. The radiation levels in the worst-hit areas of the reactor building, including the control room, have been estimated at 300 Sv/hr, providing a fatal dose in just over a minute. Compared with the global average exposure of humans to ionizing radiation of about 2.4–3 mSv per year (milli Sv), the exposure of residents who were relocated after the blast was of the order of 350 mSv. For reference, the lowest annual dose at which any increase in cancer is becoming statistically significant (but still very small) is documented at about 100 mSv. And, the recommended limit for radiation workers every 5 years is 100 mSv, i.e. 20 mSv per year. In 2009, the levels of radiation as measured in Pripyat and Chernobyl were reported to be of the order of 1 μSv per hour (one thousandth of 1 mSv per hour), corresponding to approximately 10 mSv per year.[27] This is about 3–5 times larger than the typical global average exposure of humans in the World, as mentioned above, but half the maximum dose recommended for radiation workers. Moreover, different studies have shown surprising ecological rebound in radiation-contaminated areas around Chernobyl. Elk, roe deer, wild boars, and other wildlife are apparently prospering in the radiation-contaminated preserve that is off limits to people. Researchers have found *"no evidence of a negative influence of radiation on mammal abundance"* in the Chernobyl exclusion zone straddling the Belarus-Ukraine border.[28,29] One of the

---

[27]The Chernobyl gallery, http://chernobylgallery.com/chernobyl-disaster/radiation-levels. Accessed 20 Sept. 2017.

[28]T.G. Deryabina, S.V. Kuchmel, L.L. Nagorskaya, T.G. Hinton, J.C. Beasley, A. Lerebours, and J.T. Smith, Long-term census data reveal abundant wildlife populations at Chernobyl, Current Biology 25, R811–R826, October 5 (2015).

[29]A.Pape Møller and T.A Mousseau, Are Organisms Adapting to Ionizing Radiation at Chernobyl? Trends in Ecology & Evolution 31 (4), 281–289 (2016).

co-authors of the first study, the UK environmental scientist J. Smith, summarizes the findings in the following provocative statement: "*When humans are removed, nature flourishes, even in the aftermath of the world's worst nuclear accident.*" The UK environmentalist J. Lovelock radically suggested that nuclear wastes could be the "*...perfect guardian against greedy developers...*"[30] One should however temper this controversial statement to consider the issue of doses and the evidence that some contaminated sites still show a reduction in the number of spiders and insects. Of course, radiation causes DNA damage and there is concern that these studies fail to fully assess the harm to individual animals.

Obviously, we are not implying that radiations are good or that nuclear accidents are innocuous. But the evidence suggests that the dread view, i.e. that the occurrence of an accident at a nuclear power plant would lead to the complete annihilation of the locality in a perimeter of 10–50 km from the epicenter of the catastrophe for centuries or millennia, seems simply wrong. Life, including human beings and organized society, are remarkably resilient and seems to recover much faster than imagined in the doomsday scenarios, as the above examples suggest.

The previous chapters have documented facts, which stand beyond the beliefs and fictions of the real risks associated with (present generation) nuclear energy plants. Even with the example of Fukushima, the facts paint a much less dreadful picture than the general public perception. However, it is not the public who is at fault, but likely the industry, the regulators and the policy makers who have shown a tendency for secrecy, keeping opaque the nuclear track record, thus catalyzing a climate of mistrust and doubts, and allowing an information vacuum that the sensation-loving media and anti-nuclear advocates were happy to fill.

In fact, dread was perhaps warranted in the 1980s and 1990s, given the PSA values in the late 1970s potentially as high as 10^{-3} for the core damage probability (CDP) per reactor-year. As explained later, in reaction to major eye-opening accidents, significant improvements have occurred on many fronts, which are estimated to have decreased this rather alarming early figure to a more acceptable CDP value of $\sim 10^{-5}$ per reactor-year, e.g., for retro-fitted Swiss nuclear plants in the 2010s (see Fig. 5.3). There is an expectation that Gen III reactors (and even more so for the future Gen IV reactors) could provide CDP levels below 10^{-6}. As mentioned, one should also keep in mind that the probability for core damage is not the same as the probability for a large release of radioactive material, the latter being in general estimated to be at least one order of magnitude smaller, due to the existence of proper containments.

Another factor fueling dread is the perception that a nuclear accident could pose an existential threat, mainly for small countries: a major nuclear accident with

---

[30]The full quote is: "[T]he natural world would welcome nuclear waste as the perfect guardian against greedy developers, and whatever slight harm it might represent was a small price to pay... One of the striking things about places heavily contaminated by radioactive nuclides is the richness of their wildlife... The preference of wildlife for nuclear-waste sites suggests that the best sites for its disposal are the tropical forests and other habitats in need of a reliable guardian against their destruction by hungry farmers and developers." Lovelock, James, The Revenge of Gaia, Penguin (2007).

significant release of radioactive materials close to Luxembourg, for instance, could require the evacuation over years to decades of the total population of the tiny country. It is clear that such an eventuality is unacceptable. In contrast, a large country benefits from what we could term "territorial diversification" and even a major accident does not place the whole country at risk as illustrated by the recent disaster in Japan. For small countries, or for local populations, the main way to address the threat to existence is to go towards vanishingly low risks, combined with reducing the magnitude of accidents should they occur, goals that seem possible with the novel technology and progress described in Sect. 6.2.

Thus, in the early decades of civil nuclear energy, dread could probably be rationalized as a kind of uncertainty aversion[31] or even ambiguity aversion[32] from the public, a quite healthy attitude at the dawn of the nuclear age. The challenge, taken up by this book, is to inform and develop a culture together with an awareness that can change dread[33] into a fresh ambition for a completely safe future for nuclear energy, as suggested in Chap. 7.

5.6 Progress and Framework to Make Nuclear a Viable Option

Future nuclear power reactors have to be considered and understood as a part of an integrated nuclear system that includes all elements and aspects of the fuel cycle. Furthermore, this future nuclear system must comply with sustainability criteria such as availability and preservation of resources, curbed greenhouse gas emissions during normal operation and limited amount together with safe management of wastes. Elimination of accident risk, conflict potential and resistance to it all need to be considered from cradle to grave (see Sect. 1.4).

New nuclear must be economically viable and affordable. Regulatory procedures for new reactor designs need to be developed as existing regulations may serve as unreasonable or irrelevant barriers to novel designs. And the competence of new nuclear countries needs to be assured through regulation and training. Further, the potential for different future nuclear technologies to address and be adapted to different needs, infrastructures, and objectives, need to be kept in mind.

---

[31]Uncertainty aversion refers to the dislike and avoidance of risks for which the probability of occurrence is not known, in favor of risks for which the probabilities can be assessed.

[32]Ambiguity aversion refers to the dislike and avoidance of risks that are unknown in their amplitude, in addition to their probabilities.

[33]Can the dread that many people hold towards anything related to nuclear be eased? This will require important and continuous efforts, for the reasons exposed above, which link dread to deep-seated fears of the mysteries of invisible radiations and the terror associated with the atomic bomb. After all, our brains are just only slightly more evolved than that of our ape cousins, with the rational cortex being often under the control of the emotional brain, associated with the limbic system also known as the paleo-mammalian cortex (and in particular the hypothalamus and amygdala).

From the beginning of nuclear development, the possibility of *unusual hazards from peaceful, beneficial applications of nuclear energy* was recognized at an early time. Therefore, the developers had resolved to seek an exceptionally high level of safety. This objective, which was unprecedented in industrial development and has surpassed the pattern of other engineering disciplines, has been maintained and improved upon throughout the history of the nuclear industry.[34] In other words, nuclear safety requires a continual quest for excellence.[35]

As accidents at nuclear plants could not be ruled out absolutely as recognized in the USA as early as 1952, special protection from the consequences of potential severe accidents was added to plants, i.e. certain nuclear power plants, notably light water reactors, were housed in strong concrete buildings designed to contain radio-active material and prevent its escape from the plant if an accident did happen.[36] This containment system became a cornerstone of safety strategy and provided the final barrier to fission product release in the system of *defense in depth*. Accident prevention has been regarded as the first safety priority of both designers and operators, mainly achieved through the use of highly reliable structures, compo-nents, systems, and procedures as well as through commitment to a strong safety culture. The necessary additional protection was realized by the incorporation of many safety features engineered into the plant, e.g., to ensure safe shutdown, continued core cooling and confinement of fission products, for which design parameters are defined by deterministic analysis. As already outlined in Sect. 2.2 the accidents in the spectrum requiring the most extreme design parameters for the safety feature are termed the design base accidents. Since beyond design base accidents could occur, due to their tiny yet non-zero probability, other procedural measures including technical measures were introduced, notably severe accident management, to mitigate their potentially major consequences.

By introduction of the method of probabilistic safety analysis (PSA) in the mid-1970s, the behavior of the reactor and its safety features became subject of the analysis, aiming at the simultaneous estimation of probabilities and conse-quences of the entire set of potential accidents, in particular beyond those considered in the design base. Besides this, historical advances have been achieved by use of experience feedback and extensive research programs.[37] However important safety lessons were learned from three severe accidents that have occurred and from other events that were not so severe. The accidents at Three Mile Island (TMI), Chernobyl

---

[34]INSAG-5, The Safety of Nuclear Power, IAEA, Vienna 1992.

[35]INSAG-12, IAEA, Vienna, 1999.

[36]This requirement was soon adopted for light water and heavy water reactors throughout most of the world. However, early light water reactors in the Soviet Union and Eastern Europe were provided only with partial containment buildings. It was in the mid-1970s that full containment of newer light water reactors was introduced in Eastern Europe. Lack of a tight, sturdy containment building around large gas cooled reactors (including RBMK) has been compensated for by other design features.

[37]Research to provide the necessary engineering base have spanned the past 35 years with the cost equivalent of USD 5000 billion around the world.

and Fukushima as well as the extensive electrical fire at the Browns Ferry plant triggered extensive safety reviews, sometimes including outages for inspection and retro-fits at least at plants of the same type and design. These actions have improved safety standards worldwide:

- TMI (1979) intensified research and demonstrated that severe accidents, up to that time only assumed to be possible, could indeed take place and were very costly. It confirmed the wisdom of having a tight containment building, which at TMI did its job as there was no major release of radioactivity into the environment. It was recognized human factors had not been adequately included in previous safety cases. It was also discovered that preparations for an accident were totally absent, contributing to the fear of the public. It also highlighted the value of proper PSA in revealing safety weaknesses and encouraged further development of PSA as more deficits came to light.[38] The accident prompted numerous advances in design of safety systems, containment, operating practices and the man-machine interface. Combined, all of these measures led to improvements in the safety of nuclear plants throughout the world.
- Chernobyl (1986) revealed basic design defects in the RBMK light water graphite moderated reactor under certain operational conditions that could cause excessive chain reactions (power bursts), without a means to stop them. If the reactor had been located in a full-sized containment building, the consequences might have been more benign, although it is debatable whether such a building could have been designed for a plant of this type to withstand the physical loads caused by the accident. Mistakes were also made by the personnel and training since they did not fully understand the physics and behavior of their plant. There was a lack of preparedness both on-site and off-site to be able to properly manage the accident, Since the accident, RBMK reactors have continued in service in Russia, but their safety has been significantly improved. It can be ruled that no other plant of this design will ever be built.

    Because of the large release of radioactive material, large parts of continental and northern Europe experienced radiological exposures, resulting in a tremendous increase in public fear and organized resistance against the use of nuclear power.
- The Browns Ferry (1979) electrical fire revealed deficits in the spatial separation of redundant safety systems and in the fire resistance of related cables. It led to significant improvements, in particular in fire prevention at nuclear plants in many countries.

We will now take a more in-depth look at Fukushima-Daiichi disaster, which started on 11 March 2011, and the lessons learned from it. The root cause of the disaster was flooding from a Tsunami that followed the Tōhoku earthquake. The tsunami exceeded the design basis of the sea wall, even although tsunamis of this

---

[38]Nuclear Regulatory Commission. *Severe accident risks: an assessment for five US nuclear power plants*. No. NUREG--1150-VOL. 3. Nuclear Regulatory Commission, 1991.

size were known to be possible. The disaster demonstrated the long potential duration of such an accident, difficulties encountered due to unavailability of numerous safety systems due to a common cause, and the long time period it takes for the site to be recovered. The three reactors affected at Fukushima all had primary containment structures that helped contain the radioactive core inventories but in each case the containment also failed as a result of hydrogen explosions and core meltdown. The Fukushima-Daiichi disaster also stressed the potential effect of a filtered containment venting system (which they did not have), and the influence of hostile environment on accident management measures.

The Fukushima-Daiichi disaster raised questions regarding the controllability of nuclear energy with existing reactor designs. This affirmed the special role of nuclear accidents in media and public perception, paying lesser attention to the high death toll from the earthquake and tsunami.[39] Some countries like Germany decided to rapidly phase out the operation of nuclear power plants, and continue relying on coal for electricity generation. Other countries followed a more cautious path where, for example Switzerland has declared it will not replace existing reactors once they are closed. France has announced a reduction of nuclear power in the electricity mix, opting for wind power instead. Other countries including Japan, the USA and the UK opted to continue with nuclear, even building new reactors. China, India and Russia are actively expanding their nuclear fleets. And other countries like the UAE, Iran and Saudi Arabia are diversifying into nuclear power for the first time. The international response to the Fukushima accident, therefore, was highly diverse.

Following the Fukushima-Daiichi disaster, many countries and organizations requested comprehensive safety reviews and risk assessments, leading to early safety improvements at the global scale. One of the most prominent examples is the *stress test* requested by the European Council of 24/25 March 2011. This was the first time that a multilateral exercise covering over 140 reactors in all EU countries (plus Switzerland) operating nuclear power plants was considered. The European Nuclear Safety Regulators Group (ENSRG) was invited to develop the scope and modalities for the stress test with the support of the Western European Nuclear Regulators' Association (WENRA). Consensus on the specifications was achieved only 2 months later. The work on the stress test should be carried out along two parallel tracks: A safety track to assess how European installations can withstand the consequences of various extreme external events and a security track to analyze threats and incidents due to malevolent or terrorist attacks, which are not discussed within publicly available reports.

The safety track included earthquakes, tsunamis, flooding and extreme weather conditions, loss of safety systems and severe accident management as main topics. The assessment was done in a three-step process: (1) Operators had to perform an

---

[39]On 10 March 2015, a Japanese National Policy Agency report confirmed 15,894 deaths, 6152 injured, and 2562 people missing across 20 prefectures, as well as 228,863 people living away from their home in either temporary housing or permanent relocation, due to the Great Eastern earthquake and tsunami (Wikipedia, visited 30.01.16).

assessment and make a proposal for safety improvements, following the specifications; (2) the national regulators independently reviewed the operators' assessments and issued requirements; (3) national reports were peer reviewed at the European level. The main results are as follows[40]:

- All countries have taken significant steps to improve the safety of their plants, with varying degrees of practical implementation.
- There was consistency across Europe in the identification of strong features, weaknesses and possible ways to increase plant safety. With regard to the external hazard topic, design basis events were well addressed in the country reports, although there were significant differences in national approaches. However, the assessment of margins beyond the design base has been quite diverse, and very few countries assessed cliff-edges in the manner requested by ENSREG. The situation is less satisfactory with regard to extreme weather, and especially for a combination of extreme weather phenomena.
- Significant measures to increase plant robustness have already been decided or are considered, including provisions of additional equipment to prevent or mitigate severe accidents, installation of hardened fixed equipment[41], and the improvement of severe accident management, together with appropriate staff training measures. In many cases, important modifications are being prepared for the near future.
- The importance of the containment function was emphasized by previous accidents. The expeditious implementation of improvements such as filtered venting appears to be a crucial issue, also in the light of the Fukushima accident.
- Further considerations are proposed to add bunkered equipment to prevent and manage severe accidents including instrumentation and communications, mobile equipment and emergency response centers protected against extreme natural hazards and contamination. And backup that could be rapidly mobilized to support local operators in long duration events.

Finally, the report concludes that although PSA is an essential tool for screening and prioritizing improvements, low numerical risk estimates should not be used as the basis for excluding scenarios from consideration of severe accident management, especially if the consequences are high. In this way, the report questions parts of the current basic safety concepts.

It has become obvious that there is no legal power at the European level to force national regulators to enforce safety improvements. Thus, only general aims and aspirations can be communicated at the European political level: for example, to reflect on proposed measures to increase robustness, to make improvements as soon as possible, to consider urgent implementation of the recognized measures to protect

---

[40]Stress tests performed on European nuclear power plants. ENSREG, 24.04.2012.

[41]Philosophy to provide an additional independent subset of safety related structures, systems and components, capable of withstanding earthquakes and flooding events significantly beyond design basis.

containment integrity and perform periodic reviews, at least every 10 years, as necessary, and so on. Nevertheless, the stress test can be regarded successful in improving the safety of European nuclear power plants, although major differences remain between European countries and between Europe and the rest of the World; notably with the USA where the regulator (NRC) refused to impose filtered vented containments.

Furthermore, the UN International Atomic Energy Agency (IAEA) plays an important role in ensuring the safety of nuclear power plants. Under its auspices, the Convention on Nuclear Safety came into force in 1996 and the Convention on the Safety of Spent Fuel and Radioactive Waste Management became effective in 2001. Two emergency preparedness and response conventions were established to promote adherence to and implementation of international legal agreements on nuclear safety. Early in 2015, the Vienna Declaration on Nuclear Safety was adopted to amend the existing convention by taking the significant number of efforts and initiatives after the Fukushima Daiichi accident into account. All activities of the IAEA are based upon a number of premises. First and foremost, each Member State can only be advised and not relieved of its responsibility for the safety of its own nuclear facilities and second, much can be gained by exchanging experience worldwide and learning lessons from past experience to prevent serious accidents. Therefore, the IAEA focuses its activities on offering its Member States a wide array of review services, in which an IAEA-led team of experts compares actual practices with widely accepted IAEA standards in, for example, nuclear safety and security.[42] Recommendations are made available to and discussed with the principal, often a governmental organization. Whether these recommendations are finally accepted and implemented in due course is up to them. The IAEA has no powers of enforcement.

To summarize, there is widely accepted premise that nuclear safety requires a continual quest for operational and safety excellence, and mechanisms are in place to ensure that. Among those mechanisms, learning from operating experience and severe accidents, which were regarded not possible until one happened, is most prominent. Severe accidents led to careful safety reviews, temporary shutdowns and even permanent shutdown of a few plants, mainly where improvements would be too costly or too difficult.[43] More importantly, major improvements of both engineered features and accident management procedures have taken place at the global scale, for which plant operators and countries bear responsibility. However, a comprehensive evaluation of past events including near-misses shows forerunners and precursors to most of the severe accidents and, thus, we suggest that operational

---

[42]Offered services include Integrated Regulatory Review Services, Integrated Nuclear Infrastructure Review, Operational Safety Review Team, Independent Safety Culture Assessment, and Emergency Preparedness Review (see iaea.org).

[43]As an example, the owner of the Swiss NPP Mühleberg decided on Oct. 2013 to permanently shut down the plant by 20 December 2019 due to economic reasons.

feedback could be used more sufficiently and effectively in future than has happened to date.

Moreover, as mentioned several times, severe nuclear accidents raised exceptional public concerns and fear as well as political reactions, although differing from country to country. An accident somewhere can still negatively impact the whole industry.

5.7 Cost Benefit Analysis of Retrofits

Figure 5.3 shows estimates of the evolution of the core damage probability (CDP) for the US, and for a Swiss reactor. Attributing these improvements to costly major retrofits, this allows for a simple indicative assessment of return on investment (i.e., cost versus expected cost of accidents avoided). The results of this exercise are summarized in Table 5.4, where the CDF is assumed to start at 10^{-3}, and ends at 10^{-5} (hypothetical-optimistic post-Fukushima level), corresponding to a two order-of-magnitude drops.

On this basis, TMI-related retrofits were overwhelmingly justified, delivering a benefit of near 60 Million USD per reactor-year. The post-Fukushima level, with CDF 10^{-5} per reactor-year, is a target for many units, probably only achieved for units with superior safety systems and limited natural hazard exposure.[44] It is also hypothetical, given limited operating experience post-Fukushima, and retrofits still ongoing. However, even assuming the CDF of 10^{-5} indicates that the pure reduction in risk from CDF 10^{-4} does not compensate the cost. This characterizes a diminishing return on safety investment for the current fleet, although can be reasonably justified with aversion to risk and uncertainty. Moreover, there is the reasonable expectation that Gen III and IV designs will be able to achieve substantial additional safety improvements.

These major retrofits were reactive, being triggered by major accidents and precursors. Proactive future improvements may rely upon strong regulation based on sound risk information, and accident liability for plant owners. This is important because a "tragedy of the commons"[45] may incentivize avoiding proactive safety investment: when a sensitive event takes place at any unit in the World, the whole fleet suffers disruption, costly retrofits, increased costs due to regulatory ratcheting, or a legally mandated early shutdown. Therefore, there is limited incentive to have a superior safety level when the global fleet is treated as only as strong as its weakest

---

[44]Not all plants have bunkered safety equipment and other BDBA-hardened emergency systems, and there is substantial variation in the degree of redundancy and diversity from country to country.

[45]The tragedy of the commons refers to a situation within a shared-resource system where individual users tending to maximize their own independent self-interests behave in a way that turns out to be contrary to the common good of all users, and thus of themselves in the end, by depleting or destroying that resource through their collective action. Hardin, G., The Tragedy of the Commons, Science 162 (3859), 1243–1248 (1968).

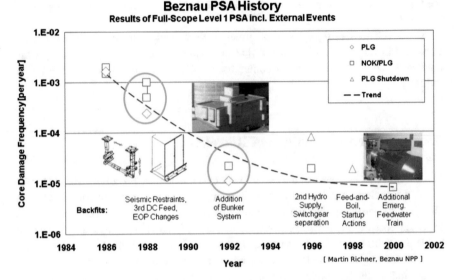

Fig. 5.3 Retrospective core damage probability (CDP) at the Swiss KKB station, suggesting a downward trend following major retrofits, courtesy of KKB. Note that KKB has undergone more extensive and costly upgrades than most units globally, resulting in higher quality, redundant, and diverse safety systems

Table 5.4 Hypothetical cost-benefit (risk reduction) analysis of retro-fits, summarized by the final column: reduction in accident risk minus cost of retrofits, measured in Millions of 2017 9 USD per reactor-year

Period	CDF/ ry	Accident Risk/ry	Externality[a]	Retrofit cost[b]	Net benefit/ ry
Pre-TMI	10^{-3}	$100 Mil./ry	1–2 ¢/kWh	NA	NA
1980–2011	10^{-4}	$10 Mil./ry	0.1–0.2 ¢/kWh	$33Mil./ry	57 Mil./ry
Post-Fukushima	10^{-5}	$1 Mil./ry	0.01–0.02 ¢/kWh	$30Mil./ry	−21 Mil./ry

The risk reduction is driven by the reduction in CDF from the previous time period (row), caused by a costly retrofit, with cost spread evenly over remaining reactor life. CDF levels are based on US (Fig. 4.6) and Swiss (Fig. 5.3) figures
[a]For a 5–10 TWh/ry output. Assuming average cost of USD 100 Billion per core damage accident (as in Sect. 5.3)
[b]Assumed retrofit costs (see A.2.5): TMI USD1B/unit over 30ry remaining life; Fukushima USD 0.3B/unit over 10ry remaining life

link. Also, operators may be tempted to opt for mandatory early shut-down if compensation is offered, transferring future liability costs to the public. This underscores the statement by the IAEA that "Nuclear power plant safety requires a continuing quest for excellence"[46] and one should add "of the whole fleet".

---

[46]INSAG, IAEA. "Basic safety principles for nuclear power plants [R]. 1999." (12).

5.8   Key Requirements for Innovative, Less Vulnerable Designs

We should not overburden the use of nuclear energy with a selective "zero risk" expectation. However, there are major barriers to make its future, potentially extended use, acceptable to the public. These barriers include risk aversion—the unequal treatment of probabilities and consequences of the extraordinarily high potential societal and economic damage in the case a severe accident happens despite its rarity—and the perceived dread of even low doses of radiation, as well as the sensitivity to instabilities within the operational environment.

As a pragmatic way out, we uphold that future nuclear power reactors should be less dependent on: (1) properly designed safety systems and security measures as well as protection against external events, either natural or malicious-intentional; (2) the adequacy of infrastructure, safety culture and operational modes; and, *last but not least,* (3) on the stability of our societies.[47] The following requirements, aiming as far as possible at a deterministic exclusion of serious conditions and states, are put forward:

1. Elimination of potential reactivity induced accidents by reactor core design or at least controllability by passive means. This can be achieved by

    (a) subcritical systems (receiving additional neutrons from accelerator driven systems)
    (b) weak, negative reactivity coefficients (graceful reaction on increasing fuel temperature, power, void fraction, burn up, where appropriate),
    (c) small reactivity surplus at startup with fresh fuel,
    (d) fail-safe design of shutdown absorber rods.

2. Forgiveness against loss of active core cooling by reactor design and inherent/ passive means. This can be achieved by

    (a) low power density and power size (to avoid exceeding critical temperature limits),
    (b) strategies to avoid high fission product inventory, e.g., by dispersed fuel,
    (c) temperature resistant fuel cladding and structural material that will not melt or burn if adverse conditions occur,
    (d) sufficient heat storage capability and inherent/passive heat transfer mechanisms in case of normal (forced) cooling/loss of coolant (depressurization)/ total loss of power,
    (e) passive decay heat removal systems.

3. Securing structural integrity to avoid geometric disorders (e.g., loosing core cooling capability) or loss of confinement of radioactive inventory. This can be obtained by

---

[47]D. Sornette, A civil super-Apollo project in nuclear R&D for a safer and prosperous world, Energy Research & Social Science 8, 60–65 (2015).

(a) low primary circuit pressure or leak/rupture proof components (notably reactor pressure vessel),
(b) radiation resistant and robust core structures,
(c) underground siting for protection against extreme external impact, including conventional weapons' attack.

4. Use of chemically non-reactive, non-toxic materials and fluids or avoid direct contact of reacting substances. This can be achieved by intermediate cycles if necessary.
5. Avoidance/incineration of long-lived radioisotopes (actinides) by fuel cycle designs allowing for reduced long-term stewardship (husbandry times). This can be achieved by

(a) a switch to fuel cycles (thorium) with drastically smaller generation of long-lived minor actinides,
(b) waste burner core designs,
(c) striving for enhanced closed fuel cycles or for long-term stable, high burn-up spent fuel as an open fuel cycle option.

6. Intrinsic proliferation resistance characteristics of the fuel, fuel cycle and related processes. Based on the discourse on proliferation issues (Sect. 3.2 and its annex), the following recommendation principles, means and strategies to enhance proliferation resistance characteristics of fuel, entire fuel cycle and related processes, can be assigned and should be applied:

(a) Avoid use of highly enriched uranium (HEU),
(b) Configure nuclear reactors to enable maximum burn up of fuel and thereby decrease the amount of plutonium in spent fuel that could be used for weapons,
(c) Avoid high-grade plutonium generation, e.g. by employed blankets,
(d) Reprocess spent fuel only if there is a clear plan to minimize the time during which weapons–grade material, notably plutonium, is in separated form and to reuse it as soon as feasible; to avoid accumulating a stockpile,
(e) Strive for online reprocessing including fuel fabrication at the reactor location and avoid transportation of sensitive material,
(f) Implement protective measures throughout the entire fuel cycle.

Annexes

A.1 The International Nuclear and Radiological Event Scale (INES)

The INES was introduced in 1990 by the International Atomic Agency (IAEA) in order to enable prompt communication of safety-significant information in case of

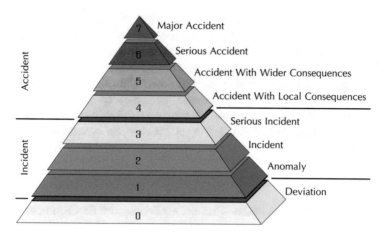

Fig. 5.4 International Nuclear Event Scale, standard pyramid representation of the 8 levels (INES, IAEA. "The international nuclear and radiological event scale user's manual 2008 edition." *IAEA and OECD/NEA* (2008))

nuclear accidents, and was refined in 1992 and extended to be applicable to any event associated with radioactive material and/or radiation, including the transport of radioactive material. The most recent user manual was published in 2008.[48] The scale is intended to be logarithmic, similar to the moment magnitude scale that is used to describe the sizes of earthquakes. Each increasing level represents an event approximately ten times more severe than the previous level. Compared to earthquakes, where the event energy can be quantitatively evaluated, the level of severity of a man-made disaster, such as a nuclear accident, is more subject to interpretation (Fig. 5.4).

The level of the scale is determined by the highest of scores from three tracks: off-site effects (also called People and Environment), on-site effects (also called Radiological Barriers and Controls), and defense in depth degradation (potential consequences). Effects include health and environmental impact; costs are not considered explicitly. Below, real events are given as examples for INES levels 3–7, scored according to the relevant track.

A.1.1 Level 7: Major Accident—Impact on People and Environment

Major release of radioactive material with widespread health and environmental effects requiring implementation of planned and extended countermeasures. There have been two such events to date:

---

[48]IAEA, "INES: the International Nuclear Event Scale User's Manual", IAEA–INES–2009.

- Chernobyl disaster (Ukraine/former Soviet Union), 26 April 1986. A power surge during a test procedure resulted in a criticality accident, leading to the release of a significant fraction of core material into the environment. See also Sect. 3.4.2.
- Fukushima Daiichi nuclear disaster, a series of events beginning on 11 March 2011, Japan. Major damage to the backup power and containment system caused by the Tohoku submarine earthquake and induced tsunamis resulted in overheating and leaking from some of the nuclear plant's reactors. Each reactor accident was rated separately; out of the six reactors, three were rated level 5, one was rated level 3, and the situation as a whole was rated level 7. See also Sect. 3.4.3.

A.1.2 Level 6: Serious Accident—Impact on People and Environment

Significant release of radioactive material likely to require implementation of planned counter-measures.

- Kyshtym disaster at Mayak Chemical Combine (former Soviet Union), 29 September 1957. A failed cooling system at a military nuclear waste reprocessing facility caused a steam explosion with a force equivalent to 70–100 tonnes (metric) of TNT. About 70–80 metric tons of highly radioactive materials were carried into the surrounding environment. The impact on local population is not fully known, but at least 22 villages were affected with doses that may lead to shortened life expectancy due to higher risk of cancer.[49]

A.1.3 Level 5: Accident with Wider Consequences

- Impact on people and environment
  Limited release of radioactive material likely to require implementation of some planned countermeasures. And/or several deaths likely to result from radiation.
- Impact on radiological barriers and control
  Severe damage to reactor core (more than 1% of inventory); release of large quantities of radioactive material with a high probability of significant public exposure. This could arise from a major criticality accident or fire.

  - Windscale fire (UK), 10 October 1957. Annealing of graphite moderator at a military air-cooled reactor caused the graphite and the metallic uranium fuel to catch fire, releasing radioactive pile material as dust into the environment.
  - Three Mile Island accident near Harrisburg (USA), 28 March 1979. A combination of design and operator errors caused a gradual loss of coolant, leading to a partial meltdown. See also Sect. 3.4.1.

---

[49]Standring, W. J., Dowdall, M., & Strand, P. (2009). Overview of dose assessment developments and the health of riverside residents close to the "Mayak" PA facilities, Russia. *International journal of environmental research and public health*, 6(1), 174–199.

- First Chalk River accident (Canada), 12 December 1952. Reactor core damaged.
- Lucens (INES level 4–5) partial core meltdown accident (Switzerland), 21 January 1969. A test reactor build in an underground cavern suffered a loss-of-coolant accident during a startup, leading to a partial core meltdown and massive radioactive contamination of the cavern, which was then sealed.
- Goiânia accident (Brazil), 13 September 1987. An unsecured cesium chloride radiation source left in an abandoned hospital was recovered by scavenger thieves unaware of its nature and sold at a scrapyard. 249 people were contaminated and 4 died.

A.1.4 Level 4: Accident with Local Consequences

- Impact on people and environment
  Minor release of radioactive material unlikely to result in implementation of planned countermeasures other than local food controls. At least one premature death from radiation.
- Impact on radiological barriers and control
  Fuel melt or damage to fuel resulting in more than 0.1% release of core inventory. Release of significant quantities of radioactive material within an installation with a high probability of significant public exposure.

- SL-1 Experimental Power Station (USA), 1961, reactor reached prompt critically, killing three operators.
- Saint-Laurent Nuclear Power Plant (France), 1969, partial core meltdown; 1980, graphite overheating.
- Buenos Aires (Argentina), 1983, criticality accident on research reactor RA-2 during fuel rod rearrangement killed one operator and injured two others.
- Tokai-mura nuclear accident (Japan), 1999, three inexperienced operators at a reprocessing facility caused a criticality accident; two of them died.

A.1.5 Level 3: Serious Incident

- Impact on people and environment
  Exposure in excess of ten times the statutory annual limit for workers; non-lethal deterministic health effect (e.g. burns) from radiation.
- Impact on radiological barriers and control
  Exposure rates of more than 1 Sv/h in an operating area. Severe contamination in an area not expected by design, but with a low probability of significant public exposure.
- Impact on defense-in-depth
  Near-accident at a nuclear power plant with no safety provisions remaining. Lost or stolen highly radioactive sealed source. Highly radioactive source in the wrong or unexpected place without adequate procedures in place to handle it.

- THORP plant, Sellafield (UK), 2005; leak of uranium and plutonium dissolved in nitric acid into stainless steel containment structure.
- Paks Nuclear Power Plant (Hungary), 2003; fuel rod damage in cleaning tank
- Vandellos Nuclear Power Plant (Spain), 1989; fire destroyed many control systems; the reactor was shut down safely.
- Davis-Besse Nuclear Power Station (USA), 2002; negligent inspections failed to record corrosion through 15.24 cm of the carbon steel reactor head leaving only 9.5 mm of stainless steel cladding intact that was holding back the high coolant pressure of 17 Mpa.

Table 5.5 Radioactive releases compared (numerical values in TeraBecquerel $= 10^{12}$ Bq, % of core inventory) for Chernobyl[a], Fukushima[b], TMI[c], and Windscale[d]. Empty values mean nothing measured

Material	Half-life	Windscale	Three Mile Island	Chernobyl	Fukushima Daiichi (atmospheric)
Iodine-131	8 days	600	0.6	50–60%, 2,900,000	120,000
Cesium-137	30 years	46	<10	20–40% 260,000	8800
Xenon-133	5.2 days	12,000	58,500	100% 6,600,000	7,300,000
Xenon-135	9.1 hours	35	11,100		
Strontium-90	29 years	0.22	<0.1	20,000	~0
Plutonium-238-240	88 to 24,100 years			3400	~0.001

[a]Guntay, S., Powers, D.A. and Devell, L. "The Chernobyl reactor accident source term: Development of the consensus view." *One Decade after Chernobyl: Summing up the Consequences of the Accident* (1996): 183–193
[b]United Nations Scientific Committee on the Effects of Atomic Radiation. "UNSCEAR 2013 Report Volume I. Report to the general assembly scientific annex A: Levels and effects of radiation exposure due to the nuclear accident after the 2011 great east-Japan earthquake and tsunami." (2014)
[c]Battist, L., and Peterson H.T. "Radiological consequences of the Three Mile Island accident." *Radiation protection* 2 (1980)
[d]Crick, M. J., and Linsley, G. S. *An assessment of the radiological impact of the Windscale reactor fire, October 1957.* No. NRPB-R--135 (ADD.). National Radiological Protection Board, 1983

A.2 Breakdown of Rough Integral Cost Estimates[50]

A.2.1 On-site

Emergency management, mitigation, capital loss, and remediation/clean-up.

TMI Unit 2 was destroyed within 1 year of operation, having cost USD 2B (Billion) to build, 7 years of downtime were caused at unit 1,[51] and substantial

---

[50]Costs are a mix of actual reported costs and estimated costs, with the aim of capturing the full social cost. All costs are in 2017 US Dollars unless stated otherwise.
[51]IAEA PRIS database, available at https://www.iaea.org/PRIS/

contamination led to nearly USD 2.5B in cleanup.[52] We find <u>USD 5-10B</u> to be reasonable, although higher estimates exits.[53]

Fukushima The six-unit site, taken into operation in the 1970s, was lost. Shutting down the near 50 reactor fleet for 7 years already led to further loss of capital. As of the end of 2017, 5 reactors have restarted and 21 are currently subject to approval to restart. TEPCO estimates for cleanup and decommissioning costs are USD 10-20B.[54] Thus, overall <u>USD 20-30B</u> is reasonable, but the more than 30 years planned for decommissioning, and ample uncertainty, leaves room for unforeseen costs.

Chernobyl In 1986 and 1987, some 440,000 recovery operation workers were employed at the Chernobyl site.[55] The Ukrainian government claims USD 20–30B on accident management and decontamination so far.[56] Unit 4, less than 4 years into its life, was lost. Despite contamination, units 1–3 (each producing about 6 TWh/year) resumed operation by the end of 1987. The new confinement structure, built over the remains of Unit 4, cost 2B USD. Also allowing for further future costs, the total on-site cost is <u>USD 25-35B</u>.

A.2.2 Life and Health

Health impact and trauma due to radiological exposure, emergency response, relocation, continued displacement, and so on. Details on late cancer fatalities in Sect. 4.3.

TMI Despite stress due to media and poor communication,[57] human consequences were relatively negligible due to successful containment. There was a minor release (Table 5.5) from the plant's auxiliary building to relieve pressure on the primary

---

[52]US NRC, Regulatory Analysis Guidelines of the US Nuclear Regulatory Commission, NUREG/BR-0058, Rev. 4 (2004).

[53]Devine, J. "Three Mile Island: The TMI-2 Accident Consequences and Costs", NEA Workshop, Paris, May 2013.

[54]Obayashi, Y. and Hamada, K. "Japan nearly doubles Fukushima disaster-related cost to $188 billion", Reuters December 9, 2016.

[55]Bennett, B., and Repacholi M. "Health effects of the Chernobyl accident and special health care programmes. Report of the UN Chernobyl Forum Expert Group Health". World Health Organization, 2006.

[56]Ministry of Ukraine of Emergencies and Affairs of population protection from the consequences of Chornobyl Catastrophe, Kyiv (Ukraine) Baloga, V.I. (Ed.). (2006). 20 years after Chernobyl Accident Future outlook, National Report of Ukraine (INIS-UA--112). Baloga, V.I. (Ed.). Ukraine: Atika.

[57]The Communications Failures Lessons of Three Mile Island; Friedman, Sharon M. "Three Mile Island, Chernobyl, and Fukushima: An analysis of traditional and new media coverage of nuclear accidents and radiation." *Bulletin of the atomic scientists* 67.5 (2011).

system, such that less than 1 additional death from cancer was expected.[58] Total payout for a class-action settlement was around USD 0.1B.[59]

Fukushima The Tōhoku earthquake and tsunami killed over 18,500 people from effects unrelated to destruction of the reactors at Fukushima Dai-ichi. Long-term displacement of more than 164,000 people, attributed to the nuclear accident is blamed for around 2000 early deaths,[60] largely effecting the elderly (90% older than 66), costing USD 2–3B[61] for years of life lost (YOLL). Attested to be of lower order are the long term expected radiological fatalities: Linear no-threshold (LNT)-based estimation allows for on the order of 100 eventual fatalities due to cancer, and therefore an expected YOLL of >1000, and a cost of about USD 1B.[62] Traumatic effects (suffering, isolation, etc.) are more difficult to estimate, with >100,000 people still displaced as of 2016, and increased health and psychological issues identified. Compensation of approx. 10 thousand USD per person per year displaced were specified[63], gives about USD 11B. This impact is the subject of overall compensation, covered in the following category. Overall here: USD 3–4B + 11B ("trauma").

Chernobyl The human impact of the major accident at Chernobyl was immense, and its assessment remains contentious. As discussed in Sect. 4.3, a range of 4000–16,000 eventual radiological fatalities have been put forth in serious studies—giving USD 3.6–11B for YOLL. Chernobyl led to more than 220,000 people being relocated. As for Fukushima, allowing 10,000 USD per person per year displaced, here limited to a 10-year period, gives 22B. Overall: USD 3.6–11B + 22B ("trauma").

A.2.3 Public Economic

Economic costs suffered by the public (incl. loss of property and income).

TMI Negligible in relative terms, and potentially already covered within the USD 0.1B payout.

---

[58]Kemeny, J.G. "Report of the President's commission on the accident at Three Mile Island." *The need for change: The Legacy of TMI* (1979).

[59]Smith, T. "The Price-Anderson Act and the Three Mile Island Accident", OECD/NEA Workshop on Nuclear Damages, Liability Issues, and Compensation Schemes, December 2013.

[60]Hayakawa, M. "Increase in disaster-related deaths: risks and social impacts of evacuation", Annals of the ICRP, Proceedings of the International Workshop on the Fukushima Dialogue Initiative (2016):123–128.

[61]"Damage caused by earthquake and tsunami", Fukushima Prefectural Government, 2017. http://www.pref.fukushima.lg.jp/site/portal-english/en03-01.html

[62]See e.g., See EPA Mortality Risk Valuation: https://www.epa.gov/environmental-economics/mortality-risk-valuation.

[63]OECD/NEA. "Japan's Compensation System for Nuclear Damage: As Related to the TEPCO Fukushima Daiichi Nuclear Accident" Legal Affairs OECD/NEA (2012).

Fukushima Compensation liabilities are currently estimated at <u>USD 50–100 Billion</u>[64], also intended to cover life and health impacts due to displacement (above). TEPCO is expected to pass these costs onto consumers in the future with higher rates. The once 1100 square km evacuation area is already less than half of its original size[65] as Cs-137 decays and is transported by rainfall to the sea.

Chernobyl A 2600 square km exclusion zone formed, the majority of which is now safe for settlement and economic activity. Total losses (not limited to this category) reported for Ukraine and Belarus are USD 210–224B. For this category, in Ukraine[66,67], there have been about USD 18B in capital losses, USD 20B in social costs (to date), and economic losses due to loss of natural resources and agriculture of USD 70B, totaling about USD 110B. A rough estimate to cover also Belarus and Russia is then <u>USD 150-250B.</u>

A.2.4 Replacement Power

The increased cost of provision of electricity, incurred by the utility, but also by the public, through higher electricity prices. According to the NRC[68], for a generic NPP unit being shut down, the replacement power costs are estimated to be approximately USD 17B, and 40B for a new unit in a pool with above average replacement energy costs—thus posing a major financial risk.

TMI Reliance on nuclear power to meet demand was high, and the accident triggered a financial crisis for the utility due to high-cost (primarily oil generated) replacement power. The cost of this replacement was later allowed to be passed onto customers through a near doubling of the rate.[69] The resulting increment in cost of electricity was up to USD 0.9B per year, for up to the 6 years of downtime of both units.[70] On this basis, a range of <u>USD 5–15B</u> can be taken, with the upper limit more in line with NRC guidelines.

Fukushima Suspending Japanese nuclear power (250 TWh/year, 30% of annual generation) required importing natural gas at an increased cost of USD 20B (1.9

---

[64]"Fukushima nuclear crisis estimated to cost ¥11 trillion: study" The Japan Times, August 27, 2014.

[65]Japanese Ministry of Environment. "Progress on Off-site Cleanup and Interim Storage Facility in Japan", August 2017.

[66]"The Human Consequences of the Chernobyl Nuclear Accident: A Strategy for Recovery", A Report Commissioned by UNDP and UNICEF with the support of UN-OCHA and WHO (2002).

[67]Belarus Foreign Ministry, "CHERNOBYL disaster", April 2009. http://chernobyl.undp.org/russian/docs/belarus_23_anniversary.pdf

[68]US NRC, Regulatory Analysis Guidelines of the US Nuclear Regulatory Commission, NUREG/BR-0058, Rev. 4 (2004).

[69]Pa. Electric Co. v. Pa. P.U.C., 78 Pa. Commw. 402, 406 (1983).

[70]EMD-80-89 Three Mile Island: The Financial Fallout.

trillion Yen[71]) per year (excl. potential health impacts). With the uncertainty about when the NPPs will be turned back on, the cost is USD 100B–150B.

Chernobyl There was sufficient need for power that operation at the (highly contaminated) site resumed within the year. The Ukrainian government claims that the Chernobyl accident led to a loss of 62 thousand TWh, much of which was not replaceable, resulting in a larger economic loss of USD 28B, giving a range of USD 10-30B.

A.2.5 Industry Response: A Cost or a Benefit?

The nuclear sector is somewhat unique in the extent to which major accidents drive increases in regulatory rigor worldwide, and costly retrofits. On the other hand, it should make the plants safer and potentially improve operational efficiency.

TMI The "TMI Action Plan"[72] retrofits implemented substantial improvements in operations, plant design, and regulation.[73] The retrofits in the US alone were roughly estimated to be between 10 and 60 Billion USD—i.e., between 100 and 600 million USD per unit.[53] The costs of delays and retrofits to the >50 operating US units and 40 units under construction, were estimated at more than USD 90B.[74,75] This indicates a total (global) cost in the range of USD 100–200B.

Fukushima This major accident strengthened scrutiny of nuclear power, triggering many safety initiatives[76] including the European "Stress Test", the IAEA "Vienna Declaration", and other activities.[77] The retrofits are estimated to cost USD 60B in Japan, USA, and France alone[78], resulting in a similar figure of USD 0.2B per

[71]Matsuo, Y., and Yamaguchi, Y. "The rise in cost of power generation in Japan after the Fukushima Daiichi accident and its impact on the finances of the electric power utilities." *The Institute of Energy Economics, Japan* 1 (2013).

[72]US Nuclear Regulatory Commission. Office of Nuclear Reactor Regulation. Division of Licensing. *Clarification of TMI action plan requirements*. Vol. 88. No. 10-737. The Commission, 1980.

[73]US Nuclear Regulatory Commission. "TMI-2 Lessons Learned Task Force Final Report." (1979).

[74]Lave, L.B., ed. *Risk assessment and management*. Vol. 5. Springer Science & Business Media, 2013.

[75]Lovering, J.R., Yip, A., and Nordhaus, T. "Historical construction costs of global nuclear power reactors." Energy Policy 91 (2016).

[76]Lokhov, A., et al. "The economics of long-term operation of nuclear power plants." OECD/NEA (2012).

[77]OECD/NEA. "Impacts of the Fukushima Daiichi Accident on Nuclear Development Policies", OECD/NEA (2017)

[78]Williams, E. "IAEA activities related to the estimation of nuclear accident costs", presentation given at NEA, 28–29 May 2013.

reactor from Japan.[79] A Swiss unit (KKM) was required to undertake > USD 0.4B of retrofitting, causing the plant to close. Extrapolating the 0.15–0.3 B globally gives a total of about USD 60–120B.

Chernobyl Major retrofits to all 19 RBMK reactors were required. The cost for this is unknown but, within a factor of two of the 0.2B number from Fukushima, provides a range of USD 2–8B.

A.2.6 Beyond/Limitations

Deeply uncertain costs and potential benefits.

TMI this accident caused an inflection point in the industry, killing growth of the nuclear industry and leading to a new regulatory-economic regime.[80] In this regard, TMI may have been the most influential nuclear accident of all.

Chernobyl As stated by the IRSN[81], such an event can durably stun/affect a nation and an economy. It is difficult to be convinced that a cost estimate is complete in such a case. In particular, the accident may well have contributed to the downfall of the USSR.[82] It also galvanized the "environmentalist movement" against nuclear power. Other things excluded are the resulting suspension of the Belorussian nuclear program[83], and contributions to the shutdown of former USSR units.

Fukushima This accident renewed and strengthened political scrutiny of nuclear power. In particular, excluded above are the potential costs of the German shutdown of 40% of the German nuclear power capacity since 2011.[84] These costs are

---

[79]Matsuo, Yuji. "Summary and evaluation of cost calculation for nuclear power generation by the cost estimation and review committee." *Institute for Energy Economics Japan (IEEJ)* (2012).

[80]EMD-80-89 Three Mile Island: The Financial Fallout.

[81]Pascucci-Cahen, L. & P. Momal. The cost of nuclear accidents. OECD/NEA International Workshop on the Full Costs of Electricity Provision. 2013.

[82]However, a detailed analysis suggests that this claim by Gorbachev may have been self-serving and used as an excuse for other mistakes or oversights in the governance of the politburo of the USSR. See D. Chernov and D. Sornette, Man-made catastrophes and risk information concealment (25 case studies of major disasters and human fallibility), Springer (2016).

[83]The nuclear program in Belarus has restarted recently, with its first nuclear power plant presently under construction and plans to have it operating from 2019. Financed by Russia, Atomstroyexport is building the 2400 MWe plant, with two VVER-1200 reactors. See world-nuclear.org (accessed 17 Dec. 2017).

[84]Another difficulty is to distinguish the responsibility for these costs, as the German decision is highly political and may be considered unwise by some. Should ill-advised political decisions be entirely put on the back of nuclear?

particularly difficult to estimate due to future uncertainties about the potential performance of future nuclear and new renewables. Official cost estimates range from tens to hundreds of billions of Euros[85], with rates expected to briefly peak at about 50% above baseline.[86]

---

[85]Neubacher, Alexander (27 July 2011). "The Latte Fallacy: German Switch to Renewables Likely to Be Expensive". Der Spiegel. Retrieved 22 September 2011.

[86]Matthes, F. "Exit economics: The relatively low cost of Germany's nuclear phase-out". Bulletin of the Atomic Scientists Volume 68 Issue 6 (2012).

Chapter 6
Candidate Features and Technology Options

Abstract Nuclear power stations have become more efficient and safer inter alia due to learning from valuable operational experience, largely in reaction to major and near accidents. However, dread of rare but possible severe accidents, reliance upon unfailing human performance at various levels, and dependence on social stability, emphasize the importance of further safety improvements, for which challenging criteria were conveyed in Sect. 2.2. To achieve them, key design features ("building blocks") are viewed and revisited and should be combined in a radically new way, to come up with "revolutionary" or even "exotic" system designs. To check whether such designs are feasible, we track most recent developments of reactor concepts focusing on differing (1) coolants, including liquid metals and molten salt, (2) neutron spectrum from thermal to fast, (3) power level, (4) fundamental design features (architecture) and purpose, and (5) ability to extend fuel reserves and "burn" waste. The designs selected are scored against the set of very stringent, highly ambitious criteria. The results show a high potential for far-reaching improvements compared to most advanced LWRs in use today. Small modular reactors emerge as being the most attractive. However, thus far, none of the candidate concepts fulfill all the criteria convincingly; avoiding criticality induced accidents and maximizing proliferation resistance appears most challenging. There is also a potential for new concept-specific risks to be introduced but this appears manageable. Although caution is warranted, a purely deterministic safety approach is tempting, in that we would like to absolutely exclude the possibility of severe accidents.

Operational experience is obtained first from individual units and national regulators, and beyond by reports to the IAEA, WANO (formed after Chernobyl), and INPO (formed after TMI), where guidelines and standards exist. Recently WANO and the NEA have signed a memorandum of understanding "to co-operate on the further development of approaches, practices and methods in order to proactively strengthen global nuclear safety" [NEA/COM(2017)5, Paris, 4 October 2017].

© Springer Nature Switzerland AG 2019
D. Sornette et al., *New Ways and Needs for Exploiting Nuclear Energy*,
https://doi.org/10.1007/978-3-319-97652-5_6

6.1   Basic Design Principles and Elements

It is the overarching goal to make the civilian use of nuclear energy safer and more acceptable by reactor and fuel cycle designs that are less dependent on reliable human performance and more robust against human failures, malicious interventions and instability of our societies. In Sect. 5.8, we have laid out the main desirable requirements. To achieve these ambitious goals and challenging requirements, key design features ("building blocks") for advanced nuclear reactors and related systems can be identified and deserve careful consideration. They include (1) neutron spectra, (2) coolants, (3) fuels, fuel claddings and core structural materials, (4) power densities and power sizes, and (5) siting issues and options. We now analyze these five key design features sequentially.

6.1.1   Neutron Spectra

A look at the fission probabilities[1] (quantified by "cross sections") of selected actinides (Table 6.1) demonstrates the attractiveness of fast neutrons compared to moderated/thermalized neutrons that stem from fission process and dominate the spectrum of today's commercial light water reactors (LWR). While thermal fission cross sections of fissile uranium-233 and -235 and plutonium-239 are significantly larger than for fast fission cross sections, the important fission-to-absorption ratio is of the same order for the isotopes mentioned above but significantly higher for all other selected isotopes, in particular atoms heavier than uranium. These so-called transuranic elements are generated from uranium or heavier atoms in the fuel of today's thermal reactors when they absorb neutrons but do not undergo fission (neutron capture). This explains that large fission-to-absorption ratios are favorable to avoid or minimize the formation of radioactive waste and minor actinides,[2] in particular. Eliminating these elements from spent nuclear fuel would reduce drastically the stewardship times of the long-lived wastes that are generated (up to a factor 100 reduction, see Sect. 2.5), which are now determined by the seven most long-lived fission products, notably iodine-129 and technetium-99.

Fast neutron spectra, together with adequate coolants, allow for high neutron economy[3] and reactor designs that are favorable to produce as much or more fissile

---

[1] A fission probability is the probability that, upon the impact of a neutron, a given nucleus will undergo a fission reaction, i.e., will split into several smaller clusters of protons and neutrons (i.e. other nuclei), including ejected neutrons. The fission probability is conveniently represented by the concept of a "cross section", which is an effective measure of the surface of capture of the incident neutron to trigger the fission reaction. The larger the cross section, the larger the probability of fission.

[2] Actinides encompass 15 metallic chemical elements with atomic numbers from 89 to 103; elements with atomic numbers from 93 onwards like Neptunium 93, Americium 95 and Curium 96 are called minor actinides.

[3] Significant extension of burn up and more efficient use of fuel.

Table 6.1 Fission probabilities of selected actinides, thermal versus fast neutrons represented by their corresponding cross sections measured in barn ($\equiv 10^{-24} cm^2$); fission % corresponds to the ratio of cross sections for fission to neutron absorption)

Isotope	Thermal Fission Cross Section	Thermal Fission %	Fast Fission Cross Section	Fast Fission %
Th-232	nil	1 (non-fissile)	0.350 bam	3 (non-fissile)
U-232	76.66 barn	59	2.370 barn	95
U-233	531.2 barn	89	2.450 barn	93
U-235	584.4 barn	81	2.056 barn	80
U-238	11.77 microbarn	1 (non-fissile)	1.136 bam	11
Np-237	0.02249 barn	3 (non-fissile)	2.247 barn	27
Pu-238	17.89 barn	7	2.721 barn	70
Pu-239	747.4 barn	63	2.338 barn	85
Pu-240	58.77 barn	1 (non-fissile)	2.253 barn	55
Pu-241	1012 barn	75	2.298 barn	87
Pu-242	.002557 barn	1 (non-fissile)	2.027 barn	53
Am-241	600.4 barn	1 (non-fissile)	0.2299 microbarn	21
Am-242 m	6409 barn	75	2.550 barn	94
Am-243	.1161 barn	1 (non-fissile)	2.140 barn	23
Cm-242	5.064 barn	1 (non-fissile)	2.907 barn	10
Cm-243	617.4 barn	78	2.500 barn	94
Cm-244	1.037 barn	4 (non-fissile)	0.08255 microbarn	33

material than consumed ("fast self-breeder" or "breeder reactor") and/or "incinerate" radioactive waste ("waste burner" or better "actinides burner").

Generally, however, the cores of reactors with a fast neutron spectrum are not in the state of highest reactivity under steady-state operational conditions; thus, changes of physical parameters, e.g., increase of coupled fuel-coolant temperature and/or void fraction may add reactivity to the core[4] and lead to disruptive power excursions. Furthermore, the margin to prompt criticality (see Sect. 2.1) is about a factor of two smaller when going from Uranium-235 to Plutonium-239, and another factor larger than two smaller when going to minor actinides, which raises additional safety concerns, narrowed down by lower excess reactivity.

---

[4]In nuclear engineering, coefficients of reactivity are used to estimate how much the reactivity of a reactor changes as physical parameters change. The void coefficient, for example, indicates the effect as voids (typically bubbles) form in the reactor moderator or coolant. If the coefficient of reactivity is positive, the core power tends to increase; if it is negative the core power tends to decrease.

6.1.2 Characteristics of Coolants

The characteristics of considered coolants, i.e., liquid metals like sodium, pure lead or lead eutectic (lead/bismuth), molten salt (ternary fluorides or chlorides) and gas (helium, CO_2) are briefly addressed here, see annexed Table 6.3 for more detailed information. All liquid metals and salts feature good heat storage and transfer capabilities and no need for pressurization for operation in a single-phase mode. But, high density and mass may lead to high static loads, notably for lead. All liquids and gas allow for core outlet temperatures of about 510 °C (molten sodium) to almost 600 °C (molten lead, molten salt) or even 650 °C (CO_2) to 850/950 °C (helium), significantly higher than for water (325 °C), resulting in relatively high thermodynamic efficiencies[5] for power production (clearly above 40% instead of 33% for modern LWR) and potential use for chemical heat applications. In detail, their characteristics differ greatly, they feature advantages and disadvantages and pose challenges of different kinds.

6.1.2.1 Sodium

Advantages

- Superior thermal hydraulic properties/heat transfer characteristics (highest thermal conductivity) allowing for high power density with low coolant volume and tight pin core lattice (albeit negative but manageable effect on natural convection);
- Excellent neutronic properties and economy[6] but positive coolant temperature coefficient of reactivity (CTC) (which is to be avoided to prevent possible instabilities);
- Good compatibility with structural materials, does not corrode steel;
- Features reasonably low melting temperatures (98 °C): in case of pool-type reactor, loss of coolant accidents can be avoided.

Disadvantages

- The relatively small margin between (low) melting and boiling point (892 °C) raises safety concerns regarding heat-up scenarios; sodium boiling leads to a significantly positive reactivity insertion;

---

[5]A basic principle of thermodynamics is that the efficiency of thermal machines is an increasing function of the difference of temperatures between the "cold" and "hot" heat sources.

[6]The neutron economy is defined as the ratio of the weighted average of the excess neutron production divided by the weighted average of the fission production. In other words, the neutron economy of a reactor results from the balance between neutrons created and the neutrons lost through absorption by non-fuel elements, resonance absorption by fuel, and leakage.

- Large coolant temperature coefficient of reactivity (CTC), and positive boiling coefficient in particular, raises questions about self-controllability;
- Low boiling point (compared with lead) limits core outlet temperatures;
- High chemical activity with water, steam and air gives rise to explosion and fire risks; need for intermediate heat transport system[7];
- High ^{24}Na isotope activity (half-life of decay: $T_{1/2} = 15$ h);
- Optical opacity (more complex maintenance and refueling).

Challenges

- Addressing sodium fire risk due to reactivity with air and water by protective measures;
- More compact configuration;
- Design of passive decay heat removal system.

6.1.2.2   Lead

Advantages

- Good natural circulation properties, excellent heat transfer and storage capability (second highest density-specific heat element);
- Superior neutronic characteristics and performance, allow (U, Pu)O_2 systems with smaller plutonium-enrichment and smallest coolant temperature reactivity coefficient (CTC);
- Chemically inactive, forms badly soluble oxides with air;
- High boiling point (1750 °C);
- High general availability leading to low cost.

Disadvantages

- High melting point of 327 °C gives rise to freezing potential;
- Erosion potential of steel giving rise to the need for a protective oxide layer or coating that is also resistant to abrasion, low limit of flow velocity[8];
- Excessive corrosion at elevated temperatures >650 °C;
- Looser lattice needed due to high viscosity and very high density, albeit a positive effect on natural circulation, resulting in lower power densities achievable;
- Optical opacity (like sodium).

---

[7]The Na risk could be mitigated with a system where the sealed reactor hall is flooded with N or CO_2.

[8]2 m/s compared to 8 m/s for sodium before turbulence occurs.

- Polonium-210 activity build up derived from Lead-210.[9]
  Lead/Bismuth Eutectic (LBE) (Compared to Lead)
- Lower melting point (123.5 °C); higher polonium-210 production.

Challenges

- Design and operational measures to prevent freezing;
- Special ballast weights necessary to immerse fuel and control rods into the coolant (to overcome buoyancy)
- Development of corrosion resistant alloys or materials, for example ceramics;
- Safety concerns regarding fast creep of hanging a vessel loaded with perhaps a thousand tonnes of molten lead with static stresses of about 30 bar.

6.1.2.3 Molten Salt

Ternary fluorides (e.g., F-Li-NaK, F-Li-Be) or chlorides (e.g., NaCl-KCl-MgCl$_2$).[10]

Advantages

- High boiling temperatures (2500 °C), voiding and associated reactivity insertion is eliminated;
- Highest density-specific heat product, capability to store decay heat during transients/accidents;
- Does not react with secondary fluids or air;
- Fuel can be dispensed in molten salt, shutdown by fuel dumping possible;
- Optical transparency, when fresh.

Disadvantages

- Need for management of positive reactivity effect of a stop of primary salt circulation;
- High melting temperature, e.g., of 396 °C for one salt composition, giving rise to freezing potential;
- Neutronically most challenging coolant for fast reactor designs;
- Very small thermal conductivity limits achievable power density, in spite of tight lattice core geometry;

---

[9]Half-life $T_{1/2}$ = 138 days, α-emitter causing operational difficulties.

[10]Chlorides are selected as being the best; fluorides feature higher melting point (450 °C) but lower boiling point (1430 °C) and smaller operational margin (Dupont, J., Consorti, C. S., & Spencer, J. (2000). Room temperature molten salts: Neoteric "green" solvents for chemical reactions and processes. Journal of the Brazilian Chemical Society, 11(4), 337–344.).

- Viscosity larger than sodium and lead, creating pumping and power constraints.

Challenges

- Design and operational measures to prevent freezing;
- Design devices that control and overcome strong positive coolant temperature coefficient of reactivity (CTC);
- Online reprocessing in case of reactors with dispersed/dissolved fuel.

6.1.3   Fuels, Fuel Claddings and Core Structural Materials

Nuclear fuel itself has by far the highest energy density (MJ/kg) of all practical energy sources.[11] Current fission reactors usually base their fuel on *metal oxide* (UO_2) rather than metals themselves, because the melting point is much higher (2850 °C) and it cannot burn, although its thermal conductivity is very low. A blend of oxide plutonium and natural or depleted uranium is called mixed oxide (MOX) fuel. It is obtained by reprocessing of commercial spent fuel and can be used in conventional LWR. *Metal* fuels including pure uranium or uranium alloys have the potential for the highest fissile atom density and the advantage of very high heat conductivity but suffer from lower melting points (1133 °C) and phase transitions in the solid phase. *Ceramic* fuels have the advantage of high heat conductivities and melting points (2700–2800 °C) but they are more prone to swelling than oxide fuels. *Uranium-carbide*, most notably in the form of crated micro particles together with ceramic (or graphite) structural material, are regarded as attractive fuel for certain future reactors. *Liquid* fuels, i.e. dissolved nuclear fuel in liquids like molten salts, offer numerous operational advantages due to inherently stable self-adjusting reactor dynamics, rapid drain ability into dump-tanks and continuous release of xenon gas that acts as a neutron absorber.

Currently dominating, UO_2 power is compacted into cylindrical pellets and sintered at high temperatures to produce pellets of high density and well-defined physical properties and chemical composition. Such pellets with uniform geometry are stacked and filled into metallic tubes. While stainless steel was used in the past, most reactors now use zirconium alloy[12] with higher melting points and lower neutron absorption as so-called cladding. The cladding prevents fission products from escaping into the coolant and must withstand internal pressure buildup and

---

[11]About 1.74 million times higher than petrol, 2.78 million times higher than coal, 4.98 million times higher than wood, 16.12 million times higher than alkaline batteries (Wikipedia/energy density).

[12]E.g., Zircaloy 4 with melting point of 1760 °C instead of 1450 °C for stainless steel (see also Fig. 2.2).

should be made of corrosion resistant material with low absorption cross section. The fuel rods are grouped into fuel assemblies of different geometry and designs depending on the reactor type.

Making fuel, fuel cladding and structural material more resistant to temperature rise and resulting core damage (see also Fig. 2.2) is a promising way to increase the robustness of nuclear reactors against potential accidents. A huge program on "accident tolerant fuel", coordinated by Westinghouse, is focused on protecting claddings from oxidation by coating[13] ceramic SiC cladding development and high density, high temperature resistant enriched U_3SiC fuel pellets.[14]

Steel alloys dominate the material for reactor (pressure) vessels; pre-stressed concrete reactor pressure vessels are technically feasible and could be "absolutely" rupture-proof.

6.1.4   Power Density and Power Level

As already mentioned, compared to other energy carriers, nuclear fission features incomparably high energy intensity and thus enables reactors with high power density and power rating. Depending on other design elements and parameters, typical power densities vary from 70 MW per cubic meter (m^3) for current commercial LWR to about 290 MW per m^3 for conceptual designs of sodium cooled fast reactors, while those of liquid lead or salt cooled fast reactors are less than half that power density. Power densities of gas-cooled reactors are smaller in principle. However, gas-cooled fast reactors show power densities slightly above water-cooled reactors while power densities of gas (helium) cooled, graphite moderated thermal reactors are significantly smaller. Power ratings follow the economy of scale and address local and regional needs. Power ratings reach into the range of a few thousands of MW thermal (MWt), with 4800 MWt (1600 MWe) of large size LWR as a reference point.[15]

The advantages of such high-power density and large size designs are mainly economic and, to a lesser extent, environmental as they require less space and land per installed MWe. However, in principle, high power density and power rating make the reactors more susceptible to loss of coolant and decay heat removal accidents, assuming all the other parameters are the same. In other words, limiting the power densities and power rating—together with other means, notably increasing

---

[13]with Ti_2AlC or functionally graded TiAlN/TiN to protect cladding from oxidation with H_2-formation under heat/steam conditions in case of unprotected loss of coolant accident.

[14]S. Ray, P. Xu, E. Lahoda, L. Hallstadius, F. Boylan, S. Johnson, Westinghouse accident tolerant fuel program—current results & future plans, Top Fuel 2015, American Nuclear Society, Zurich (Switzerland), Sept. 13–17, 2015.

[15]It is important to distinguish the energy generated in the form of thermal energy from the smaller energy ultimately obtained in the form of usable electric energy. The MWt unit refers to the former while the MWe unit refers to the later.

the heat storage capacity of the coolant and/or core structural material—provide flexibility to increase the robustness of nuclear reactors against loss of heat removal accidents and to stay away from "critical" temperatures. Proposed small-medium reactors (SMR) and the small modular high temperature reactor (HTR),[16] in particular, may serve as ground-breaking examples. Furthermore, as the inventory of fission products is proportional to the power level, in principle, a smaller amount of fission products would be released into the environment by smaller-sized reactors under loss of confinement conditions.

6.1.5 Siting Issues and Options

In the past, the main siting criterion was to choose comparatively remote or rural areas to minimize the number of people at risk in the event of an escape of radioactivity,[17] in particular in case of low probability severe accidents. Now, the selection process has evolved and includes health, safety and security factors and a set of general and specific requirements, the latter established by IAEA and adopted by member states.[18] Accordingly, the main safety objective in site evaluation is to protect the public and the environment from radiological consequences of accidental radioactive releases.

The evaluation of the suitability of a site shall consider:

- the effects of external events occurring in the region, either of natural origin, such as earthquakes and surface faulting, extreme meteorological events, flooding and water waves, or human induced impacts, such as aircraft crashes and chemical explosions[19];
- the characteristics that could influence the transfer to persons and to the environment of released radioactive material;
- the density and distribution of the population and other characteristics of the surrounding zone in so far as they could affect the possibility of implementing emergency response actions.

As a general rule, site deficiencies can be compensated by design features of the plant, measures for site protection (e.g., dykes) or administrative procedures; if not, the site shall be deemed unsuitable. A strategy for protective actions during an emergency needs to be preplanned and emergency planning zones (EPZ) are defined

---

[16]High temperature, helium cooled, graphite moderated reactor of small size (200–300 MWt), low power density (8 MWt/m^3), fully ceramic core with high inertia and inherent capability to avoid fission product release under accident conditions, see also Sect. 6.2.4.

[17]IAEA, "Managing Siting Activities for Nuclear Power Plants", *IAEA Nuclear Energy Series No. NG-T-3.7, Nuclear Energy Series Technical Reports* (2012).

[18]IAEA, "Site Evaluation for Nuclear Installations", *IAEA Safety requirements no. NS-R-3, Safety Standards Series* (2003).

[19]Willful actions by third parties are outside the scope of these requirements.

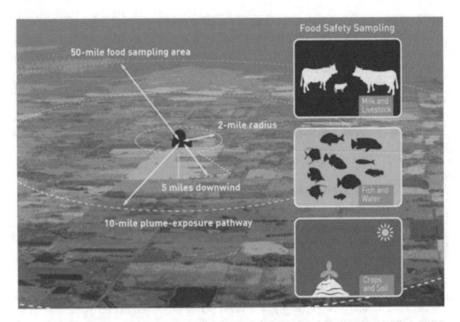

Fig. 6.1 US NRC Emergency Planning Zones (United States Nuclear Regulatory Commission (U.S. NRC), Emergency Planning Zones, 1.10.2016, http://www.nrc.gov/about-nrc/emerg-pre paredness/about-emerg-preparedness/planning-zones.html). A 2-mile ring around the plant is identified for evacuation, along with a 5-mile zone downwind of the projected release path

to facilitate their implementation in case of an extremely unlikely event. The exact size and shape of each zone is the result of detailed information. Figure 6.1 illustrates these siting considerations.

The foreseeable evolution of natural and human-made factors in the region that could have a bearing on safety shall be evaluated for the projected lifetime of the nuclear installations, e.g., 60 years for new builds. This requirement entails restrictions for spatial planning and is subject to uncertainties.

Needless to say, site characteristics are relevant for ensuring that societal risks due to severe nuclear accidents are acceptably low. Remote sites with absence of hospitals, schools, and other large-scale social infrastructure are deemed to be most suitable. However, driven by scarcity of actual remote sites and aspired use of nuclear reactors beyond power production, for example making use of waste heat, requires that sites close to consumer centers and large industrial facilities may have to be permitted.[18] Electricity transmission losses are also minimized with sites close to urban centres. The technical solution is to developed safety characteristics of the reactor so that severe accidents with release of radioactive substances beyond permissible values are precluded and emergency planning zones are no longer necessary from a technical point of view.

Accordingly, the combination of small, inherently "super-safe" reactors and of underground siting has been proposed, the latter allows to protect the plant against extreme external impacts including weapons attack. The concept of an underground, small-sized modular high temperature gas-cooled reactor (HTGR) provides an example[20,21] Besides installing the nuclear unit in a rock cavern, the so-called cut-and-cover design is favored, i.e. erecting the reactor and other sensitive parts of the plant in such an open excavated pit below ground, then cover it by soil to absorb the energy of a potential external force (missile) and put a concrete slab on top to prevent its penetration. The construction of sufficiently large pits in soil (about 60 m in diameter and 30–60 m deep) proved to be feasible, while the construction costs are expected to increase by roughly 10 to 20%.

6.2 Examining Candidate Technologies

The following sections aim to provide information about reactor and related fuel cycle concepts and associated states of development by looking into the R&D pipeline and referring to past experience. Thereby, we hope to provide a first answer to the question whether such "radically new" or even "exotic" system designs may achieve the challenging requirements outlined in Sect. 5.8 or, at least, to pinpoint directions of future research, all without anticipating or questioning all potential outcomes.

6.2.1 Sodium-Cooled Reactors

Sodium cooled reactors (SR) use abundant sodium[22] as a coolant, which is a soft highly reactive metal, with a melting point of 98 °C and a boiling point of 892 °C. Therefore, sodium allows reasonably high core outlet temperatures (about 550 °C) and thermal efficiencies (up to 40%), the latter also depends upon the power converting process employed. Due to the superior thermal hydraulic properties of sodium, sodium cooled reactors feature extremely high power density cores. More-over, due to its moderating capability and excellent neutron economy, most sodium cooled reactor concepts operate with a fast neutron spectrum. Such sodium cooled fast reactors (SFR) enable more efficient use of resources[23] and a broader field of applications, but tend to positive coolant temperature/sodium void coefficients.

Most sodium-cooled fast reactor core designs are arranged in a pool of sodium, circulated by internal primary pumps ("pool type"). The unpressurized primary

---

[20]X. Wang, "Study on Plume Emergency Planning Zone Determination for CAP 200 SMR", IAEA Int. Conference on Topical Issues in Nuclear Installation Safety, 6–9 June, Vienna, 2017.

[21]W. Kröger, "Verbrauchernahe Kernkraftwerke aus sicherheitstechnischer Sicht" (Nuclear power plants near to consumers from a safety-engineering point of view), Jül-2103, 1986.

[22]$_{11}Na^{23}$, as the only stable isotope, needs to be prepared from its compounds.

[23]Factor 60 compared to current LWR with UO_2 fuel.

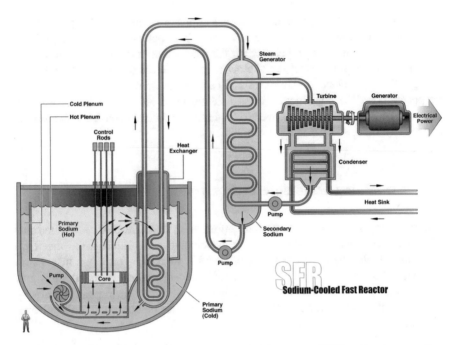

Fig. 6.2 Schematic view of a pool type sodium cooled fast reactor (SFR), with primary sodium outlet temperature typically more than 500 °C (Generation IV International Forum, A Technology Roadmap for Generation IV Nuclear Energy Systems (2002))

circuit is sealed; a secondary intermediate circuit is needed to cope with the high chemical reactivity of sodium with air and water, see Fig. 6.2 for a schematic view.

In the past, the mission of sodium cooled fast reactors was to breed more fuel as fast as possible, while nowadays the mission has shifted to broader objectives including increased resource utilization and management of plutonium and high-level waste, notably minor actinides. The fuel can be uranium or thorium based, with relatively high-enriched transuranic elements like plutonium. The fractions of minor actinides, depending on the purpose, can be either metal alloys or oxides. As part of the prospect of a closed fuel cycle, spent fuel from sodium-cooled fast reactors requires reprocessing.

The sodium-cooled fast reactor technology can rely on historical experience of design, construction, test, operation, inspection and repair of past or existing demonstration and/or prototype SFR,[24] such as the Prototype Fast Reactor (PFR, Dounreay, UK), Phénix, Superphénix (France), BN-350 (Kazakhstan, former Soviet Union) and Monju (Japan) and on numerous experimental facilities as forerunners. Several projects of sodium cooled fast reactor prototype developments were cancelled by mostly politically motivated government decisions at the times, e.g. Superphénix and the SNR-300 in Germany.

---

[24]K. Aoto, P. Dufour, Y. Hongyi et al., A summary of sodium-cooled fast reactor development, Progress in Nuclear Energy 77, 247–265 (2014).

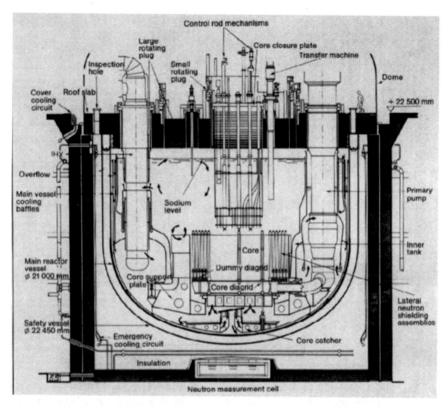

Fig. 6.3 Superphénix with a nominal power of 1200 MWe, cross-section of the reactor pool and primary circuit (K. Aoto, P. Dufour, Y. Hongyi et al., A summary of sodium-cooled fast reactor development, Progress in Nuclear Energy 77, 247-265 (2014))

Russia is operating commercially the BN-600 (600 MWe), which emerged out of the BN-350, and the BN-800 since 1980 and November 2016, respectively, while the development of the more advanced larger BN-1200 is on hold. India began its sodium fast breeder reactor program with a loop-type test reactor (FBTR: Fast Breeder Test Reactor), which has been in operation for 30 years since 1985. The construction of the 500 MWe prototype fast breeder reactor (PFBR) has been completed and commissioning is in the advanced stage in India. China's fast reactor program includes an experimental reactor (CEFR: China Experimental Fast Reactor) with a power of 65 MWt, which was connected to the grid by mid 2011, and a demo plant (China Demonstration Fast Reactor) with a preliminary designed power of 2100 MWt/870 MWe, which is planned to be constructed.[25]

Past experience has also demonstrated huge problems in mastering the highly complex sodium technology and plant layout (see Figs. 6.3 and 6.4) with the occurrence of a series of leaks, mostly on welds in the non-radioactive secondary

---

[25]I. Pioro, ed. *Handbook of generation IV nuclear reactors.* Woodhead Publishing, 2016.

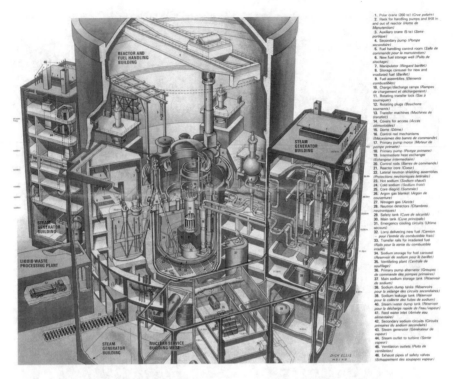

1. Polar crane (350 te) (Grue polaire)
2. Flask for handling pumps and IHX in and out of reactor (Hotte de Manutention)
3. Auxiliary crane (5 te) (Semi-portique)
4. Secondary pump (Pompe secondaire)
5. Fuel handling control room (Salle de commande pour la manutention)
6. New fuel storage well (Puits de stockage)
7. Manipulator (Ringard barillet)
8. Storage carousel for new and irradiated fuel (Barillet)
9. Fuel assemblies (Éléments combustibles)
10. Charge/discharge ramps (Rampes de chargement et déchargement)
11. Rotating transfer lock (Sas à tournquet)
12. Rotating plugs (Bouchons tournants)
13. Transfer machines (Machines de transfert)
14. Covers for access (Accès démontables)
15. Dome (Dôme)
16. Control rod mechanisms (Mécanismes des barres de commande)
17. Primary pump motor (Moteur de pompe primaire)
18. Primary pump (Pompe primaire)
19. Intermediate heat exchanger (Échangeur intermédiaire)
20. Control rods (Barres de commande)
21. Reactor core (Coeur)
22. Lateral neutron shielding assemblies (Protections neutroniques latérales)
23. Hot sodium (Sodium chaud)
24. Cold sodium (Sodium froid)
25. Core diagrid (Sommier)
26. Argon gas blanket (Argon de couverture)
27. Nitrogen gas (Azote)
28. Neutron detectors (Chambres neutroniques)
29. Safety tank (Cuve de sécurité)
30. Main tank (Cuve principale)
31. Emergency cooling circuits (Ultima secours)
32. Lorry delivering new fuel (Camion pour l'entrée du combustible frais)
33. Transfer rails for irradiated fuel (Rails pour la sortie du combustible irradié)
34. Sodium storage for fuel carousel (Réservoir de sodium pour le barillet)
35. Ventilating plant (Centrale de soufflage)
36. Primary pump alternator (Groupes de commande des pompes primaires)
37. Main sodium storage tank (Réservoir de sodium)
38. Sodium dump tanks (Réservoirs pour la vidange des circuits secondaires)
39. Sodium leakage tank (Réservoir pour la collecte des fuites de sodium)
40. Steam/water dump tank (Réservoir pour la décharge rapide de l'eau/vapeur)
41. Feed water inlet (Arrivée eau alimentaire)
42. Secondary sodium circuits (Circuits primaires du sodium secondaire)
43. Steam generator (Générateur de vapeur)
44. Steam outlet to turbine (Sortie vapeur)
45. Ventilation outlets (Puits de ventilation)
46. Exhaust pipes of safety valves (Échappement des soupapes vapeur)

Fig. 6.4 Global view of the Superphénix reactor at Creys-Malville sited by the River Rhone, in the foothills of the Alps, about 20 km west of Lyon. Note the size of the reactor core compared with the lorry on the lower left (Creys-Malville Super Phenix. The Worlds Reactors No 73, Nuclear Engineering International 23–272: 57–58 (1978))

sodium circuit,[26] of unreliable steam generators operation and fuel failures (BN350), of drop of In-Vessel Transfer Machine (Monju, 2010), and so on. Achieved availabilities were small, in general, with an average capacity factor[27] of 41.5% (1987–1996). In contrast, a maximum value of 57% (8/1995–12/1996) for Superphénix and more recently (1990–2015) an average value of 74% for BN-600 serve as positive examples.

Sodium cooled fast reactors are one of six technologies selected for collaborative R&D within the Generation IV framework (GIF), with China, EU-Euratom, France, Japan, South Korea, Russia, and the USA as participating members, working on

---

[26]The most serious leak occurred in the secondary circuit of the Monju plant, possibly at a weld point damaged by intense vibration, on December 1995; the leaked out sodium amounted to 640 kg and caused high temperature fumes.

[27]The capacity factor of a power plant is defined as the ratio of its actual output over a period of time, to its potential output obtained by operating it at full capacity continuously over the same period of time.

system arrangement, and five national driving projects as backbones and beneficiaries.[28]

Basically, three options are considered by GIF in its System Research Plan:

- A large size (600–1500 MWe) loop-type reactor with mixed uranium-plutonium oxide fuel and potentially minor actinides, supported by a fuel cycle based upon advanced aqueous re-processing at a central location serving a number of reactors;
- An intermediate-to-large size (300–1500 MWe) pool-type reactor with oxide or metal fuel, and
- A small size (50–150 MWe) modular-type reactor with uranium-plutonium-minor actinide-zirconium metal alloy fuel, supported by a fuel cycle based on pyro-metallurgical processing in facilities integrated with the reactor.

The core outlet temperatures are approximately 510–550 °C, and about 390 °C when returned to the cold pool; the envisaged deployment time is later than 2022.

Sodium cooled fast reactor development benefits from existing experience and knowledge gained from experiments for code validation, fuel behavior, components performance, and so on, as well as from operations and tests of real reactors.[29] However, the developments of the sodium cooled fast reactor concept call for intensive R&D, namely (according to GIF):

- Safety-related activities on phenomenological model development and experimental programs, conceptual studies in support of the safety provisions and preliminary assessment of safety systems;
- Development of high burn up of minor actinides bearing fuels as well as of claddings and wrappers capable of withstanding high neutron doses and temperatures including remote fuel fabrication;
- Handling techniques and system performance under irradiation;
- Alternate fuel forms and targets for heterogeneous recycling and suitable steels for core materials;
- Fuel handling system improvements to reduce outage times, improved instrumentation for sodium leaks, in-service inspection, diagnostics and repair capabilities;
- Design against seismic and other severe events.

Finally, core design studies and accident analyses need to be performed to optimize core parameters such as neutron energy spectrum, reactivity coefficients and burn up swing as well as lattice tightness and passive heat transfer mechanisms.

Available designs of future sodium cooled fast reactor are at a conceptual stage. A framework and methods for analysis of safety architecture are the subject of

---

[28]J.E. Kelly, Generation IV International Forum: A decade of progress through international cooperation, Progress in Nuclear Energy 77, 240–246 (2014).

[29]Phénix end-of-life tests, restart of Monju, lifetime extension of BN-600 and start-up of China Experimental Fast Reactor (CEFR).

development, while detailed deterministic and probabilistic safety analyses are missing. Nevertheless, comparative studies indicate vulnerability of sodium cooled fast reactors during loss-of-flow and loss-of-heat sink accidents. In view of the fuel temperature coefficient and in the presence of axial fuel expansion, positive reactivity may occur, in particular by coolant temperature increase and when natural convection is insufficient. In worst-case scenarios, sodium might start boiling, resulting in a significant reactivity insertion with power increase, leading to more boiling and voiding until fuel melts. Hejzlar et al.[30] points also this potentially severe problem but states "that it should be feasible to design a large sodium-cooled fast reactor core with self-controllability features". Furthermore, the well-known chemical reaction of sodium with water, steam and air may raise major safety concerns.

As indicated above, the safety concept is subject to further considerations and development. It will, for sure, include elements deviating from standard approaches and current regulatory praxis. For instance, decay heat removal from the core could be done by natural convection; auxiliary safety grade outside cooling of the reactor vessel could be implemented, again provided via natural air circulation (RVACS).[31]

6.2.2 Lead-Cooled Reactors

Lead cooled reactors use molten lead or lead-bismuth eutectic (LBE) as a coolant, which are low-pressure, chemically stable liquids with excellent thermodynamic properties. Lead has a high boiling point (1750 °C) and melting point (327 °C), offering a wide operational margin with potentially high outlet temperatures (typically in the range of up to 500 °C, potentially ranging over 800 °C), enabling high thermodynamic efficiency. Lead stands out for a large margin to boiling with the potential for a benign end state under severe accident conditions, though the relatively high melting point creates freezing potential, rendering the reactor inoperable and calling for specific design and operational measures. By leaking and solidifying, the coolant may damage the reactor.

Lead-bismuth (LBE) offers a lower melting point at about 125 °C combined with a lower boiling point of 1670 °C, compared to lead. While lead is abundant and quite cheap, bismuth is expensive and rare. Furthermore, polonium-210 builds up in the Pb coolant via neutron capture, causing operational issues. Lead, lead alloys and

---

[30]P. Hejzlar, Todreas, N.E., Shwageraus, E., Nikiforova, A., Petroski, R., Driscoll, M.J., Cross Comparison of fast reactor concepts with various coolants, Nuclear Engineering and Design 239, 2672–2691 (2009).

[31]As foreseen in the ANL design based in the former S-PRISM 1000 MWt design; the core design was extrapolated to 2400 MWt and studied by MIT (P. Hejzlar, Todreas, N.E., Shwageraus, E., Nikiforova, A., Petroski, R., Driscoll, M.J., Cross comparison of fast reactor concepts with various coolants, Nuclear Engineering and Design 239, 2672–2691 (2009)).

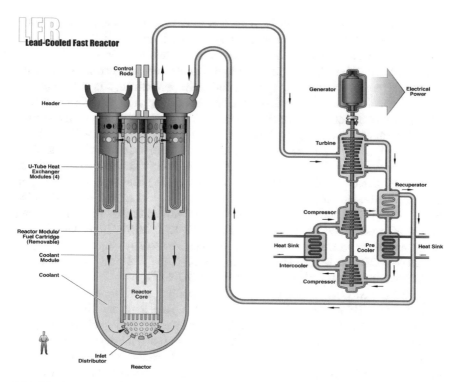

Fig. 6.5 Schematic diagram of a pool-type lead cooled fast reactor (Generation IV International Forum, A Technology Roadmap for Generation IV Nuclear Energy Systems (2002)). The reactor vessel is kept as compact as possible to reduce the total coolant inventory and mass; however, depending on reactor size, the weight of the coolant is in the range of hundreds to thousands of tonnes

lead/bismuth are chemically inactive but do interact thermally significantly with water, vapor and air, therefore there is no need for an intermediate circuit as with sodium-cooled reactors. Primary pumps transfer the heat directly to the heat exchangers and to the secondary power conversion cycle, respectively (see Fig. 6.5). However, lead and LBE are more corrosive to steel and this puts an upper limit to coolant flow velocity through the reactor.

Lead-cooled reactors feature a fast neutron spectrum (LFR) and use a closed fuel cycle enhanced by the fertile conversion of uranium; they can also be used as a "burner" to consume plutonium and minor actinides from spent LWR fuel and are flexible to the thorium fuel cycle. Instead of refueling, the whole core can be replaced after 5 to 8 years of operation. Due to its neutronic properties,[32] lead cooled reactors exhibit only a small positive coolant temperature coefficient[33] and may even prevent temperature feedback.

---

[32]Small thermal expansion coefficient, low absorption cross-section, low slowing down power and high scattering cross section.

The use of heavy liquid metals as coolant in fast reactors is not widely known about, however, there is considerable history with such reactors.[33] The most significant industrial and operational experience with such reactors was gained by Russia (former Soviet Union) in their program to design and operate lead-bismuth (LBE) cooled reactors for submarine propulsion during the period from mid 1960s until the 1990s,[34] which represents about 80 reactor-years of operating experience. Nowadays, two concepts are under development, both in Russia, with near term construction plans underway. First, the modular LBE cooled SVBR-100 with increased power (280 MWt) and increased coolant outlet temperature (490 instead of 345 °C) compared to the submarine reactors. The fuel is uranium oxide enriched to 16% U-235 and refueling times of 7 to 8 years; a site license was issued early in 2015 and start-up of the pilot unit is planned for 2019. And second, the lead cooled BREST-OD-300, a 300 MWe demo fast reactor with an on-site fuel reprocessing facility as a precursor for a commercial BREST-1200 MWe. Both development efforts are active with near-term construction plans underway—the construction of the BREST-OD-300 was approved in August 2016, the fuel plant is scheduled to begin operating in 2017 (Chapter 12.4 of Pioro (2016)[25]).

Within the GIF framework, two reactor-size options are being considered: (1) a small 50–150 MWe transportable system with a very long core life, and (2) a medium 300–600 MWe system; the coolant may either be lead (preferred) or lead-bismuth eutectic (p 47 of Pioro (2016)[25]). In Western Europe, initial efforts, starting with the fifth Framework Program (FWP) in 1997, which were related to LFR development, concentrated on Accelerator Driven Subcritical Systems (ADS, see also Sect. 6.2.5) for the transmutation of plutonium and minor actinides. Key initiatives include EFIT[35] and MYRRHA[36]: while EFIT served as a starting point for the design of later critical reactor systems cooled by lead, MYRRHA continues as a major project intended to demonstrate both subcritical and critical operation of a multi-purpose irradiation facility cooled by LBE. It is planned that MYRRHA will replace the BR2 reactor at the Belgian nuclear research center in Mol (see also Sect. 6.2.6). Considerable effort has been spent to exploit to the greatest degree possible the beneficial characteristics of lead and LBE as a coolant and to make safety design a primary goal from the beginning. In 2010, based on experience gained from previous projects, two new projects were launched as part of the seventh FWP: (1) the CDT-FASTEF project conducting conceptual design of the MYRRHA facility and (2) the LEADER project concentrating its activities on the development of an enhanced concept for an industrial-sized critical reactor, called the European

---

[33] A. Alemberti, Smirnov V et al., Overview of lead-cooled fast reactor activities, Progress in Nuclear Energy 77, 300–307 (2014).

[34] A total of 12 reactors and 15 reactor cores, 155 MWt each, were built and deployed including two reactors/three of the cores were operated onshore.

[35] European Facility for Industrial Transmutation.

[36] Multi-purpose hybrid research reactor for high technology application, 50–100 MWt, 55% lead and 45% bismuth eutectic as coolant, full power operation around 2025, total investment costs Euro 960 million (as of 2009), visit www.myrrha.scken.be

Lead-cooled Fast Reactor (ELFR). In parallel, design activities for a smaller (120 MWe) LFR demonstrator, called ALFRED, are being undertaken in Romania as an EU-project of national interest.

Japan and South Korea have conducted significant research into LFR since at least the 1990s while China only recently started its activities in this field. As an example, Japan proposed the LBE-cooled long-life simple small portable reactor, fabricated in an energy park and transported to its operating site with a sealed vessel that would not be opened at the site for refueling to ensure proliferation resistance. Work on LFR concepts and technology in the USA has been carried out from 1997 to the present by various research organizations. Currently the development of a small, secure, transportable, autonomous reactor (SSTAR), carried out by Argonne National Laboratory (ANL) and other organizations, is of particular relevance.

The 20 MWe SSTAR, the BREST-OD-300 MWe and the 600 MWe ELFR are reference concepts that form the basis of the LFR Provisional System Steering Committee established within the GenIV framework with the EU, Japan and the Russian Federation as active partners in R&D projects. Figure 6.6 provides an overview sketch of the BREST-OD-300 that is regarded to be the most mature LFR concept.

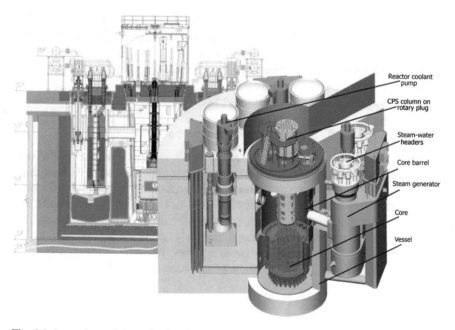

Fig. 6.6 Lower layer: Schematic view (symmetric) of the BREST-OD-300 MWe/700 MWt lead-cooled system with uranium-plutonium nitride fuel, conversion ratio 1, coolant temperature of 420/540 °C and core diameter/height of 2.6/1.1 m. Note that the implied efficiency is an impressive 43% (A total of 12 reactors and 15 reactor cores, 155 MWt each, were built and deployed including two reactors/three of the cores were operated onshore). Front layer: Labeled three-dimensional image of the system (Rosatom, "BREST-OD-300 natural-safety reactor facility for the NPP pilot & demonstration complex with an on-site fuel cycle", http://www.nikiet.ru/eng/images/stories/NIKIET/Areas/brest_eng.pdf)

LFR are in an early stage of development. Although they can profit from considerable experience and knowledge gained so far and from various ongoing research activities. However, major R&D challenges remain related to[37]:

- The high melting point of lead where solidification would render the reactor inoperable;
- The accumulation of radioactive polonium-210 in the lead coolant;
- Optical opacity of the coolant with challenges related to inspection and monitoring as well as repair of reactor core components and to re-fueling;
- High density and mass/weight (hundreds to thousands of tonnes) that require careful consideration of foundations, structural and seismic design;
- Corrosion control of structural steels at high temperatures and flow rates;
- Management of fuel in molten lead.

Needless to say, designs are in an early, conceptual stage and detailed concept-specific analyses are missing. Therefore, safety assessments are premature and somewhat speculative. However, comparative assessments carried out by MIT and JRC Petten for large LFR self-breeding cores (around 2400 MWt, conversion ratio of 1, low pressure drop) allow for first estimates:

- Regarding unprotected loss-of-flow and loss-of-heat-sink accidents as well as total loss-of-power accidents (with some restrictions), LFR show remarkable positive behavior due to excellent natural circulation, massive heat storage capability and high boiling temperatures of lead/LBE; severe overheating appears virtually impossible.
- The chemical inactivity of lead excludes strongly exothermic reactions with air, water and water vapor.
- Reactivity-induced accidents can be eliminated by design.

Tuček et al.[38] show advantages of lead-cooled fast neutron reactors (LFR) over sodium cooled fast reactors (SFR) regarding behavior in severe accident conditions, concluding that a LFR is a very robust system when it comes to safety. Kelly[41] notes a high potential of self-controllability of LFR due to reactivity feedbacks that inherently shut down the reactor before temperature limits on reactor structures are exceeded during accidents. This occurs passively as decay heat is removed by convection, obviating the need for external power. However, accident sequences leading to temperatures below freezing of the coolant as well as severe seismic events still deserve special attention.

In short, smaller capacity LFR seem to have a high potential to achieve stringent safety requirements. As in case of sodium-cooled (fast) reactors, the safety concept

---

[37]Kelly J.E., Generation IV International Forum: A decade of progress through international cooperation, Progress in Nuclear Energy 77, 240–246 (2014).

[38]K. Tuček, J. Carlsson, H. Wider, Comparison of sodium and lead-cooled fast reactors regarding reactor physics aspects, severe safety and economical issues, Nuclear Engineering and Design 236, 1589–1598 (2006).

and measures are under development and will for sure deviate from standard approaches and current regulatory praxis, requiring adaptation.

6.2.3 Molten Salt-Cooled Reactors

Molten salt cooled reactors (MSR) use a molten salt mixture as coolant, which exhibit high volumetric heat capacity, high viscosity and small thermal conductivity and can corrode metallic alloys. A range of such mixtures exists that have different physical properties. Molten fluoride salt with a melting point of 300 °C and a boiling point of more than 1400 °C is one popular option. Alternatively and regarded to be better are chlorides with higher melting and boiling points, allowing operating temperatures[39] of up to 750–900 °C, much higher than for current LWR. Accordingly, MSR facilitate higher thermal efficiency of the steam cycle or advanced power conversion process of up to 44%. The nuclear fuel may be solid or, mostly favored, dissolved in the molten salt, e.g. as uranium or thorium tetra-fluoride. The salt fuel flows into an intermediate heat exchanger, where the heat is transferred to a secondary molten salt coolant.

MSR are highly flexible to the fuel composition. The fuel can be uranium or thorium based, with U-235 and Pu-239 in case of the uranium cycle or U-233 in case of the thorium cycle. The reactor can be designed as burner or breeder; any other waste, in particular minor actinides from other reactor lines, can be added and consumed, turning a MSR into a "waste burner". MSR can be operated with semi-thermal or fast neutron spectra, depending on the purpose of operation; graphite may serve as moderator or structural material if solid fuel is used. As the power production is controlled by increasing or decreasing the fuel fraction, control rods can be largely avoided. Moreover, due to its highly favorable neutron balance and flexibility regarding the fuel composition and continuous addition, the reactor does not need excess reactivity, which is also positive from a safety point of view. Gaseous and volatile fission products can be continuously separated and then conditioned and stored, thus reducing significantly the radioactive inventory and requirements to decay heat removal from the core, compared to current LWR designs and other innovative designs with solid fuel.

The reactor core is embedded in an unpressurized vessel with some control rod drives on top. The reactor vessel is surrounded by containers/safety tanks, which enclose all significant components including the reactor vessel, primary heat exchanger, chemical processing and degassing system as well as dump or overflow tank. The concepts follow the defense in depth/multi-barrier approach (see Fig. 6.7) and provide an overflow system as a key safety feature. This system is equipped with

---

[39]Following specific numbers are given for the Seaborg waste-burner (SWaB) pilot plant early design.

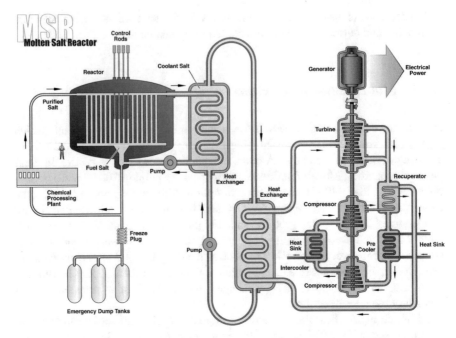

Fig. 6.7 Molten salt reactor scheme with fuel dissolved in the molten salt and online chemical reprocessing (Generation IV International Forum, A Technology Roadmap for Generation IV Nuclear Energy Systems (2002)). Fission occurs within the fuel salt, heat is then transferred to a secondary liquid-salt coolant circuit, which also separates fuel and fission products from the steam cycle

a dump tank, separated from the reactor by a freeze plug,[40] which allows one to passively dump the fuel to the tank where the sub-critical salt is cooled by passive means in most crucial failure scenarios, notably core heat-up. Solidified salt can be electrically heated and pumped back into the reactor to restart the operation.

The MSR concept is not new and goes back to experiments and design proposals, mainly by ORNL[41] in the 1960s and 1970s. Finally, although successful, it was given up in favor of other competing programs. The interest has been renewed within the GIF international collaboration effort to develop next generation Gen IV technologies.

As one of the six technologies pursued within the GIF framework, the molten salt reactor concept (MSR) benefits from R&D efforts since 2005, coordinated by a

---

[40]The salt plug is actively cooled to melting temperature and opens in case of loss of power or overheated fuel.

[41]Oak Ridge National Laboratory molten salt reactor experiment (MSRE) with a 7,4 MWt test reactor with LiF-BeF_2-ZrF_4-$4F_4$ fuel salt, graphite core moderated, 650 °C outlet temperature, running for 4 years; the molten salt breeder (MSBR) design project with LiF-BeF_2-ThF_4-UF_4 fuel salt, graphite moderated, 705 °C peak temperature, closed down in the early 1970s.

"Provisional System Steering Committee". In general, three lines of R&D are in focus[42]:

- The EU, France and Russia are focusing on the development of a fast neutron spectrum molten salt reactor, without any solid moderator in the core, capable either of breeding (MSFR)[43] or trans-mutation of actinides from spent nuclear fuel (MOSART).
- The USA is focusing on thermal fluoride-salt-cooled High-Temperature Reactors (FHR) with solid coated particle fuel as a nearer term MSR option.
- China launched in January 2011 a pilot project of the "Thorium Molten Salt Reactor" (TMSR) aiming at developing solid and (afterwards) liquid fueled MSR and utilization of thorium energy within 20 to 30 years, with two test reactors, a solid-fueled and a liquid fueled one, as near term goals.

However, the overall effort is actually quite small in view of the tremendous research and development needs and compared to other development lines; a legitimate deployment date is mostly missing. Nevertheless, young private enterprises have taken up this technology and have come up with promising concepts,[44] see Fig. 6.8.

To capitalize on the advantages, i.e. extended resource utilization and waste minimization, MSR requires a fast neutron spectrum and a closed fuel cycle including reprocessing, which may restrict public acceptance. Besides corrosion issues, related to the type of salt, severe problems still need to be solved concerning the composition of fuel and secondary salt, the material for vessels and heat exchangers and other components as well as miscellaneous reliability issues[45]:

- Stability of salts, their resistance to high neutron fluxes, avoidance of deposits of nuclear material, liquid salt chemistry and properties, in general;
- Material resistance against mechanical abrasion and chemical corrosion by fast moving salt;
- Oxygen at high temperature on the external surface;
- Valve and pump reliability over many years of operation that need to work in an adverse environment.

Furthermore, combined uranium and thorium cycles need chemical separation and reprocessing units that do not yet exist. These issues place molten salt reactors in

---

[42]J. Serp, M. Allibert, O. Beneš, S. Delpech, O. Feynberg, V. Ghetta, D. Heuer, D. Holcomb, V. Ignatiev, J.L. Kloosterman, L. Luzzi, E. Merle-Lucotte, J. Uhlíř, R. Yoshioka, D. Zhimin, The molten salt reactor (MSR) in generation IV: overview and perspectives, Progress in Nuclear Energy 77, 308–319 (2014).

[43]The MSFR is a 3000 MWt reactor with a total fuel salt volume of 18 m^3 (reactor power density 300 MW/m^3) composed of lithium fluoride and thorium fluoride, operated at maximum fuel salt temperature of 750 °C.

[44]For example, Transatomic Power Corporation (TAP), USA; Seaborg Technologies, Denmark.

[45]E. Mearns, *Molten Salt Fast Reactor Technology – An Overview*, (Online) http://euanmearns. com/molten-salt-fast-reactor-technology-an-overview/

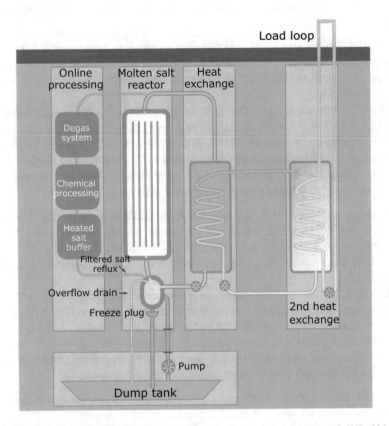

Fig. 6.8 Provisional schematic of the Seaborg Waste-Burner molten salt reactor (SaWB, 50 MWt) with primary and secondary circuit (fuel salt: 78LiF + 22ThF₄; secondary coolant salt: 46SliF + 11.5NaF + 42KF), online reprocessing, extraction of volatile fission products and fuel handling as well as an overflow system (Seaborg, *Tentative schematic of the SWaB reactor*, (Online Image) 2.10.16, http://seaborg.co/ppp2015.pdf, p. 6)

the domain of theoretical research, calling for a substantial increase above the present existing funding.

Available designs of MSR, including the SWaB, are at an early stage. Therefore, sufficiently secure assessments of potential failures and failure scenarios and associated risks are not yet possible. However,

– overheating or even meltdown scenarios and station blackout scenarios are reduced significantly or potentially eliminated due to physical properties and plant design;
– prompt criticality scenarios, e.g. triggered by unintentionally or intentionally flawed fuel addition, or faulty chemical reprocessing deserve more research attention;

– core vessel meltdown as a result of failure to dump molten salt fuel resulting in a release of radioactivity outside of the first barrier is imaginable with a measurable probability and stresses the importance of incorporating a second barrier.

Other MSR-design specific scenarios need careful consideration as soon as the layout has become sufficiently detailed; the likelihood of operating incidents or accidents may increase due to the in-line reprocessing and high complexity of the overall plant layout.

More importantly, the classical safety approach and analysis methods cannot be applied to MSR due to their unique safety-relevant characteristics,[46] which are fundamentally different from current LWRs. As for other innovative technologies, new safety schemes, principles and requirements need to be developed and adequate knowledge and regulatory competence needs to be accumulated. This requires a strong political will that today does not seem to exist within the OECD, creating uncertainties about the future of what may be regarded as a vital future energy source for humanity.

Finally, it is worth mentioning that all thorium fueled MSRs will use either U-235 or Pu-239 to "ignite" the fission chain reaction since thorium-232 is fertile but not fissile by itself and requires other fissile fuel to initiate the fission cycle, which lowers the advantage with regard to low contents of minor actinides of pure thorium-232/uranium-233 cycles. Cost estimates are not available and would be premature.

6.2.4 High-temperature Gas-cooled Reactors

To date, high-temperature Gas-cooled Reactors (HTGR) have used helium as coolant, an inert gas that does not react with any material or substance. It is immune to neutron radiation and thus cannot become radioactive. The core outlet temperatures of "conservative" designs (called near term demonstrators) lie in the range of 750–800 °C, enabling high thermodynamic efficiency of the water-steam cycle (about 40%). The water-steam cycle is coupled to the reactor via a heat exchanger; the operating pressure of the reactor is roughly 70/80 bar (70 to 80 times the atmospheric pressure). Further developments like the VHTR[47] strive for outlet temperatures up to 1000 °C for process heat applications like hydrogen production that could be used in vehicle transport.

HTGR operate with thermal neutron spectra and graphite as the moderator and core structural material. Ceramic fuel pellets or pebbles comprise a U oxide or U carbide fuel kernel, a porous buffer of pyrolytic graphite (PyC) to accommodate fuel swelling and fission gases, a dense PyC buffer and a dense silicon carbide (SiC) layer

---

[46]Notably, fuel molten during normal operation and absence of cladding, strong negative feedback coefficients, no separation of reactor and processing/recycling plant, etc.

[47]The Very High Temperature Reactor is one of the six concepts within the Generation IV development framework, see also p. 55 of Pioro (2016)[26].

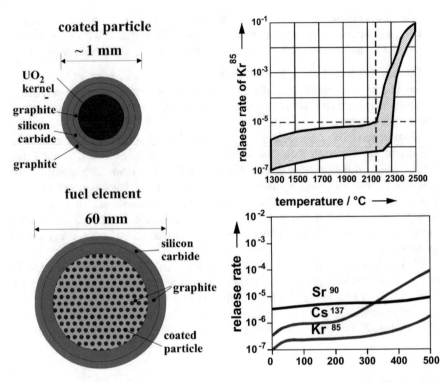

Fig. 6.9 Bottom left shows a cross section of a ceramic fuel pellet (TRISO coated). Thousands of these pellets are packed into cylindrical compacts (top left). Note that ceramic-coated pellets are packed into ceramic-coated containers providing two barriers between spent fuel and the environment. Right shows the results of heating experiments conducted on ceramic fuel designed to test their retention. For [85]Kr, the fuel did not fail until a temperature of ~2200 °C was reached (top right). Bottom right shows heating at 1600 °C over a period of ~21 days (K. Kugeler et al., *Safety of HTR—state of knowledge and necessary research*. In: Proceedings of the Workshop on Advanced Nuclear Reactor Safety Issues and Research Needs, 18–20 February 2002, Paris, France)

as a barrier against fission product escape and a final PyC layer for better bonding with the graphite matrix into which the coated particles are integrated (see Fig. 6.9). These "TRISO coated particles", with a diameter of approximately 1 mm each, are the basis for all modern HTGR-designs. Up to almost 10,000 particles, baked into a graphite matrix, are given a macroscopic shape, usually in the form of thumb-sized cylinders ("compacts"), placed in a hexagonal graphite block, or in the form of a spherical elements ("pebbles"), assembled to constitute a fully ceramic reactor core contained in a pressure vessel[48] (see Fig. 6.10, also annexed Table 6.4 for further details).

---

[48]M.A. Fütterer, L. Fu, C. Sink, S. de Groot, M. Pouchon, Y. Wan Kim, F. Carré and Y.Tachibana, Status of the very high temperature reactor system, Progress in Nuclear Energy 77, 266–281 (2014).

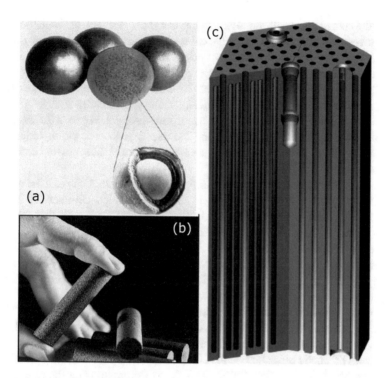

Fig. 6.10 Macroscopic forms of HTGR fuel. Pebbles (**a**) are assembled into a pebble bed core as illustrated in Fig. 6.9 or cylindrical compacts (**b**) are placed in hexagonal graphite blocks (**c**)

HTGR use uranium fuel with low enriched U-235, (8.5–16.5%), in the form of uranium dioxide (UO_2) or uranium carbide (UC) but has potential to also use the thorium fuel cycle. When the fuel depletes in fissile isotopes, on-line refueling is done by simply adding fresh pebbles to the bed. This provides a means of micromanaging the fission reaction, ensures criticality while avoiding excess reactivity, which is a positive characteristic from a safety point of view. HTGRs aim at high burn-up[49]; they are based on an open fuel cycle with direct geologic disposal of spent fuel elements, which are regarded extremely resistant to leaking in a final repository. That is because the spent fuel is encapsulated in a hard, ceramic pellet of pebble. For an actinide-burning alternative, specific plutonium-based driver fuel and transmutation fuel containing minor actinides would have to be developed.[48] The retention capability of the dense silicon carbide diffusion barrier of TRISO has been confirmed by intensive testing and, therefore, fuel characteristics serve as the key inherent feature of the safety concept of modern HTGR designs (see Fig. 6.10). Together with the comparatively low power density,[50] this limits the reactor power to the range of

---

[49] 100 GWd/tHM and beyond, double compared to modern LWR; approaching 200 GWd/tHM as a visionary goal.

[50] About 8 MW/m^3 compared to 100 MW/m^3 for LWR.

250–300 MWt and a small modular design. For these designs, the metallic reactor pressure vessel is connected to an adjacent steam generator through a duct designed to carry very hot gas (He). Robustness against loss-of-coolant[51] and against loss-of-active-core-cooling accidents has been demonstrated in prototype reactors. Decay heat is removed sufficiently rapidly solely by conduction, convection and radiation. Under incident conditions, this is sufficient to keep maximum fuel temperatures below a critical value, presently 1600 °C. Efforts are ongoing to raise the critical tolerable value to 1800 °C by improving the heat tolerance of the materials used. This would allow one to increase important parameters like power, core outlet temperature and/or fuel burn up.

Gas-cooled reactors were deployed originally for their simplicity and the quest for high thermodynamic efficiencies. Twenty six CO_2-cooled commercial Magnox reactors were built in the UK between 1956 and 1971 with a further two units built in Italy and Japan. All have since closed down. Fourteen Advanced Gas-cooled Reactors (AGR) are still operated, all in the UK, with a temperature limitation due to CO_2 as coolant. Various development steps have been taken towards using He instead of CO_2. In parallel, full ceramic fuel development of coated particle fuel took place in 1957–1961 by the UK atomic energy authority (UKAEA) and Battelle, a US based research organization. There followed the development of TRISO particles, accompanied by extensive conceptual work and testing of HTGR designs. Finally, two different types of reactors were designed, and built.[52] An experimental HTGR with prismatic block core in the UK (Dragon, 21.5 MWt, 1966–74), followed by Fort St. Vrain Station (842 MWt, 1976–89) in Colorado. Over the same period, Germany pursued the pebble bed design[53] and built an experimental reactor, the AVR, 46 MWt, operated from 1966–88 and a 300 MWe (750 MWth) prototype commercial power reactor, aimed at using thorium fuel, the THTR 300 that operated at full power from 2/1987 to 9/1989 after a construction time from 1970 to 1983. These reactors established the technical feasibility of HTGR even though they experienced problems including power fluctuations, jamming of control rods into the core, and leakage of moisture (Fort St. Vrain). The THTR 300 operated in Germany had technical difficulties with the pebble withdrawal system, hot gas channel insulation and so on, and thus operated for only 2 years. These problems finally caused early closure and decommissioning, mainly for induced economic and political reasons.

However, in the 1980s, the industry in Germany led the development of the 200 MWt HTR-Module, a simplified design concept with a power size and power rating to enhance passive and inherent safety features as mentioned before. The HTR-Module was the basis for the HTR-10 (10 MWt) and HTR-PM reactors in

---

[51]Physically, helium cannot get lost in case of leaks but heat transfer capability will be reduced significantly.

[52]Beck et al., 2010. High Temperature Gas-Cooled Reactors Lessons Learned Applicable to the Next Generation Nuclear Plant (INL/EXT-10-19,329). Idaho National Laboratory, Idaho Falls, ID.

[53]Developed and pushed by Prof. R Schulten.

Fig. 6.11 Schematic view of the HTR-PM (250 MWt/ 100 MWe) core design with (1) a steel reactor pressure vessel, (2) a pebble bed core, (3) graphite reflector, (4) continuous fuel feeding and (5) withdrawal and (6) helium flowing from the top to the bottom before (7) reaching the high temperature He outlet; (8) control rods move in drilled holes in the reflectors (Zhang, Zuoyi, et al., Current status and technical description of Chinese 2× 250 MWth HTR-PM demonstration plant, Nuclear Engineering and Design 239 (7), 1212–1219 (2009))

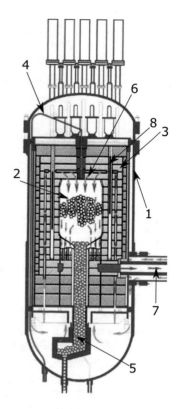

China and for projects in other countries, notably in the Republic of South Africa, featuring an upgraded modular HTGR and power generation via a gas turbine. Past experience with experimental and prototype HTGRs have demonstrated technical viability and excellent safety characteristics[54] but were not given the time to prove their economic competitiveness in comparison with LWR or other energy sources. For example, the HTGR development project in South Africa was given up after a duration of 10 years and US$1.4 billion spent, because of lack of commercial interest. The South African project was initiated in 1994 at a time when nuclear power had fallen totally out of favor with the international community in the wake of the Chernobyl disaster in 1986. Some argue that the HTGR development was too fast, in particular the extension into power production.

Current HTGR programs are found mainly in China, which has announced it is to scale up the pebble bed HTR-10 test reactor (first criticality December 2000, full power operation January 2003) to a demonstration plant with two 250 MWt modules, each with a helium core outlet temperature of 750 °C, driving a single steam turbine generator set (HTR-PM, see Fig. 6.11). Construction started at the beginning

---

[54]In particular, the AVR demonstrated passive safety performance and survived a water ingress accident provoked by a steam generator leak.

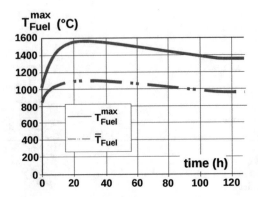

Fig. 6.12 Safety experiments and analyses for a pebble bed modular HTGR (300 MWthermal) demonstrate excellent inherent safety features (K. Kugeler et al., Safety of HTR—state of knowledge and necessary research. In: Proceedings of the Workshop on Advanced Nuclear Reactor Safety Issues and Research Needs, 18–20 February 2002, Paris, France): automatic removal of residual heat in the case where active heat removal fails. In addition, it was assumed that the shutdown system failed too

of 2014 (it was delayed following the Fukushima Daiichi accident); all buildings are already constructed and the grid connection is planned for 2020. In 2006, the Chinese State Council declared that the small HTR is second of two high-priority National Major Science and Technology projects for the next 15 years.[55] Further research, development and demonstration (RD&D) efforts aim at helium outlet temperature of 950 °C and more ambitious power conversion systems and heat delivery for chemical and industrial purposes. Activities in Japan concentrate on the restart of the HTR test facility and further safety tests, on innovative designs including oxidation-resistant graphite. This is in response to an air or water ingress accident and incineration of surplus plutonium from reprocessing of LWR fuel and on hydrogen production technologies.

Safety assessments and risk estimates are based on key design features and knowledge gained from analytical and experimental work in the past and at present, and point clearly to the following conclusions:

– If optimally designed, i.e. by limiting the power rate and power density, HTGR are not vulnerable to failure of the active heat removal and first shutdown systems. Temperatures stay below critical values with no fuel melt or fission products release (see Fig. 6.12, also 6.10).
– Due to inherent passive safety features, the sensitivity to station blackout scenarios is eliminated (at least for the short and midterm).
– As graphite at high to very high temperatures is susceptible to oxidation (>700 °C), although slowly developing, water and air ingress scenarios deserve special attention. Both can be counteracted by making graphite more resistant to

---

[55]Private communication, Sept. 2017.

oxidation (e.g. surface coating of the fuel elements) or limiting the total amount of air and water, which might ingress into the core in case of a leak by means of smart design.[56]

– To make 'super-safe' modular HTGR robust against extreme external events, including bomb attacks, underground siting has been proposed.[57]

When restricted to conservative design parameters, notably core outlet temperatures around 750 °C, small modular HTGRs are sufficiently well developed to be regarded as deployable. Further qualification of graphite and coated particles requires further research. Going for temperatures up to 1000 °C calls for certification of fuel, graphite environments and of pressure vessel materials as well as the steam turbine generator and auxiliary heat utilization systems as major RD&D challenges.[58] Replacing steel pressure vessels by pre-stressed concrete vessels is often proposed as an ultimately safe design option.

Cost estimates for electricity produced from HTGRs that are sufficiently reliable are not available. However, cost constraints and lack of competitiveness became obvious in the course of various pilot projects in Germany and South Africa.

Regulatory issues arise as major barriers for the cost and time effective deployment of HTGR because both the safety characteristics (inherent-passive instead of active) and the safety approach (coating as primary confinement of radioactive material under all accident conditions) differ significantly from standards set by and for current LWR.

6.2.5 Accelerator-Driven Systems

An accelerator-driven system (ADS) is a reactor without a self-sustaining chain reaction. This kind of device is called a subcritical reactor where additional neutrons are provided from an external source. Most current ADS designs propose to use a high-intensity proton accelerator where the required proton energy is about 1 GeV and the beam current is 2 mA. Protons are then directed to a high atomic number target able to produce neutrons by spallation.[59] Up to one neutron can be produced per 25 MeV of the incident proton beam, thus a 1 GeV beam will create 20–30 spallation neutrons per proton.

---

[56]Filling the containment building with inert gas; safe steam generator arrangement/placement or replacing the steam-turbine cycle by a gas-turbine cycle.

[57]IAEA, Advanced nuclear plant design options to cope with external events, Vienna, 2006.

[58]Kelly J.E., Generation IV International Forum: A decade of progress through international cooperation, Progress in Nuclear Energy 77, 240–246 (2014).

[59]Spallation is the process where nucleons are ejected from a heavy nucleus being hit by a high-energy particle. In this case, a high-energy proton beam directed at a heavy target, e.g. depleted uranium, thorium, lead/lead-bismuth, expels a number of spallation particles, including neutrons.

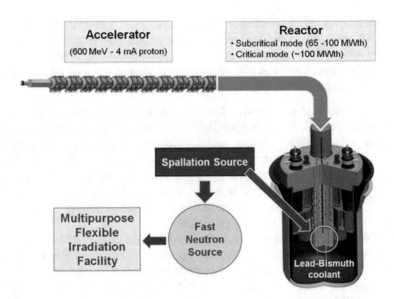

Fig. 6.13 Schematic view of an accelerator driven system, by way of example the R&D project MYRRHA in Mol, Belgium (*MYRRHA: an Accelerator Driven System*, (Online Image), 3.10.16, http://myrrha.sckcen.be/en/MYRRHA/ADS)

The spallation neutrons have only a very small probability of causing additional fission events in the target. However, the target still needs to be cooled due to heating caused by the accelerator beam. When the spallation target is located in the heart of the reactor (see Fig. 6.13), it is surrounded by a blanket assembly of nuclear fuel, such as fissile isotopes of uranium or plutonium or thorium-232 that can breed to uranium-233. There is then a possibility of sustaining a fission reaction where only a few percent of the neutrons stem from the spallation source and the bulk of neutrons arise from fission events in the subcritical reactor.[60]

An ADS can only run with neutrons supplied to it and the nuclear reactor can be turned off simply by switching off the proton beam, as opposed to the need to insert control rods to absorb neutrons and make the fuel assembly subcritical in a conventional reactor. Therefore, reactivity induced accidents are eliminated e.g. due to the ejection or incorrect handling of control rods. This is seen as major safety advantage. However, as fission products accumulate in ADS cores, the requirement to remove decay heat reliably and sufficiently remains.

Current ADS use lead/bismuth eutectic (LBE) as the spallation target, target coolant and primary coolant that circulates by natural convection around the core, similar to lead-cooled fast conventional reactors. For a description of Pb-Bi cooled reactors and the challenges they face, see Sect. 6.2.2.

---

[60]World Nuclear Association (WNA), *Accelerator-driven Nuclear Energy* (Online), 3.10.16, http://www.world-nuclear.org/information-library/current-and-future-generation/accelerator-driven-nuclear-energy.aspx

ADS are suited for transmutation (burning) of long-lived waste from used fuel from commercial reactors, i.e. fission products such as Tc-99 and I-129 and minor actinides in particular, due to their neutronic and design properties. Even pure minor actinide blankets are an option that avoids the building up of new long-lived wastes as occurs when using uranium or plutonium. The need for reprocessing remains, as fission products must be separated.

There are technical difficulties to overcome before ADS can become economic and eventually be integrated into future nuclear waste management. The accelerator must provide a high intensity and be highly reliable. There are concerns about the window separating the protons from the spallation target, which will be exposed to stress under extreme conditions. However, recent experience with the MEGAPIE[61] Pb-Bi neutron spallation source tested at the Paul Scherrer Institut in Switzerland has demonstrated a working beam window under a proton beam power of 0.78 MW. The chemical separation of the transuranic elements and the fuel manufacturing, as well as the provision of structure materials, are important issues. Finally, the lack of nuclear data at high neutron energies limits the efficiency of the design.[62]

Some laboratory experiments and many theoretical studies have demonstrated the theoretical possibility of ADS. Carlo Rubbia, a nuclear physicist and Nobel laureate, was one of the first to conceive a design of a subcritical reactor called the energy amplifier[63,64]

What was claimed to be the world's first ADS experiment began in March 2009 at the Kyoto University Research Reactor Institute (KURRI). The experiment irradiates a high-energy proton beam (100 MeV) from the accelerator onto a heavy metal target set within the critical assembly, after which the neutrons produced by spallation are bombarded into a subcritical fuel core.

The Spallation Neutron Source (SNS) at the Oak Ridge National Lab, USA, produces neutrons with an ADS that delivers short (microseconds) proton pulses to a target/moderator system, where neutrons are produced by a spallation process.

The Indian Atomic Energy Commission is proceeding with design studies for a 200 MWe PHWR accelerator-driven system fueled by natural uranium and thorium.

In March 2016, a strategic cooperation agreement to develop accelerator-driven advanced nuclear energy systems (ADANES) was signed between China General Nuclear and the Chinese Academy of Sciences, which has a major R&D program on thorium molten salt reactors, including a 2 MWe accelerator-driven, sub-critical liquid fuel prototype designed to demonstrate the thorium cycle. This Chinese project is the most ambitious project in the world. The project is aiming at a complete

---

[61]Megawatt Pilot Experiment, www.psi.ch/info

[62]Wikipedia, *Subcritical Reactor* (Online), 3.10.16, https://en.wikipedia.org/wiki/Subcritical_reactor

[63]C. Rubbia et al., *Conceptual Design of a Fast Neutron Operated High Power Energy Amplifier*, CERN report 289,551, Geneva, 1995.

[64]1500 MWt lead-cooled, fast hybrid reactor with thorium based closed fuel cycle (see PSI Nr. 96–17, 1996 for detailed assessment).

energy system, integrating nuclear waste transmutation, nuclear fuel multiplication and energy production, at 1000 MWe. A site was chosen. As far as the authors of this book know, the project also aims at innovating in the spallation target technology.

The Belgian Nuclear Research Centre (SCK.CEN) has initiated the pre-licensing process and is planning to begin construction work on the MYRRHA research reactor at Mol. The decision of the Belgian Government to support that project is expected by early 2018. Initially, it will be a 65–100 MWt ADS, consisting of a proton accelerator delivering a 600 MeV, 4 mA proton beam to a liquid lead-bismuth (Pb-Bi) spallation target that in turn couples to a Pb-Bi cooled, subcritical fast nuclear core (see also Sect. 6.2.2). The total costs are estimated to be 1.6 billion Euro, shared by Belgium (40%) and EURATOM. The first R&D facility including the high-power accelerator, to be developed, is scheduled to become operable around 2024, the whole ADS not earlier than 2036.

In mid-2014, a conditional license was issued for the construction of the European Spallation Source (ESS) facility at Lund in Sweden, to be used for material and life sciences research. It is designed around a linear accelerator, producing intense pulses of neutrons from a heavy metal target. These pulses of neutrons are led through beamlines to experimental stations. The European spallation source will provide neutron beams up to 30 times brighter than any current neutron source. Completion is scheduled by 2019 and the facility should be fully operational by 2025. Funding for the project involves 13 European countries. Construction costs of the ESS facility are estimated at about €1.8 billion, with annual operating costs of some €140 million.

ADS concepts[65] compete with "conventional" fast reactor designs, in particular LFR (lead cooled fast reactor), and face partially the same and in part even stronger technical challenges. There are some potential show stoppers, such as the need for more powerful proton and neutron sources. ADS are flexible regarding blanket composition and are best suited for minor actinide incineration. They are in general at very early stage of development. Therefore, reliable safety assessments are currently impossible. However, there is high potential to rule out reactivity induced accidents while unprotected loss-of-adequate-decay-removal accidents seem to be controllable by passive means.

6.2.6 Small Modular and Medium-Sized Reactors

While the size of reactor units has grown to more than 1600 MWe with corresponding economies of scale in operation and increasing complexity, there is

---

[65]See also contributions on developments in accelerators in: Thorium Energy for the World, Revol, J. P., Bourquin, M., Kadi, Y., Lillestol, E., de Mestral, J. C., & Samec, K. (2016). Thorium Energy for the World. Thorium Energy for the World, by J.-P. Revol. ISBN 978–3–319-26,540-7. Springer International Publishing Switzerland, 2016.

a revival of interest in small and simpler units for electricity production and other purposes. The incentive to develop small, often modular, and medium-sized reactors (SMR)[66] comes from different sources. There is a strong belief[67,68] that SMR:

- Would open additional market sectors, for example process heat production or seawater desalination and, based on enhanced safety characteristics, allow for sites close to consumers, which are not accessible to large reactors;
- Would better adapt to low growth rates of energy demand and provide power away from large grid systems, often found in small or developing countries, and lower requirement for access to cooling water and land, the latter comparing favorably with respect to wind and solar energy;
- Would reduce the impact of capital cost and lead to easier financing and earlier revenues;
- Would better meet specific user requirements, mostly in relation to safety by design incorporating reduced core inventory, and high level of passive and/or inherent safety features that would hopefully help to improve public acceptance;
- Would allow for greater simplicity of design and modularization, enabling series production largely in factories, shipping to utilities and site-assembling, going along with potential usage of domestic resources, implementation of higher quality standards and shorter construction times;
- Would add ability to remove reactor modules or facilitate in-situ decommissioning at the end of lifetime.

Some argue that SMR facilitate a wider spread of nuclear energy, i.e. to new-comer States, and make easier the build up a domestic nuclear infrastructure and capabilities.

All types of large reactors currently in use or being developed are represented in the SMR lines, including thermal neutron spectrum water-cooled reactors, various kinds of fast neutron spectrum reactors (FR) including liquid metal and molten salt cooled reactors, and gas-cooled, graphite moderated high temperature reactors (HTR-M). See previous Sects. 6.2.1 to 6.2.4. The first LWR-based SMR has the lowest technological risk, while some fast SMR concepts enable longer operation before refueling, which is regarded advantageous, where appropriate.

Safety standards for SMR have been issued in many countries, and have much in common with the safety requirements that we recommend in Sect. 5.8, in particular:

---

[66]The IAEA defines "small" as under 300 MWe, and "medium" as under 600MWe. "SMR" is often used as an acronym for "small modular reactor"; a subcategory of very small SMR is proposed for units under 15 MWe.

[67]OECD-NEA, "Nuclear Energy Market Potential for Near-term Deployment", *NEA Nuclear Development Publication* (2016).

[68]World Nuclear Association (WNA), *Small Nuclear Power Reactors* (Online), 3.10.16, http://www.world-nuclear.org/information-library/nuclear-fuel-cycle/nuclear-power-reactors/small-nuclear-power-reactors.aspx

- A simpler and more rugged design, e.g. by avoiding active systems or processes for normal operation and safety functions;
- Increased safety margins to avoid the need for early action by the operator to rectify abnormal situations;
- Lowered or eliminated risk of core damage triggered by induced reactivity or loss of active heat removal incidents;
- Exclusion of any possibility of radioactive releases to the environment, potentially implying significant favorable consequences for the population and the environment and/or at least a significant reduction of emergency planning zones (EPZ).

In general, SMR seem to achieve most of these requirements by design optimized for safety. Certain innate features, such as the small power rating and proportionally reduced inventory of fission products, reduce the impact of accidents, should they occur. Some SMR designs have lower power density and higher heat storage capacity alongside inherent and passive safety features. Most of the proposed SMR concepts make use of the potential for below-ground or even underwater locations providing greater protection from natural hazards and malicious attacks. Several countries are pioneering in the development and application of transportable plants, including floating and sea-based SMR.[69] A recent report[70] on standardizing SMR licensing and harmonization of regulatory requirements also sees an enormous potential of SMR based on the factors detailed above.

However, in pursuing individual SMR concepts with regards to operation, safety and security requirements, the IAEA points out that the necessary size of emergency planning zones will lead to modifications of the licensing process and partially new legal and regulatory frameworks, The IAEA regards this as the main challenge for SMR deployment. The question whether the current regulatory framework needs to be adapted to better take SMR safety characteristics into consideration has been raised by industry, regulators, and academia. There seems to be a growing consensus that water-cooled SMR can be licensed according to the current, "slightly relaxed" regulatory requirements.[71] However, most of the innovative/exotic designs with novel safety features call for the regulatory framework to be rethought.[72]

The projected timelines of readiness for commercial deployment of SMR generally range from the present to 2020–2030, depending on the novelty and maturity of the concepts.

Currently, there are more than 45 small modular designs under development for various purposes and applications (see Fig. 6.14 and for overview and

---

[69]For example, Russian designers offer a new class of water cooled floating SMR for electricity and heat production to remotely located areas or nuclear newcomer states.

[70]WNA, "Facilitating International Licensing of Small Modular Reactors", *Report No. 2015/004* (2015).

[71]S. Magruder, "SMR Regulators' Forum", IAEA (2016).

[72]W. Kröger, Small-sized reactors of different types: Regulatory framework to be re-thought, Modern Science and Engineering (ISSN 2333–2581), Academic Star Publishing Company, 2017.

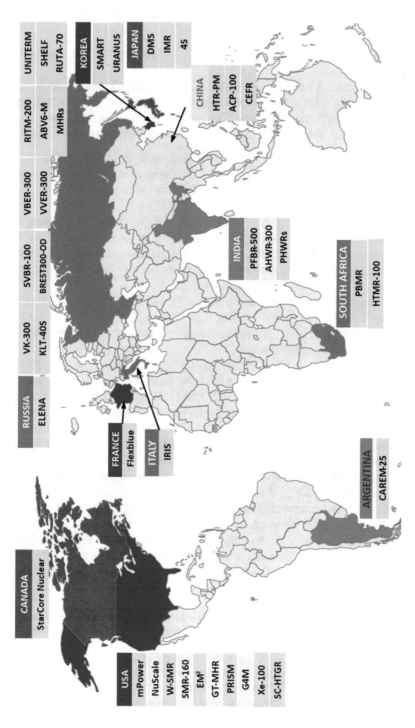

Fig. 6.14 Worldwide development of small and medium sized modular reactors (SMR) (source: https://www.iaea.org/NuclearPower/SMR/)

p. 661–693 of Pioro (2016)[25] for further details). Three designs are under construction[73,74]:

- CAREM 25 MWe/100 MWt integral PWR in Argentina with passive safety systems;
- 2 KLT-405, 38.5 MWe/150 MWt each, floating PWR (based on marine/ice-breaker plant) in the Russian Federation with passive and active safety systems;
- 2 HTR-PM, each module with 100 MWe/250 MWt, pebble bed high-temperature, gas-cooled reactor in China with inherent/passive safety concept.

In China, the ACP 100 (125 MWe/385 MWt), with the entire nuclear steam generation system integrated in the reactor pressure vessel and passive safety systems, has been granted permission for construction, with construction expected to start by the end of 2017.

The IAEA is coordinating the efforts of its Member States to facilitate the development of SMR of various types by taking a systematic approach to the identification and development of key enabling technologies to achieve competitiveness and reliable performance of such reactors, and by addressing common infrastructure issues that could facilitate their deployment.

The governments of countries other than those mentioned above are pushing the development of market-ready SMR concepts. In Europe, The UK has strong interest in SMR concepts based on proven light-water reactor technology and recently awarded R&D funding to Rolls Royce.

In the USA, the Department of Energy (DOE) announced $452 Million in funding to support the realization of the licensing processes in order to support the commercial operation of one SMR by 2022. The DOE plans to finance in particular an SMR demonstration project with maximum power of 300 MWe that is factory produced and transported to the site. The DOE will finance at most 50% of the costs over 5 years, while private enterprises have to fund the other half. The funding must be paid back, if the project is not completed. Babcock and Wilcox (B&W) won the first funding round in November 2012 with its concept mPower (see Fig. 6.15). Babcock and Wilcox (B&W) also won the second funding round in March 2013 with NuScale. In the meantime, the funding of the SMR projects was reduced because of high costs for development and licensing in the US. B&W announced in April 2014 that they want to reduce their financing for the mPower to $15 Million per year.[75] Furthermore, GE Hitachi is promoting its PRISM concept, a small, safe, sodium-cooled fast reactor. In South Korea, in cooperation with the Kingdom of

---

[73]For overview and further details, see also the IAEA booklet on Advances in SMR /www.iaea.org/ .../, IAEA 'Status of small and medium sized reactor designs', 9/2012 (http://aris.iaea.org) and IAEA "Advances in Small Reactors Technology Development", 2014.

[74]See also S. Buchholz, A. Krüssenberg and A. Schaffrath, Safety and International Development of Small Modular Reactors (SMR) – A Study of GRS, atw 60 (11), (2015), with a summary of SMR under development and characteristics of safety systems.

[75]World Nuclear News (WNN), *Funding for mPower reduced* (Online). 3.10.16, http://www.world-nuclear-news.org/C-Funding-for-mPower-reduced-1404141.html

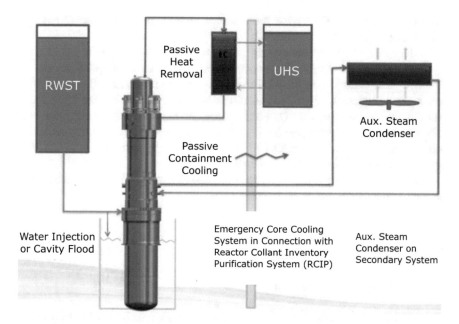

Fig. 6.15 Design features and decay heat removal strategy of the mPower integral pressurized water reactor. It has 530 MWt, forced primary circulation, 14.1 MPa system pressure, UO_2 fuel, internal once-through steam generator, pressurizer and control rod drive mechanism (IAEA, "Advances in small modular reactor technology developments", 2014.). Acronyms: reactor water storage tank (RWST), emergency cooler (EC), ultimate heat sink (UHS)

Saudi Arabia, the development of the integral SMART concept is ongoing, with a thermal power of 365 MW and extensive use of passive systems to enhance safety, heat removal and containment the coolant. All the concepts mentioned in this section are light-water cooled and based on proven technology.

Despite a lot of enthusiasm and hope for big success, there are those who warn that these small reactors may still be too expensive to build and operate in a competitive market.[76] Others point to obstacles to the development and deployment of nuclear power that continue to exist, such as skeptical public opinion, concerns over safety and security of nuclear waste as well as concerns about proliferation besides high economic costs.[77] A more recent study carried out by the OECD-NEA concluded that "if SMRs are produced in series in factory conditions, they are likely to be cheaper to build than advanced light-water reactors in terms of both absolute and per kWe total construction costs, although they will have higher variable costs.

---

[76]M. V. Ramana, S. Saini, "A Radioactive Money Pit - The hidden risks of small-scale nuclear reactors", Harpers Magazine, 2/2016.

[77]International Risk Governance Council (IRGC), "Preserving the nuclear option - Overcoming the institutional challenges facing small modular reactors", *Opinion Piece* (2015).

In economics terms, SMR costs are therefore situated between those of coal and large nuclear plants."[78]

SMR concepts are at different stages of development and employ different safety systems and concepts that are reviewed and described in numerous publications.[79,80] Reliable deterministic and probabilistic safety assessments are still missing or in some cases underway. However, although basic statements, on the behavior of various reactor types and designs under accident conditions, are mostly for large reactors, they still apply to the SMRs. But notably, further stringent and ambitious safety requirements and advanced safety features can be more easily achieved by small to very small sized power reactors.

Large scale funding is necessary to overcome technological challenges and risks as well as regulatory and licensing obstacles.

6.2.7 Special Types

6.2.7.1 Travelling Wave Reactor

A travelling wave reactor[81] (TWR) is a pool-type fast reactor, cooled by sodium, which can convert fertile material into usable (fissionable) fuel. In contrast to sodium cooled fast reactors (SFR, see Sect. 6.2.1), TWR use fuel efficiently without uranium enrichment or reprocessing, instead directly operating with depleted uranium, natural uranium, thorium, spent fuel removed from light reactors, or some combination of these materials. Only a small amount (~10%) of enriched uranium-235 or other fissile fuel is needed to initiate the self-sustained reaction, which could theoretically run continuously for decades (i.e. 40 years or more) without refueling or removing spent fuel.[82]

Initially, the core is loaded with fertile material, with a few rods of fissile fuel concentrated in the central region. After the reactor is started, four zones form within the core: (1) the depleted zone, which contains mostly fission products and leftover fuel; (2) the fission zone, where fission of bred fuel takes place; (3) the breeding zone, where fissile material is created by neutron capture; and (4) the fresh zone, which contains unreacted fertile material. The energy-generating fission zone

[78]Small Modular Reactors: Nuclear Energy Market Potential for Near-term Deployment, NEA No.7213, OECD, 2016.

[79]S.Buchholz et al., "Studie zur Sicherheit und zu internationalen Entwicklungen von Small Modular Reactors", GRS 376, 2015.

[80]D.T. Ingersoll, Deliberately small reactors and the second nuclear era, Progress in Nuclear Energy 51, 589–603 (2009).

[81]The name refers to the fact that fission remains to a boundary zone in the reactor core (rather than involving the whole core), which slowly advances over time. TWR were first proposed in the 1950s and have been studied intermittently.

[82]Wikipedia, *Traveling Wave Reactor* (Online), 20.10.16, https://en.wikipedia.org/wiki/Traveling_wave_reactor

Fig. 6.16 Schematic view of the 1150 MWthermal/600 MWelectric Travelling Wave Reactor (TWR) concept, developed by Terra Power, with (1) containment dome, housing non-radioactive parts; (2) reactor vessel, containing (3) the reactor core and its components, supported from (8) the reactor head; (4) in-vessel fuel shuffling machine; (5) control and safety rods; (6) the primary sodium pool; (7) intermediate heat exchangers (TerraPower, *TerraPower TWR*, (Online Image), 20.10.16, http://terrapower.com/uploads/img/twr_cross_section_cropped.png)

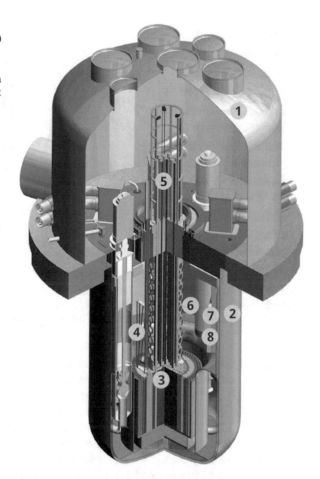

steadily advances through the core, effectively consuming fertile material in front of it and leaving spent fuel behind. Meanwhile, the heat released by fission is absorbed by the molten sodium and subsequently transferred into a closed-cycle aqueous loop, where electric power is generated by steam turbines.[83]

In the late 2000s, Terra Power launched a conceptual design effort of a 1200 MWt/600 MWe, first of a kind demonstration TWR power plant (TP-1), using the cylindrical standing wave concept, achieved by periodically shuffling fuel within the core.[84] Characteristics given here refer to this concept,[85] and are

---

[83]Wald M., *TR10: Traveling-Wave Reactor*, MIT Technology Review (Online), 20.10.16, http://www2.technologyreview.com/news/412188/tr10-traveling-wave-reactor, 2009.

[84]The breed burn wave does not move from one end of the reactor, travel through the reactor, but gradually from the center out; the fuel rods themselves are moved through a largely stationary burn wave.

[85]Ahlfeld C. et al., "Conceptual Design of a 500MWe Traveling Wave Demonstration Reactor Plant", ICAPP 2011, Nice, France, May 2–5, 2011.

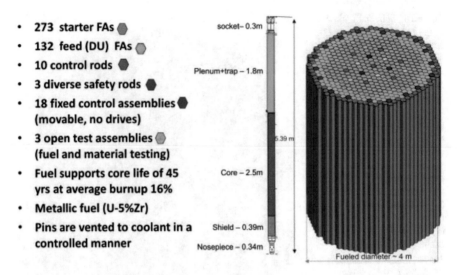

- **273 starter FAs** ⬡
- **132 feed (DU) FAs** ⬡
- **10 control rods** ●
- **3 diverse safety rods** ●
- **18 fixed control assemblies** ●
  (movable, no drives)
- **3 open test assemblies** ◐
  (fuel and material testing)
- **Fuel supports core life of 45**
  yrs at average burnup 16%
- **Metallic fuel (U-5%Zr)**
- **Pins are vented to coolant in a**
  controlled manner

socket– 0.3m
Plenum+trap – 1.8m
5.39 m
Core – 2.5m
Shield – 0.39m
Nosepiece – 0.34m
Fueled diameter ~ 4 m

Fig. 6.17 TP-1 fuel pin and color coded core assembly layout linked to the legend on the left. A full reactor fuel load comprises 405 fuel assemblies and each assembly contains 169 fuel pins (Ahlfeld C. et al., "Conceptual Design of a 500MWe Traveling Wave Demonstration Reactor Plant", ICAPP 2011, Nice, France, May 2–5, 2011)

schematically shown in Fig. 6.16. The reactor core and all primary cooling system components are contained in the unpressurized reactor vessel with the addition of a number of in-vessel structures. The reactor core is composed of two fuel assembly types. As shown in Fig. 6.17, fuel assemblies in the active control zone have (metallic) uranium alloy fuel slugs with beginning-of-life enrichment less than 20% U-235. Fuel assemblies in the fixed control zone have uranium alloy fuel pins that contain only depleted uranium at beginning-of-life.

The two types of fuel assemblies are initially distributed to achieve criticality and to start the breed-burn process. Excess reactivity in the core increases monotonically and control rods are slowly inserted to offset this increase. At a pre-determined point, the reactor is shut down and higher burn up assemblies are moved to the periphery of the core and depleted uranium fuel assemblies replace the high burn up fuel. This process, referred to as "shuffling", occurs about every 18 to 24 months. Each fuel pin has a vent assembly to continuously release gaseous fission products in a controlled manner. These products are removed from the sodium and stored in specific systems, thereby reducing the core radioactive inventory, but become distributed core inventory.[86]

The control function is accomplished with control rods that are ducted clusters of clad B_4C absorber pins. In the active control zone, these control rods are inserted or withdrawn as needed for core operation. In the fixed control zone, their purpose is to control reactivity and power in the peripheral region of the core as high burn up fuel

---

[86]Like in molten salt reactors with dispersed fuel.

is moved into this region. In-vessel handling machines can relocate the assemblies during a shuffling shutdown to optimize core performance. A dedicated group of three safety rods with diverse system design are used to rapidly shut down the reactor during transient conditions.

The sodium coolant is heated to a nominal 510 °C as it exits the core and flows through intermediate heat exchanges (IHX) that are cooled by non-radioactive sodium with higher pressure than on the primary side. By use of advanced steam generator technology, the gross electrical conversion efficiency yields 42%.

According to Terra Power, the TP-1 design uses proven technologies for most of the plant. Notable exceptions are the venting fuel pins, the innovative design of IHX and the sodium technology in general. Furthermore, lifetime of major systems/ components might be limited due to relatively high core temperatures and concentration of power production; high material flow and neutron flux may cause additional problems. TerraPower aims to achieve startup of its prototype TWR in the mid-2020s.

The reactor safety concept is based on strategies "that proved successful for previously operated fast reactors combined with state-of-the-art analytic methods"[87]; "classical" IAEA safety series requirements have been adopted.

Similar to other pool type reactors, loss-of-coolant accidents (LOCA) are eliminated from the design base, while loss of coolant flow or heat sink together with transient overpower events dominate design basis accidents. Decay heat removal is accomplished by active or passive means, normally by circulating primary and intermediate sodium coolant with pump pony motors connected to emergency diesel generators. In the case they are unavailable, natural circulation is sufficient to keep the reactor within design temperature limits. The heat is dissipated through the steam generators using a large dedicated inventory of feed water. If this system is disabled, a passive system[88] draws ambient air down coaxial chimneys and circulates the air through a baffle structure surrounding the guard vessel (see Fig. 6.16). Its capacity is sufficient to remove the peak decay heat that occurs about 1 day after shutdown.

A level 1 probabilistic safety analysis (PSA) is in progress, based on the conceptual design. Sequences leading to core damage and late (rather than early) large release have been identified showing the dominating role of seismic events, disabling all decay heat removal capability and causing bulk primary sodium boiling approximately one day after the initiating event.

Unprotected overpower transients, e.g., caused by loss of flow followed by a failure to scram the reactor or excessive control rod removal, are challenging because of the positive sodium void reactivity coefficient and potential power excursions. Such events are made extremely rare by some design features. However, complete

---

[87]Ahlfeld, Charles, T. Burke, T. Ellis, P. Hejzlar, K. Weaver, C. Whitmer, J. Gilleland et al. "Conceptual design of a 500 MWe traveling wave demonstrator reactor plant.", Proceedings of ICAPP 2011, Societe Francaise d'Energie Nucleaire – SFEN (2011).

[88]Reactor Vessel Air Cooling System (PVACS) without moving parts and required operator actions.

self-controllability of reactivity induced accidents and extreme loss of decay heat sequences appear not to be a given within the current TWR TP-1 design.

As reprocessing of spent fuel is not foreseen and the need for enrichment of fresh fuel (besides a small amount to initiate the nuclear reaction) is eliminated, the proliferation risk is deemed minimal. Fissile fuel is both produced and then consumed in the reactor, greatly improving fuel efficiency and availability, while easing waste management and disposal.

6.2.7.2 Fusion-Fission Hybrid Systems

A fusion-fission hybrid is defined as a subcritical reactor, like the accelerator driven system, here consisting of a fusion[89] core surrounded by a fission blanket, thus combining the neutron-rich characteristics of a fusion system with the energy-rich characteristics of a fission system. The fusion core provides an independent source of neutrons, which allows the fission blanket to operate subcritically. Hybrid advocates suggest that this extra design freedom and fuel cycle flexibility (see below) would be beneficial to achieving practical sustainability of nuclear power related to long-term fuel supply, environment friendly power production and proper waste management,[90] potentially better than pure fission solutions. The main applications of hybrids are (a) nuclear waste management by means of burning long-lived radioactive elements (actinides), (b) the simultaneous production of energy and management of existing and new nuclear waste by very high burn-up ("deep-burn") fuel cycles, and (c) the breeding of new fissile fuel by substantially increasing the utilization efficiency of uranium-238, and (d) alternatively, converting to a thorium-232 fuel cycle. Because of proliferation concerns, applications (a) and (c) are the main trends currently followed.

The idea of the hybrid can be traced back about 50 years[91] and resurfaced every decade or so, but there has never been a serious experimental research program on hybrid applications, because of lack of sufficient motivation, the perceived complexity and the unproven fusion core. The problem of controlling fusion still exists despite significant research in plasma physics.

---

[89]In nuclear fusion, two or more atomic nuclei merge to form one or more different small atomic nuclei and subatomic particles. The primary source of solar energy and of similar size stars is the fusion of hydrogen to form helium (the proton-proton chain reaction), which occurs at a solar-core temperature of 14 million kelvins. The International Thermonuclear Experimental Reactor (ITER) is under construction, next to the Cadarache facility in Saint-Paul-lès-Durance, in southern France, to become the world's largest magnetic confinement plasma physics experiment (a tokamak nuclear fusion reactor). It aims to demonstrate the controlled principle of producing more thermal power from the fusion process than is used to heat the plasma.

[90]U.S. Department of Energy, Research Needs for Fusion-Fission Hybrid System, Report of the Research Needs Workshop, Gaithersburg, Maryland, Sept. 30–Oct. 2, 2009.

[91]L. M. Lidsky, Fission-fusion systems – Hybrid, symbiotic and Augean, Nuclear Fusion 15, 151–173 (1975); H. Bethe, The fusion hybrid, Physics Today 32, 44–51 (May 1979).

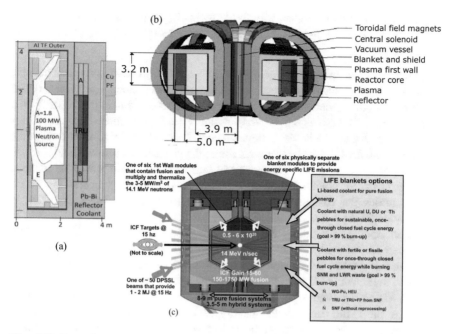

Fig. 6.18 Recently proposed hybrid systems based on the following fusion drivers: (**a**) the spherical tokamak, (**b**) the standard tokamak, and (**c**) laser-driven confinement (USDOE Research Needs for Fusion-Fission Hybrid Systems, ReNeW, Sept 30–Oct 30, 2009)

The fusion core, independent from its design details, is simply the source of neutrons that are supplied to the surrounding fission blanket to help accomplish its primary application of interest. These neutrons are especially efficient for transmutations, since their energy is 14 MeV, to be compared to the 1 MeV of those provided by fast fission reactors. Fusion drivers proposed for the most recent hybrid systems, i.e. the spherical and standard tokamak or laser-driven inertial confinement, are shown in Fig. 6.18. These ideas appear to have attractive features, but they require various levels of advances in plasma sciences and nuclear technology. Namely:

– A waste transmuter based on leading magnetic fusion and fast burner reactor technology, e.g. combining leading ITER physics and technology with sodium-cooled fast burner reactor technology, plus associated fuel fabrication and reprocessing technologies.

– A waste transmuter with a removable fusion core based on spherical tokamak concept that employs a compact replaceable fusion core, extractable as a single unit from the fission reactor to achieve minimal electromagnetic and mechanical coupling between both systems.

– Once-through burn-and-bury-energy producers with a very deep-burn fuel cycle based on laser fusion, almost completely burning nuclear fuel without requiring enrichment and, if deep burn were successful, no reprocessing would be necessary.

- Efficient LWR fuel breeders with fissile fuel being produced in a flowing target; the fissile fuel is removed on line in order to suppress its subsequent fission in the hybrid system.

Many proposals exist for fusion drivers, a few proposals have conceptual designs at various stages of elaboration, but most of them are in a pre-conceptual stage or not much more than a general idea. As an example, the reversed-field pinch (RFP) as a neutron source for fusion-fission hybrid systems has been proposed and a community has been formed to boost its development.[92] The RFP is similar to a tokamak with a toroidal field[93] that changes sign within the plasma. The toroidal field is much smaller than in conventional tokamaks and thus external magnets are also smaller. This plasma has the potential of ohmic heating to thermonuclear temperatures, which also decreases the cost and complexity of the machine. However, the RFP is probably on a longer timeline than the other tokamaks.[94]

Present scaling shows that the RFP might provide the necessary fusion core. This could be checked with a machine only twice as big as the present largest RFP (RFX-mod in Padua, Italy: large radius of 2 m and small radius of 0.46 m). All proposed systems are likely to require substantial long-term research and development work and present significant development risks. As an extreme point of reference, the ITER experimental facility is designed to achieve first plasma by 2020 and to start deuterium-tritium operations by 2027. More specifically, its goal is to achieve a duty factor of 25% for burn periods greater than 10 minutes, and to operate continuously for periods of 12 consecutive days.

However, one should not confuse the pure fusion and the hybrid fusion-fission reactors. Whilst it is true that the former still has a remote prospect, the fusion core of the latter requires only a nuclear reactivity, which was already obtained in the JET tokamak during the 1997 D-T operation. One does not need a machine as big as ITER for the development of fusion-fission hybrid system, which makes them more viable. There is evidence that the Chinese are working actively and secretly on the hybrid fusion-fission. An indication of their progress is the present engineering design of the CFETR (China Fusion Engineering Test Reactor) tokamak, which can provide the necessary fusion core for a hybrid reactor.[95] As any advanced waste management strategy, using either novel pure fission technology or fusion-fission hybrid technology will still require a long-term geological repository for the final remaining long-lived waste, though significantly less per unit of energy produced compared to current nuclear technologies.

A fusion reactor is generally considered "safe" and a subcritical fission reactor is arguably "safer" than its critical counterpart. As in the case of accelerator-driven

---

[92]R. Piovan et al. The RFP as neutron source for fusion-fission hybrid systems, second Int. Conf. on Fusion-Fission Sub-critical Systems for Waste Management and Safety.

[93]i.e. having the shape of a torus

[94]U.S. Department of Energy, Research Needs for Fusion-Fission Hybrid Systems, Report of the Research Needs Workshop, Gaithersburg, Sept 30–Oct. 2, Maryland, 2009, p. 41.

[95]D. Escande, private communication (October 2017).

systems, reactivity-induced accidents are practically eliminated but it is worth noting that the main safety issues are normally associated with other types of accidents. Mixing fusion and fission together in one facility will likely introduce new safety issues including potential new accident initiators, which deserve careful concept-specific evaluation. In the end, hybrids may have attractive safety features, but this is not a given. Hybrids produce significant quantities of fissile materials generally not retained in individually accountable fuel rods. This raises considerable proliferation concerns, which depend on the particular concept and associated applications:

- A transmuter of plutonium and minor actinides does not avoid the reprocessing step and the associated diversion risks;
- the once-through deep burn-and-bury concept requires a large flow of fuel spheres through the system, which together with the associated flow of damaged spheres for replacement, also represents a significant diversion risk;
- the production of LWR fuel in a fission-suppressed blanket entails large flows of liquid fuels that are difficult to account for, also representing a significant diversion risk.[96]

Because of the early stage of the development of these concept and lack of sufficiently mature designs, the appraisal of safety features and associated risks, including proliferation, are preliminary and subject to large uncertainties.

Projections on economics for a hybrid are speculative at best, although it is generally agreed that, due to their higher technological complexity and lower overall power density, hybrids will almost certainly be more expensive than an equivalent LWR.

The time scale for the development and testing of fusion-fission hybrids is subject to many technological, energy-political and financial (funding) factors and remains speculative as well; the deployment times could be expected to be of the order of roughly 50 years from now, if the present rate of development is not increased. On the other hand, enthusiasts focus on the happy marriage between fusion and fission promised by the neutron energy of 14 MeV of the fusion part that only requires nuclear reactivity, which is particularly favorable for transmutations, and claim more optimistic goals of 20 years or less if significant larger R&D efforts are undertaken.

6.3   Comparative Assessment

In Table 6.2 as well as the annexed Table 6.5, we have aggregated the information on characteristics and key parameters of the main candidate concepts highlighted above. Furthermore, we have systematically assessed those concepts by considering various aspects such as key design features, fuel cycle/waste management issues, reactor safety, states of development, costs/construction times and overarching

---

[96]Wald M., *TR10: Traveling-Wave Reactor*, MIT Technology Review (Online), 20.10.16, http://www2.technologyreview.com/news/412188/tr10-traveling-wave-reactor, 2009.

contextual issues. This detailed compilation (see annexed Table 6.5) demonstrates that all potential future concepts enable high power densities (compared to all other energy sources) and low operating pressures, allow for enhanced safety by inherent or, at least, passive means and for a significant reduction of husbandry times by avoiding or burning long lived actinides, strive for elevated fuel burn-ups and fuel cycle lengths and possess some other positive features. However, the developments include "show stoppers" and need new regulatory approaches and modernized pathways for licensing approval, all of course to different degrees. All large reactor types include smaller sized, mostly modular lines that turn out to be beneficial from a reactor safety point of view.

In the following sections, the concepts, viewed before (besides special types), are contrasted and assessed in greater detail before being evaluated against the key requirements set up in Sect. 5.8.

6.3.1 Concept by Concept Comparison

6.3.1.1 Liquid Metal and Salt Cooled Critical and Subcritical Reactors (Mainly based on Hejzlar[97] and Tuczek[98,99])

General Remarks

- Liquid metal (sodium, lead, lead/bismuth) and molten salt coolants—with heat transfer properties superior to water—allow for high power density and, due to low moderation capability and good neutron economy, for reactors with fast neutron spectrum.
- High boiling points at ambient pressure allow for non-pressurized reactor vessel, exclusion of loss of coolant accidents and high shut down temperatures.
- Such fast neutron spectrum reactors can be designed for various purposes—as consumer of fissile material, as breeder of as much or more fissile material than consumed, or as waste burners.[100]

---

[97]P. Hejzlar, Todreas, N.E., Shwageraus, E., Nikiforova, A., Petroski, R., Driscoll, M.J., Cross. "Comparison of fast reactor concepts with various coolants", Nuclear Engineering and Design 239, 2672–2691 (2009).

[98]K. Tuček et al., "Comparison of sodium and lead-cooled fast reactors regarding reactor physics aspects, severe safety and economic issues", Nuclear Engineering and Design 236, 1589–98 (2006).

[99]K.Tuček et al., "Comparative study of minor actinide transmutation in sodium and lead-cooled fast reactor cores", Progress in Nuclear Energy 50, 382–388 (2008).

[100]Heterogeneous and/or homogeneous incineration of minor actinides (MA).

– Such fast neutron spectrum reactors, with their multiple critical masses and positive coolant temperature coefficient, have the potential for disruptive power excursion under severe accident conditions.
– Comparative investigations have been made for various mid-sized concepts (1000 to 1400 MWt) with extrapolations to uprated power levels (2400 MWt), all at a conceptual stage; accident analyses include reactivity transients and "unprotected" severe accident scenarios.[101]
– Such fast neutron spectrum reactors are flexible to fuel cycles, either uranium or thorium, and different kinds of fuel assembly.[102] They apply solid fuel pins with different core designs, e.g., tight or loose lattice, while molten-salt cooled fast reactors (MSFR) will have fuel particles dispersed in the salt.
– Conventional fast neutron spectrum reactors are devices with sustained chain reactions, called critical reactors, while accelerator-driven system (ADS) work with fast subcritical reactors and additional neutrons from an external source, eliminating reactivity-induced power excursion accidents by design, while challenges with respect to sufficient decay heat removal remain.

More Specific Remarks

– All investigated fast neutron spectrum reactor cores[103] can utilize uranium resources effectively, can manage their own waste and reduce the long-term stewardship burden by depleting transuranic elements (TRU) from spent LWR fuel.[104]
– All fast neutron spectrum reactors follow a closed fuel cycle including reprocessing and use fuel with highly enriched transuranic elements (e.g. plutonium 15–20%). Thermal power production is in the range of 70 MWd/kg compared to roughly 45–50 MWd/kg in today's LWR, corresponding to an almost doubled efficiency of fuel burn-up.
– Fast neutron spectrum reactors have difficulties in achieving the self-controllability/inherent shutdown goal, as coolant temperature coefficients ought to be positive, although to different degrees (MSFR worst, SFR worse, LFR best/small) and compensation by negative fuel-temperature reactivity coefficient is not fully sufficient, all subject to the specific fuel and reactor design. In particular, self-breeder fast neutron spectrum reactors feature low excess reactivity to compensate fuel burn-up ("burn-up swing"), limiting the consequences of accidentally inserted reactivity.

---

[101]Loss-Of-Flow (LOF), Loss-Of-Heat-Sink (LOHS), total Loss-Of-Power/Station Blackout (LOP).

[102]Metallic (e.g., U-TRU-Zr) or oxide (U/Th-TRU)O$_2$.

[103]Liquid sodium cooled (SFR), lead cooled (lead or lead/bismuth cooled (LFR)), molten salt cooled (MSFR), all fast.

[104]Stewardship burden ("husbandry times") could be determined by long-lived fission products (J-129, Tc-99), no longer by minor actinides – depending on separation and incineration effectiveness.

- LFR and MSFR can be designed to be capable of removing decay heat in case of unprotected accidents[105] passively without excessive decay heat removal systems and to keep peak temperatures below given limits, with differences stemming from different coolant temperature reactivity coefficients and the efficiency of natural convection (MSFR ranked best). SFR (ranked worst) may have difficulties to transit to new asymptotic power levels without sodium boiling due to net reactivity increase, limited natural convection and low boiling temperature to meet the self-controlled heat removal criterion.
- All liquid metal and molten salt cooled reactors have a large capacity to store decay heat and, therefore, exhibit milder transient response with significantly longer times to reach peak cladding temperatures in accidents with loss of heat sink, compared to today's LWRs.
- Some liquid metal and molten salt cooled reactors are susceptible to scenarios with primary system overcooling and coolant freezing, especially those with high melting points (SFR less likely, followed by lead-bismuth and lead-alloy and the MSFR). Due to the high chemical reactivity of sodium, leaks in the primary circuit of SFR, potentially followed by fire or even explosion, present a high level risk and deserve special attention.
- As mentioned before, liquid metal and molten-salt coolants do not need to be pressurized and enable fairly high core outlet temperatures. Therefore, together with advanced conversion systems, thermodynamic efficiencies of about 42% can be achieved.
- All concepts face various operational problems and show a high degree of complexity; sodium-cooled reactors require an intermediate heat transport system.

6.3.1.2 Accelerator-Driven Systems (ADS) by Comparison to Critical Fast Reactors

General Remarks

Accelerator-driven Systems (ADS) use lead/bismuth eutectic as spallation target and coolant and compete with "conventional" fast reactors, in particular LFR; stated coolant-related characteristics apply.

- ADS are flexible to all kinds of fuel and purposes; following a closed fuel cycle, they are especially suited to burn long-lived waste from present commercial reactors.
- Neutrons, necessary to reach criticality of the reactor, are provided from an external source, namely a neutron accelerator, which further increases the technical complexity and technological risk.

---

[105]SFR cores require special, albeit passive devices, large SFR face more difficulties than small; LFR and SFR call for slight moderation (softer spectrum), albeit with potential negative impact on "burn-up swings" (see Table 6.5).

– As the proton beam and the external neutron beam can be simply switched off, reactivity induced accidents are practically eliminated. Sufficient and reliable decay heat removal remains a challenge.

More Specific Remarks

– ADS have to be designed to be capable of removing decay heat passively in case of unprotected accidents.
– ADS are under development (with MYRRHA as the flagship) and face even stronger technical challenges and show-stoppers than conventional LFR. The required high-energy proton accelerator is still under development.

6.3.1.3 Present High-Temperature Gas-Cooled Reactors (HTGR)

General Remarks

– HTGR use helium as coolant, which is chemically and physically non-reactive but with heat transfer properties inferior to water, liquid metal or molten salt, allowing only for relatively low power density reactors (about 8 MW/m^3 compared to 100 MW/m^3 of current LWR). On the other hand, low power density can be viewed as a positive safety feature.
– HTGR use graphite as the moderator and core structural material and therefore operate with a thermal neutron spectrum.
– HTGR has a fully ceramic core that cannot melt and, if properly designed, reactivity induced accidents and loss of active core cooling accidents are controlled by passive/inherent means.
– High core outlet temperatures enable high thermodynamic efficiency and/or process heat application, e.g., hydrogen production. Due to chemical reactivity of graphite at high temperature (>700 °C), air and water ingress accidents deserve special attention.
– HTGR use low enriched uranium-235 (roughly 4 times higher enrichment than LWR) but are flexible to the uranium-thorium fuel cycle. They aim at high burn-ups and direct storage of the spent fuel. Continuous refueling (pebble bed design) allows for small excess reactivity.

More Specific Remarks (See Also Remarks on Small Sized, Modular HTGR)

– Investigations have shown that small modular HTGR, like the 250 MWt High-Temperature Reactor with Pebble-bed fuel (HTR-PM), are not sensitive to failure

of active heat removal and failure of primary shutdown systems. Fission products are confined inherently inside two coatings of highly resistant ceramic material.
– Air and water ingress events can become controllable by means of limiting the respective amounts or by coating the graphite to make it oxidation resistant.
– Small modular HTGR with conservative design parameters, including restriction of core outlet temperature to 750 °C, are ready be deployed but sufficiently reliable cost and construction time estimates are not yet available.
– Helium losses of about 0.5% per day due to diffusion through the steel of the reactor pressure vessel need to be handled/compensated.

6.3.1.4 Small Sized, Modular Reactors (SMR)

General Remarks

– There is a revival of interest in small-sized (under 300 MWe) modular reactors (SMR) as they allow for greater simplicity and adaptability to special needs and conditions, enjoy enhanced safety by design, enable factory production and shipping to the construction site, are easier to finance, and present other favorable features.
– All types of large reactors, presently operated or subject of development, are represented in SMR lines including the HTR-PM. Concepts using light water as a coolant serve as a bridging technology.
– Lower power size and increased safety margins and special design characteristics (like a fully ceramic core) can lower or even eliminate risk of core damage and radioactive releases, triggered by reactivity induced or loss of heat removal incidents. Active systems and the essential need for early operator interventions can largely be avoided.

More Specific Remarks

– SMR concepts are at different stages of development. Reliable safety cases and probabilistic safety assessments are largely pending.
– SMRs follow different fuel cycle concepts including high burn-up and direct storage of spent fuel, extended refueling time periods, and so on.
– Some of the SMR concepts, if properly designed, have the potential to achieve ambitious safety requirements if pertinent basic design principles are followed (as set up in Sect. 5.8).
– Very small floating PWR, factory fabricated, transported to the operation site and back to factory for refueling, might be an attractive option for special conditions, also minimizing proliferation risks.
– Many innovative SMR concepts have been developed in the past, but have been given up for economic reasons. Technological and regulatory risks are lowest for PWR-based concepts.

6.3.2 Summary and Conclusions

We have presented a systematic evaluation of the major nuclear plant concepts using the key requirements of Sect. 5.8, developed to make the civilian use of nuclear energy safer and more robust against human interventions and less dependent on the stability of our societies. The results, summarized in Table 6.2 and further detailed in annexed Table 6.5, indicate a high potential for far-reaching improvements compared to the most advanced existing (Generation III+) large light water reactors

Table 6.2 Ranking of Gen IV reactors against key safety criteria detailed in Sects. 5.8 and 6.1 compared with the Gen III+ large EPR. Ranking from excellent (++) to neutral (~), to very poor (−−)

Key requirements	Candidate reactor concepts—varying coolant, selected designs in brackets				
	Water-thermal (large EPR)	Sodium-fast (PRISM)	Molten Salt-fast (SaWB)	Helium-thermal (HTR-PM)	Lead-fast (BREST-OD-300)
Elimination of Reactivity Induced Accidents	+	−	−	++	−/~
Resistance to Loss of Active Core Cooling	−−	−	~	++	−/~
– avoid exceeding critical temperatures	−−	n.a.	n.a.	++	
– avoid high fission product inventory	−−	+[a]	++[b]	+[a]	+[a]
– provide sufficient heat storage & transfer capacity	+	++	+	+	++
Structural Integrity	−	+	+	++	+
– avoid high operating pressure [suitability of underground siting]	−− [−]	+[c] [?]	++ [++][d]	+ [++][d]	+[c] [+]
Use Non-chemically Reactive/ Non Toxic Materials	+	−−[e]	−[e] (non-stable)	++	+
Avoid Long-lived Radioisotopes	−	+	++	+	++
Enhance Proliferation Resistance	+	−	−	~	−
– avoid high enriched uranium	++	−[f]	−[f]	−/~	−[f]

[a]Due to small power size
[b]In case of dispersed fuel & due to small power size
[c]Not pressurized but high static load
[d]Foreseen
[e]Intermediate cycle (IHX) foreseen
[f]Close to HEU lower limit

serving as benchmark, and to finally achieve very stringent, highly ambitious safety criteria.

However, none of the best versions, i.e. small sized in general, of the candidate concepts fulfills completely all requirements convincingly, yet. Thermal helium cooled reactors (HTR-PM) come closest, promising inherent robustness against classical severe accidents and largely avoiding long-lived radioisotopes when using thorium fue. Currently, they are however not capable of burning waste. With respect to burning waste, molten salt fast reactors (MSFR) promise to do best but appear most susceptible to reactivity-induced accidents as all liquid metal cooled fast reactors are, albeit to different degrees. There is also a potential of new concept-specific accidents, such as overcooling/freezing of coolant, chemical reactions following coolant outflows after leaks or air/water ingress into hot graphite cores, which deserve special attention.

Thus, future research and development appears necessary, aiming at further improving some essential characteristics and features and/or combining design elements in a radically new and innovative way.

All concepts seem to have limited capabilities to achieve the goal of reducing proliferation risk or even to maintain the current level, mainly due to partially elevated enrichment and/or significantly heightened by the need for reprocessing as part of the pursued strategy of a fully closed fuel cycle. We should also mention that there could be non-technical political means to minimize or even remove proliferation risks, For instance, the UAE has a deal with South Korea to manage fuel issues for them, with legislation making wastes return to the country of origin, idem between Russia and some of its clients. Moreover, the use of thorium instead of uranium, as foreseen for future HTR-PM, would improve proliferation resistance significantly.

Finally, a purely deterministic safety approach, i.e. without taking probabilities into account, to fulfill the safety requirements is deemed attractive but hard to achieve.

Also important is that, for revolutionary designs and technologies, experience from Gen-II units is of limited relevance, partially leading to a "starting from scratch" situation. New designs often tend to represent a jump in complexity, which may make them more difficult to understand and control. This also presents regulatory barriers since the regulator needs to assess unfamiliar technologies. This calls for unprecedented excellence in research, development, design, demonstration, and operation. For instance, exceedingly rare events will need to be rigorously accounted for.[106] Further, lack of experience and greater complexity are potential weaknesses that should be considered when evaluating the safety of new designs. Above all, the development of innovative, partially exotic technology is moving at a snail's pace, due mainly to a lack of funding.

---

[106]IAEA. Low Level Event and Near Miss Process for Nuclear Power Plants: Best Practices. IAEA, Safety Reports Series No. 73 (2012).

Annex

Table 6.3 Main characteristics of fast reactor concepts, all with 2400 MWthermal reactors, compared to the 4800 MW thermal European Pressurized Water Reactor (EPR) as a benchmark; SFR refers to sodium cooled fast reactors

	EPR	SFR	LFR	MSFR[a]
Operating characteristics				
Power density (kW/l)	70	290	112	130
Cycle length (years)	3–5	3.5	5.4	5.8
Coolant parameters				
Coolant	Water	Sodium	Lead	NaCl-KCl-MgCl$_2$ salt
Primary system pressure (MPa)	15.51	0.1	0.1	0.1
Core inlet temperature (°C)	295.2	371	479	496
Core outlet temperature (°C)	330	510	573	581
Core flow rate (kg/s)	7865	13,580	173,600	29,000
Fuel and its characteristics				
Fuel	UO$_2$	U-TRU-Zr	U-TRU-Zr	U-TRU-Zr
TRU enrichment (wt)	4.95%	15.2%	16.7%	15.7%
Discharge burn up (MWd/kg)	65	72	77	67
Specific power (kW/kg)	33	64.8	44.7	35.0
Geometry and dimensions				
Lattice P/D	1.3	1.08	1.3	1.19
Core active height (m)	4.2	1.02	1.3	1.3
Equivalent core diameter (m)	3.87	3.2	4.6	4.25
Active core volume (m^3)	32.6	8.7	21.5	18.5
Vessel outer diameter (m)	4.72	10.2	10.2	10.2
Number of assemblies	241	360	349	451
Number of pins per assembly	265	271	441/416	390/372
Other				
Cladding	M5	HT-9	T-91	T-91
Peak cladding temperature (°C)	450	550	614	650

(continued)

Table 6.3 (continued)

	EPR	SFR	LFR	MSFR[a]
Decay heat removal	Active CHRS	Passive RVACS	Passive RVACS + ACS	Passive RVACS + ACS
Power conversion system	Water/ steam	Water-steam/ S-CO$_2$	Indirect S-CO$_2$	Indirect S-CO$_2$
Reactor vessel	Pool type	Pool type	Pool type	Pool type

LFR refers to Lead-cooled reactors that feature a fast neutron spectrum. Data for fast reactors developed by K. Tuček et al.[b] Glossary: Transuranics (TRU), Zirconium (Zr), Reactor Vessel Air Cooling System (RVACS), Auxiliary Cooling System (ACS), Supercritical Carbon Dioxide (S-CO$_2$)

[a]The MSFR is a 3000 MWt reactor with a total fuel salt volume of 18 m^3 (reactor power density 300 MW/m^3) composed of lithium fluoride and thorium fluoride, operated at maximum fuel salt temperature of 750 °C

[b]K. Tuček et al., Comparison of sodium and lead-cooled fast reactors regarding reactor physics aspects, severe safety and economical issues, Nuclear Engineering and Design 236, 1589–1598 (2006)

Table 6.4 Examples of nominal characteristics of ceramic fuel used in HTGRs: (1) German AVR GLE-4 particles and pebbles and (2) US NGNP particles and compacts (M.A. Fütterer, L. Fu, C. Sink et al., Status of the very high temperature reactor system, Progress in Nuclear Energy 77, 266–281 (2014))

Coated particle	AVR pebble	NCNP compact
Kernel composition	UO$_2$	UCO
Kernel diameter (μm)	502	425
Enrichment (U-235 wt.%)	16.76	14
Thickness of coatings (μm)		
Buffer	92	100
Inner PyC	40	40
SiC	35	35
Outer PyC	40	40
Particle diameter (μm)	916	855
Fuel element (pebble or compact)		
Dimensions	Ø60 mm (spherical)	Ø12.3 × 25 mm (cylindrical)
Heavy metal loading (g/FE)	6.0	1.27
U-235 content (g/FE)	1.00	0.18
Number of coated particles per FE	9560	3175
Volume packing fraction (%)	6.2	35
Defective SiC coatings	7.8×10^{-6}	$<1.2 \times 10^{-5}$
Matrix type	A3–3	A3–3
Matrix density (kg/m^3)	1750	1600
Temperature at final heat treatment (°C)	1900	1850

Table 6.5 Comparative survey of concepts with potentials to achieve key requirements (as outlined in Sect. 6.3.1)

ASPECTS	Pressurized Water Reactor (PWR)	Sodium cooled (SR)	Molten Salt cooled (MSR)	High Temperature, Gas-cooled (HTGR)	Lead/Lead Bismuth (LR/LBR)	Accelerator-Driven Systems (ADS)
REACTOR LINES						
Key Reactor Design Features						
Neutron spectrum, moderator; reactivity coefficients; excess reactivity	thermal, light water; all negative; needed	fast, -; pos. CTC²) small	semi-thermal to fast *b) -; most challenging, CTC positive to different degrees; small	thermal, graphite; all negative; not needed	fast; -; CTC small positive/negative; small	subcritical fast reactor, -; additional neutrons from outside source; not needed
Coolant; primary pressure (Mpa) i. melting/boiling temperature (°C) ii. core outlet temperature (°C) iii. neutronic characteristics iv. chemical characteristics v. heat transfer/storage capabilities	light water; high (15,5) i. reasonable (0/345²) ii. moderate (330) iii. absorptive, scattering iv. stable, low corrosive, non-toxic v. good	sodium; unpressurized i. low (98) / high (892) ii. moderate (510) iii. excellent iv. highly reactive with water and air v. excellent	molten salt fluorides*a) or chlorides*b); unpressurized i. reasonable (fluorides: 450/chlorides 396) ii. high to very high (750ᵃ to 900) iii. neutral iv. non-stable; corrode metallic alloys v. good, high storage capability	helium; moderate (7) i. (single phase) ii. high (750 -1000 envisaged) iii. neutral iv. inert gas v. poor	Pb/Pb-Bi ; unpressurized i. reasonable/high (327/1737, 125/1670) ii. high (573) iii. neutral / Po-210 build-up iv. chemically inactive v. excellent thermodynamic properties	Pb-Bi used as spallation target, as target and reactor coolant i–v: see LBR
Fuel; enrichment and cladding	UO₂, (U, Pu)O₂, solid; low (3,7); zircaloy	(U, TRU)O₂ or metal (U-Pu-Zr), solid; elevated (15,2); high-strength stainless steel	U, Th, TRU, dissolved/dispersed in salt; elevated (15,7ᵃ); -	UO₂ or ThO₂, solid; elevated (8,5); ceramic coating	(U,TRU)O₂ or metal, solid; elevated (16,7); stabilized stainless steel	see LBR
Power density (MW/m³), size(MWt); burnup (GWd/tHM)	high (80), large (4800); moderate (45-50)	extremely high (290), small (200) to large (2400); elevated (72)	high (70) to extremely high (300ᵃ), small (50) to large (2400ᵇ); elevated (67ᵇ)	small (8), small (250); high (100)	high(130), small (70) to large (2400); elevated (77)	high, small (100 critical mode);
General layout; structural materials; siting	loop type; metallics; above ground	pool type; metallics; n.a.	pool type with overflow tank; metallics (special alloys); above ground/underground	loop type; ceramic core, steel vessel; underground	pool or loop type; metallics - steel or prestressed concrete vessel; above ground/ n.a.	pool type;
Start up; **refueling**, cycle length (years)	no startup fuel; periodical, 3-5	no startup fuel; periodical, 3,5	startup fuel; continuous, 5-6ᵇ)	fuel continuously added; continuous, 15 cycles	needed in case of Th fuel; periodical, 5.4 to 8	see LBR
Fuel Cycle/ Waste Management						
Reactor characteristics; Basic fuel cycle **concept**	fuel burner; open or partially closed (MOX fuel)	fuel breeder, waste burner; closed cycle	fuel burner, breederᵃ), **waste burner**ᵇ); fully closed	fuel burner (power & heat production); open	self/fuel **breeder**, waste burner; fully closed	waste burner; fully closed
Use of energy content; amount of waste	Small (2/3); high	efficient; minimized	efficient; minimized	small-moderate; high volume	efficient; minimized	efficient; minimized
Radiotoxicity of disposed waste; husbandry times	high; extremely long³)	lowered⁶), reduced to historical scale (as waste burner)	lowered; reduced to historical scale (as waste burner)	high; extremely long (but leach proof coated particle)	lowered; reduced to historical scale (as waste burner)	see LBR
Proliferation resistance	moderate (smaller in case of reprocessing)	concerns due to elevated, almost high enrichment and reprocessing (reduced in case of thorium)	defense in depth; no target values; passive incl. dumping system (all premature)	moderate to small (due to elevated enrichment)	concerns due to elevated enrichment and reprocessing (reduced in case of Th fuel)	see LBR
Reactor Safety						
Basic approach; target values for accidental release; decay heat removal	defense in depth; CDF⁴) 2*10⁻⁴/a; passive (premature)	defense in depth; ?; passive (premature)		inherently safe design; no target values; inherent/passive	defense in depth, inherently safe design; no target values; inherent/passive (all premature)	see LBR
Core fission product inventory (Bq)	extremely high (10¹⁸ ... 10¹⁹)	moderate to extremely **high** (depending on scale)	low (continuous extraction of some fission products)	**moderate** due to power size (about 10¹⁷)	moderate to high (depending on scale)	see LBR

(continued)

Table 6.5 (continued)

ASPECTS	Pressurized Water Reactor (PWR)	Sodium cooled (SR)	Molten Salt cooled (MSR)	High Temperature, Gas-cooled (HTGR)	Lead/Lead Bismuth (LR/LBR)	Accelerator-Driven Systems (ADS)
Sensitivity to unprotected scenarios i. reactivity induced accidents; ii. loss of cooling accidents; iii. others (concept specific)	i. small; ii. high; iii. non-full power / startup: notable	i. notable, self-controllability might be ensured; ii. high; iii. non-full power / startup: ensure self-control; iii. sodium fire/explosion following leaks	i. distinctive, self-controllability hard to ensure; ii. narrowed down, decay heat removal by passive means; iii. flawed fuel addition/chemical reprocessing, overcooling/freezing	i. very small, self-controllability; ii. eliminated; iii. potential air/water ingress scenarios	i. very small, self-controllability: ii. narrowed; down, decay heat removal by passive means; iii. overcooling/freezing (LBE)	i. eliminated; ii-iii: see LBR
Importance of containment/protection against external impact	high/high	high/high	minor/n.a.	minor/minor	minor/n.a.	(see LBR)
Necessity of off-site protective measures	not accounted for	n.a.	n.a./probably not	regarded not necessary	n.a./probably not	(see LBR)
State of Development						
Design state/maturity	commercially deployed	feasibility partly proven, use of real experiments/demos	early research state	feasibility proven, **at advanced stage of construction**	early design state of new concepts, reactors operated in Russia	early design state, **under development**
Technology approach; development risk	proven; small	renewed; non-standard; high	exotic; high	non-standard/innovative; moderate	evolutionary; high	revolutionary; extremely high
Show stoppers; major barriers	none; economics, lack of acceptance	mastering sodium technology, passive system design; public resistance	insufficient stability of salts and material resistance; long term reliability of components	no/fuel and materials qualification for very high temperatures; economic	insufficient corrosion control of steels; fast creep of hanging, heavy loaded vessel	lack of high-intensity pro09ton accelerator
Regulatory issues	certified, licensed	partially new approach and methods needed, can base on experience	new approach and methods needed	certified, **licensed in PR China**	new approach and methods needed, can base on experience	new approach and methods needed
Costs/Construction Times						
Capital/Investment costs	high (8.5-10.5 billion €)	presumably high	n.a.	lowered[6] (modular design)	n.a.	high (about **1 billion €**)
Construction; construction time (years)	on-site; long (8-11 Flamanville, 14 Olkiluoto)	on-site; presumably long	n.a.	on-site; lowered	n.a.	on-site; long (>10)
Electricity generating costs	competitive	n.a.	n.a.	probably not competitive	n.a.	n.a.
Overarching, contextual issues						
Complexity; novelty	moderate, evolutionary standard	high, innovative (can base on experience)	high, highly innovative (but past and ongoing R&D)	small, innovative	high, innovative	very high, innovative
Dependence on socio-political stability; safety culture	up to major (waste); to be ensured	high - lowered for minimized waste; to be ensured	n.a. - lowered for minimized waste; n.a.	less[6], less	n.a. - lowered for minimized waste; n.a.	(see LBR)
Need of intricate infrastructure	moderate (depending on fuel cycle)	moderate, secondary sodium cycle	high, on-line processing, secondary salt coolant cycle	small	small, no intermediate cooling cycle	high, high intensity proton accelerator and spallation source
Overall uncertainties	moderate (mainly management, cost/construction time overruns)	high - public resistance	high, barely assessable	moderate (first of its kind problems)	high-moderate, barely assessable	very high, barely assessable

Concrete benchmarks by column: PWR: European Pressurized Water Reactor (EPR); SR: **bold numbers** for MIT Sodium Fast Reactor concept; MSR: **bold numbers** for Seaborg Waste Burner (SaWB), a) for Eur. Molten Salt Fast Reactor concept (MSR), b) for MIT Molten Salt Fast Reactor concept; HTGR: **bold numbers** for Chinese pebble-bed, modular HTR (HTR-PM); LR/LBR: **bold numbers** for MYRRHA, Mol/Belgium ADS: **bold numbers** for MIT Lead/LBE cooled Fast Reactor concept.
Annotations: [1] indicating loss of core cooling, loss of heat sink (transients), station blackout; [2] at system pressure; [3] beyond historical time frame; [4] core damage frequency; [5] coolant temperature coefficient; [6] compared to EPR benchmark; "n.a.": not assessable due to premature design status.
Please note: All concepts conform with the need for deep geological disposal of remaining high level wastes. Systems with closed cycles need to overcome political resistance against reprocessing.
All lines include Small Modular Reactors (SMR) that reduce challenges and increase the possibility to achieve the stated requirements, compared to large reactors.

Chapter 7
Potentials and Vision for the Future of Nuclear Energy

Abstract International treaties (Kyoto and Paris), EU and national legislation have created a growing demand for a larger share, and overall larger amount of concentrated de-carbonized electricity. With such high stakes, to rely on wind and solar as the only feasible solutions is a strategic error. To address the existential need for more and more energy of our growing and wealthier societies across the world, we argue for keeping the nuclear option open, supported by future revolutionary safe and clean nuclear technologies, which would be acceptable to an otherwise presently mostly nuclear-averse society. This proposal is further supported when one acknowledges the real problem of stewardship of already existing high-grade nuclear waste over time scales eclipsing that of stable societies. To realize this vision, substantial ongoing national and international R&D programs exist, although funding is at historically low levels, and the sufficiency of current policies and activities to meet expected energy demand at an acceptable level of emissions has been questioned. Moreover, in the nuclear industry, there is the risk of stagnation of essential human-capital and know-how. The promising concepts and designs presented in Chap. 6 provide the impulse to get us over the existing hurdles, but the scope is ambitious, and time delay from R&D to commercial deployment in general is too long, stemming in part from regulatory inertia. Therefore, we call for an urgent increase in government and international R&D funding by two orders of magnitude—i.e., of the order of hundreds of billions of USD per year, for an international civilian "super-Apollo" program. We emphasize that such a large-scale public program is not unprecedented in size, and experience indicates that such investments in fundamental technology are not only of immense public benefit but also enable revolutionary innovations to be spun out that would not otherwise ever have been attained.

7.1 Paradox of Human Societies and Irreversibility

The development of nuclear technology has already created about 250,000 tonnes of high level radioactive waste, currently in temporary storage in secure water tank facilities while the radioactivity decays with time (with a 1000-fold decrease after 40 years). By weight, these high-level wastes represent only 3% of the total while

© Springer Nature Switzerland AG 2019

D. Sornette et al., *New Ways and Needs for Exploiting Nuclear Energy*,

https://doi.org/10.1007/978-3-319-97652-5_7

containing 95% of the radioactivity, 7% are intermediate-level, and the remaining 90% are low-level wastes mostly generated by non-energy related nuclear technology. Most of the wastes have their radioactivity decaying to levels similar to the background level over one or two generations. But, some by-products of reactor operations last for a hundred years e.g. Cesium-137 with half-life of 30 years; to hundreds of thousands of years, e.g. Plutonium-239 with a half-life of 24,000 years; to even millions of years, e.g. Technetium-99 with a half-life of 211,000 years. Technetium-99 is important and a problem since it constitutes 6% of the waste produced from fission of uranium-235 and has the longest half-life. Note that, as a rule of thumb, once five half-lives have passed, radioactivity will have declined close to background levels. Thus, to protect the biosphere and in particular to protect human health, it is already essential that the artificially concentrated and man-made nuclear materials are not entering the biological cycles, over a broad range of time scales. Therefore, we are already "stuck" with the existence of concentrated radioactive matter, and substantial efforts are ongoing to improve processing and storage of radioactive waste. Given this, future nuclear would benefit from economies of scale, and the issue of dealing with the current waste could be better served through innovation of the future nuclear community, for example through the development of waste burning reactors.[1]

Given the above, a troubling reality emphasizes the severity of the current dilemma. The upper bound of the million-year time scale of required human husbandry stands in stark contrast with all other activities involving time scales of decades to centuries at most, even in the worst cases of chemical pollution.[2] Even the long-time scales invoked for CO_2 driven global warming are dwarfed by those resulting from human nuclear activities. In this backdrop, we need to recognize that, without exception, societies are in continuous evolution, formation, aggregation, fusion, consolidation, disaggregation and collapse. They are punctuated by transitions taking the form of revolutions, civil wars, conflicts, ethnic collisions,

---

[1]The amount of existing wastes is known (about 20,000 cubic meters) and the current rate of production is also known (http://www.world-nuclear.org/information-library/nuclear-fuel-cycle/ nuclear-wastes/radioactive-waste-management.aspx). As it stands, is there enough waste in all countries to economically justify the (currently) necessary long term deep geological disposal? It could be argued that producing more waste would actually result in a more proper handling of the existing wastes by requiring the continuity of the required technological expertise. Also, this gives force to the goal of developing new technology to burn these wastes (see Chap. 6).

[2]The management of heavy metal wastes also involves very long time scales, see e.g. Raymond A. Wuana and Felix E. Okieimen, Heavy Metals in Contaminated Soils: A Review of Sources, Chemistry, Risks and Best Available Strategies for Remediation, International Scholarly Research Network, ISRN Ecology, Volume 2011, Article ID 402647, 20 pages, https://doi.org/10.5402/ 2011/402647 (2011).

instabilities, and so on, at times scales of decades to centuries, at best.[3,4] A large body of evidence on geopolitical dynamics shows that these evolutions often occur as sudden mutations,[5,6] rather than through smooth transitions.[7] This is in contrast with the claim of "The End of History" by Fukuyama[8] (see Box 7.1). How is it possible to ensure that teams of skilled technicians will then professionally continue their routine maintenance of key nuclear facilities and waste storage sites in the presence of a local revolution, conflict or war threatening their families? Even worse, during severe conflicts between nations, nuclear power plants (NPP) and other critical infrastructures may become prime targets in the goal of crippling the enemy.[9] This necessitates a deep reassessment of mankind's choices with respect to the nuclear industry.

Box 7.1 Fukuyama's "The End of History" Rebuked: The Pervasiveness of Instability in Human Society

In 1992, Fukuyama argued that "What we may be witnessing is not just the end of the Cold War, or the passing of a particular period of post-war history, but the end of history as such: that is, the end point of mankind's ideological evolution and the universalization of Western liberal democracy as the final form of human government." In the 25 years since publication—less than half of the lifetime of civilian nuclear power—many of the theses of Fukuyama seem to fall in the domain of faulty/hasty/inductive fallacies, having been influenced by the vivid recent historical events (the fall of the Berlin wall) over-interpreted as representative of a new order. Many have even argued that the military nuclear threat of mutually assured destruction has been the very mechanism of stability in the western world.

(continued)

---

[3]Chien-Chih Chen, Chih-Yuan Tseng, Luciano Telesca, Sung-Ching Chi and Li-Chung Sun, Collective Weibull behavior of social atoms: Application of the rank-ordering statistics to historical extreme events, Europhysics Letters 97, 48010 (2012).

[4]Cherif, A. and, Barley K., Cliophysics: Socio-Political Reliability Theory, Polity Duration and African Political (In)stabilities, PLoS ONE 5(12): e15169 (doi:https://doi.org/10.1371/journal.pone.0015169) (2010).

[5]Diamond J, Robinson JA, editors. Natural experiments of history. Belknap Press (2011).

[6]D. Sornette and P. Cauwels, Managing risks in a creepy world, Journal of Risk Management in Financial Institutions (JRMFI) 8(1), 83–108 (2015) (http://ssrn.com/abstract=2388739).

[7]Scheffer M., Critical transitions in nature and society (Princeton studies in complexity). Princeton University Press (2009).

[8]Francis Fukuyama, The End of History and the Last Man, Free Press (1992).

[9]As a recent vivid illustration, during the Ukrainian civil war, there was active social media activity concerning the calls to attack the Zaropozhskay NPP (the largest NPP in Europe and the fifth largest in the world), which is 200 km from the war zone. In February 2014, operatives of the Right Sector were arrested by guards of the NPP when trying to infiltrate it, forcing NATO nuclear specialists to check that all Ukrainian NPPs have adequate protection measures. One can also recall the example of Saddam Hussein lighting the Iraqi oil fields on fire, when he realized that all was lost for his regime.

Box 7.1 (continued)

While there is general complacency in the Western world view that wars and instabilities are a thing of the past, after the two devastating World Wars (WWI and WWII) and the Cold War ending in 1990, consider the crisis in Europe in the early 1990s, leading to the breakup of Yugoslavia following a series of political upheavals and inter-ethnic and religious Yugoslav wars. Consider also the so-called Arab Spring[10] started in December 2010, which destabilized regimes that had been stable for decades, albeit authoritarian dictatorships. Its consequences and long-term impact are still ongoing. Reflecting on the situation in many countries and regions, the political and social world is very far from stable, even on a time scale of years to decades. And what about the growing inequalities in the Western world in the last decades, characterised by a progressive but steady relative impoverishment of the bottom 99% of the population[11]? This may breed again the spirit of revolution with new large scale instabilities with uncertain and perhaps extraordinary consequences?

7.2 Need for Keeping the Nuclear Option Open

Global oil, gas and coal production and consumption data show that mankind's greenhouse gas emissions, in particular carbon dioxide, have accelerated in recent decades[12,13] notwithstanding the series of conventions and treaties since 1990[14] that commit state parties to reduce greenhouse gas emissions. There is growing but not necessarily consistent overall political will[15] to transition to a low-carbon economy that is less reliant on coal, oil and gas. The desire for decarbonisation of the energy sector, and of the electric power sector in particular, seems to be getting support internationally, as illustrated by the Paris Climate Agreement within the United Nations Framework Convention on Climate Change

---

[10]which should arguably be renamed "Arab winter"!

[11]Piketty T. Capital in the 21st century. Cambridge, MA/London, UK: The Belknap Press of Harvard University Press (2014).

[12]BP, BP Statistical Review of World Energy, 66th Edition (June 2017).

[13]Andreas D. Hüsler and Didier Sornette, Human population and atmospheric carbon dioxide growth dynamics: diagnostics for the future Eur. Phys. J. Special Topics 223, 2065–2085 (2014).

[14]United Nation General Assembly Resolution 45/212 Negotiating mandate (1990), the 1992 United Nations Framework Convention on Climate Change (UNFCCC) in Rio de Janeiro, COP 1 (the Berlin mandate in 1995), the 1997 Kyoto protocol (COP 3), COP 7 (the Marrakesh accords in 2001), the entry into force of the Kyoto protocol (2005), COP 15 (the Copenhagen accord in 2009), the Doha amendment to the Kyoto protocol (2012), COP 21 (the Paris protocol, 2015)...

[15]Barack Obama, The irreversible momentum of clean energy, Science 355(6321), 126–129 (2017).

(UNFCCC) adopted by consensus by 196 parties/nations.[16] The Paris Climate Agreement plans to implement measures in the year 2020 to mitigate greenhouse gas emissions. Intermediate measures include energy saving and improvement of the efficiency of electricity generation and use. Improving energy efficiency in the transport sector and transitioning towards electric cars is also a priority.[17] Throughout the OECD countries, in electricity generation, the focus is on replacing fossil fuels by new intermittent renewable energy sources mainly wind and solar photovoltaics. Russia and China are favouring more the expansion of nuclear fission energy, including new fast breeder reactors in Russia. In the much longer term, controlled nuclear fusion[18] remains a goal. The commitment to decarbonisation needs to be questioned in some countries like Germany that is actually closing down nuclear power stations and replacing them with unabated coal and gas. And France too has announced closure of nuclear power stations to be replaced by wind and solar.

Taking decarbonising of the energy sector, and the power sector in particular, as a long-term imperative,[19] nuclear energy qualifies rationally as the most viable option. Continued use of fossil fuels requires the addition of carbon capture and sequestration (CCS), a quest which is dying because of the huge costs.[20] Nuclear power is

---

[16]"Paris Agreement" (12 December 2015). United Nations Treaty Collection. Entry into force and registration 4 Nov. 2016 (https://treaties.un.org/pages/ViewDetails.aspx?src=TREATY&mtdsg_no=XXVII-7-d&chapter=27&clang=_en, accessed 6 Aug. 2017).

[17]One should point out that priorities are not always well thought of. Consider that one car driving 15,000 km a year emit approximately 101 grammes of Sulphur oxide gases (or SOx) in that time. The world's largest ships' diesel engines that typically operate for about 280 days a year generate roughly 5200 tonnes of SOx, the equivalent of 52 million cars. Thus, based on engine size and the quality of fuel typically used by ships and cars shows that just 15 of the world's biggest ships may now emit as much pollution as all the world's 760 million cars. Low-grade ship bunker fuel (or fuel oil) has up to 2000 times the sulphur content of diesel fuel used in US and European automobiles. And there are about 90,000 cargo ships in operation. Europe, which has some of the busiest shipping lanes in the world, has dramatically cleaned up sulphur and nitrogen emissions from land-based transport in the past 20 years but has resisted imposing tight laws on the shipping industry, even though the technology exists to remove emissions. The corresponding added cost to shipping that would follow if higher grade fuel oil was imposed are feared to stymie global commerce and economic growth. Pollution from the world's 90,000 cargo ships is estimated to lead to 60,000 deaths a year and costs up to $330 billion per year in health costs from lung and heart diseases. (Ellycia Harrould-Kolieb, Shipping impacts on climate: a source with solutions, Oceana (July 2008)) (http://usa.oceana.org/sites/default/files/reports/Oceana_Shipping_Report1.pdf) and John Vidal), Health risks of shipping pollution have been 'underestimated', The Guardian, Thursday 9 April 2009.

[18]World Nuclear Association, Nuclear fusion power, http://www.world-nuclear.org/information-library/current-and-future-generation/nuclear-fusion-power.aspx, accessed 6 Aug. 2017.

[19]"Paris Agreement" (12 December 2015). United Nations Treaty Collection. Entry into force and registration 4 November 2016 (https://treaties.un.org/pages/ViewDetails.aspx?src=TREATY&mtdsg_no=XXVII-7-d&chapter=27&clang=_en, accessed 6 Aug. 2017).

[20]Gerard Wynn, IEEFA Europe: The Carbon-Capture Dream Is Dying, July 20, 2017 (http://ieefa.org/ieefa-europe-carbon-capture-dream-dying, accessed 6 Aug. 2017). Here are a few extracts: "In 2007, EU leaders endorsed a European Commission plan for up to 12 carbon capture and storage

actually lower in CO_2 emissions than solar, wind, and biomass (based on results of Life Cycle Analysis, see Sect. 1.4).[21] Due to the incomparable energy density of uranium, nuclear power is by far the most concentrated source of energy available, by several million-fold compared with other sources, see Table 1.3. It thus seems tailor-made for the on-going inevitable concentration of human societies in urban centers and mega-cities, via the logic that concentrated needs are best met by concentrated sources.[22]

Nuclear energy is easily scalable from small to mid to large sized reactors, for example from a 50MWe SMR to the 3200MWe twin EPR, and can adapt to specific needs. It features firm, dispatchable and reliable capacity rather than an intermittent and stochastic energy source like wind and solar. At present, intermittent wind and solar can only be dispatched with the assistance of load-following gas or coal generation with perhaps localised assistance from biomass and hydroelectric power.[23]

Nuclear power is deployable and has proven affordable and competitive against other low CO_2 emitting technologies. France runs on it, has one of the most advanced electric rail services in the world and one of the lowest electricity CO_2 intensities in Europe. It enjoys the ability to provide base-load or function in load-following mode and has proven ability to power a large grid on its own.

Nuclear reactors can be designed as burners of abundantly available fissionable or fertile uranium and thorium, which are useless for other purposes. They can also be

---

(CCS) demonstration power plants by 2015. There are no such plants, nor plans." However, we should mention that, in June 2017, the first commercial plant for capturing carbon dioxide directly from the air opened in the Climeworks AG facility near Zurich (www.climeworks.com). The company says that the plant will capture about 900 tons of CO_2 annually and direct the gas to help grow vegetables. Gerard Wynn continues: "CCS has also had big backing from the International Energy Agency and the Intergovernmental Panel on Climate Change, both of which have promoted the technology as the cheapest way to transition quickly to a low-carbon economy, because in theory it allows us to keep using—rather than writing off—existing fossil fuel infrastructure. The core problem with CCS is that it is so hugely expensive up front, even with large subsidies."

[21]In its 2011 report (Toxic Air, The Case for Cleaning Up Coal-fired Power Plants, March 2011), The American Lung Association estimated that smoke from coal-fired plants kills about 13,000 people every year in the US alone. Similarly, the "Europe's dark cloud report" prepared by the WWF European Policy Office, Sandbag, CAN Europe and HEAL in Brussels, Belgium, found that EU's currently operational coal-fired power plants were responsible for about 22,900 premature deaths in 2013. In the study (P. R. Epstein et al., Full cost accounting for the life cycle of coal, Ann. N.Y. Acad. Sci. 1219, 73–98 (2011)), coal costs USD500 billion per year in large part due to health-induced problems. Disaggregating this global US cost figure, the health costs of cancer, lung disease, and respiratory illnesses connected to pollutant emissions total over USD185 billion per year.

[22]There are technological and economic logics to put these concentrated sources close to consumers. On the other hand, safety considerations suggest to put these concentrated power sources far from consumers (see however Chap. 6).

[23]One of the greatest ironies of the modern era is that OECD energy policies based around wind and solar are in fact locking us into a long-term dependency on gas and coal to provide back-up and load following capability.

designed as breeders that burn long-lived isotopes of minor actinides found in nuclear waste. In nuclear power, only about 2% of the total cost is buying yellow cake. The costs are concentrated in capital equipment in both fuel manufacture and power stations. Nuclear power, therefore, is not sensitive to variations in the uranium price.

Radioactivity remains perceived as dangerous (see Sect. 5.5 on dread) by a public ill-equipped to understand the real risks. Despite their rarity, severe accidents with large consequences including high associated costs remain a possibility in the present fleet of mostly Gen II nuclear plants. When an accident does happen, the public fear, enormously exacerbated by media sensationalism, and the resulting backlash are huge. This form of scare-mongering causes deaths, for instance due to over-reactions such as exaggerated evacuations and continued population displacement as occurred following the Fukushima disaster,[24] and puts the future existence of nuclear power at risk, as evidenced by the German decision to phase out of nuclear energy entirely. Given that future prosperity of human society may depend upon nuclear power, this situation can only be described as grave. Society must reclaim proper scientific evaluation of facts and evidence to guide policy decisions and reject media sensationalism that is often founded on propaganda distributed by Green NGOs without solid scientific foundations.

Next-generation nuclear reactors are expected to be much safer than their predecessors, less dependent on their environment and promise "to take fear out of nuclear". Radical design advances have potential to achieve the most stringent safety requirements. While most are still on the drawing board, some are under development and some of them are even close to deployment, notably small to medium sized, modular reactors (SMR) of different lines such as the high temperature gas cooled pebble bed (see Sect. 6.2.4). While past experience should warn us against claims of zero risk, there is a positive trend toward achieving such tiny risk levels that they become essentially irrelevant compared with other types of risks and in

---

[24]There are of course significant risks to health due to radioactivity but these tend to be overestimated in the minds of the public and the politicians thus over-react to cater to the general perception. If we take one of the earliest accidents in 1957 at Windscale, UK, that caught fire and burned for 3 days, creating an INES 5 accident, there was no evacuation. There were no immediate deaths but between zero (K J Bunch, T J Vincent, R J Black, M S Pearce, R J Q McNally, P A McKinney, L Parker, A W Craft and M F G Murphy, Updated investigations of cancer excesses in individuals born or resident in the vicinity of Sellafield and Dounreay, British Journal of Cancer 111, 1814–1823, doi: https://doi.org/10.1038/bjc.2014.357 (2014)) and perhaps 500 early deaths from cancer (J.A. Garland and R. Wakeford, Atmospheric emissions from the Windscale accident of October 1957, Atmospheric Environment 41(18), 3904–3920 (2007)) did occur many years later. This needs to be contrasted with Fukushima in 2011 where the evacuation of ~300,000 people led to immense trauma and a conservative estimate of 573 deaths from suicide since the accident. One estimate places the number of deaths resulting from the evacuation as high as 1600, mainly from the elderly population. Had the evacuation not taken place, all these people would likely still be alive today. However, avoided latent delayed fatalities due to reduced exposure must also be credited. These examples are offered as matter for thought to neither fall in the extreme of over- nor under-reacting. We call for rational scientific approaches to such evacuation decisions, which may have large scale consequences for the public.

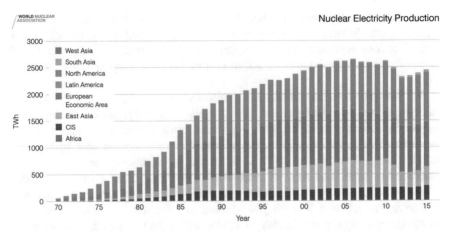

Fig. 7.1 Nuclear electricity production by region from 1970 to 2015 in TWh. Reproduced from the World Nuclear Association (World Nuclear Association, Report No. 2016/001: "World Nuclear Performance Report", *WNA* (2016)). Note the resumption of growth of nuclear power over the past 3 years, which in part reflects reactors in Japan coming back on line and the burgeoning of China's surge in nuclear energy

view of the benefits that flow from having reliable, cheap, low carbon electricity. Further, we suggest that nuclear energy should not just be judged on its turbulent early past history but rather on its maturation towards much higher achievable levels of future safety and efficiency.[25]

As of 2017, there are 446 nuclear power reactors in operation worldwide with a total net installed capacity of approximately 390 GWe.[26] Nuclear power plants have provided a significant electric output (see Fig. 7.1), contributing appreciably to the global energy system over recent decades.[27] Within the context of overall growing energy consumption and the need to meet decarbonization targets, in the foreseeable future, there is a clear case for a big increase in nuclear energy production.[28]

However, the development of nuclear energy in the USA and in Europe is hindered by long construction times and high capital costs for new builds.[29] In the

---

[25]Gen-II nuclear plants, based on technology from the 1970–1980s, generate significant radioactive wastes, and have been made safer through substantial retrofitting rather than by initial design. Anchoring on the technology and safety record of the early days, and ignoring both the measurable improvements, and that present and future technology can be a game-changer, would lead to excluding what can be considered as one of the most promising energy sources for the twenty-first century and beyond.

[26]https://www.iaea.org/pris/

[27]In 2015, the global nuclear electricity generation reached 2441 billion kWh, which amounts to 11.5% of the total, as calculated by the World Nuclear Association (http://www.world-nuclear.org/information-library/facts-and-figures/world-nuclear-power-reactors-and-uranium-requireme.aspx).

[28]The European Atomic Forum (FORATOM), Position Paper January 17, 2017.

[29]For Hinkley Point C with two 1600 MW electrical EPR units, capital costs are 20.9 billion Euro with a 20 years construction time. However, some argue that nuclear new build lifecycle cost is

USA and Europe, nuclear new-build also faces public and often political opposition in addition to lengthy and laborious design approval and extensive regulation. It is noteworthy that these problems do not seem to exist for recently built nuclear plants in South Korea, China, and Russia. Furthermore, the use of nuclear power requires extensive infrastructure including education, training and technical support.

Again, SMR pledge to remedy to the high costs and long construction times, and may add flexibility to siting. Due to the high specific energy of fissionable material[30] and the high-power density of reactor cores,[31] nuclear power demonstrates a potential to reduce human demands on land use, to decouple development from nature and to power dense human settlements and large growing economies.[32] The historic growth of megacities demonstrates the need for concentrated forms of electricity generation (see Sect. 1.2.3).[33]

An energy controversy is raging between those who argue that 100% renewable energy systems are the way forward and those who reason for diversity of supply that has nuclear power at its core. Serious outlooks and road maps support for a doubling of nuclear capacity by 2050 to satisfy needs, and if international commitments to reduce CO_2 emissions are to be met.

7.3   Ongoing Research and Development on a National and Global Scale

Most current commercially operated nuclear power plants are of "Generation II–III", while new and advanced reactor designs, generally falling into the evolutionary Generation III+ and revolutionary Generation IV categories, are being researched and developed worldwide. Considering the time needed to reach the construction and then the commercial exploitation stage from the start of basic research and conceptual work on the designs, it is understandable that different countries have already started their respective efforts by launching and continuing several high-profile research programs. These programs are not just conducted on a national scale, but are also often seen as a joint venture between countries and interest groups, including private industries, towards potentially more efficient energy generating reactor models, as is the case with the Gen IV International Forum (GIF).

---

close to that of on-shore wind (FORATOM, https://www.foratom.org/facts-figures). Moreover, new designs of SMRs integrating inherent super-safety measures could be much cheaper and allow for earlier return on initial investment.

[30]Compared to petrol/coal, the specific energy of fissionable material is 1.74–2.78 million times higher respectively, and 16.1 million times higher than (Li-Ion) battery.

[31]About 80 MW thermal per cubic meter for current light water reactors.

[32]An Ecomodernist Manifesto, 2015, www.ecomodernism.org

[33]See also: Allianz Risk Pulse The megacity state: The world's biggest cities shaping our future, Nov. 2015.

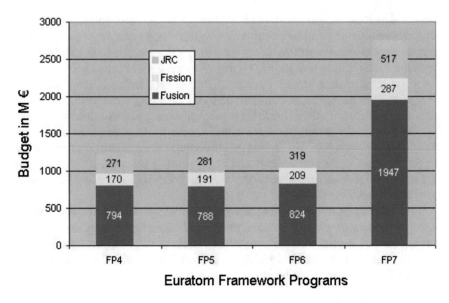

Fig. 7.2 Budget comparison of EURATOM Framework Programs FP4 to FP7 (EURATOM, http://ec.europa.eu/research/energy/euratom/index_en.cfm?pg=faq#4). Note that FP7 (from 2012 to 2016) had a budget of 2.7 billion € funding split the Joint Research Center (JRC), fission and radiation protection research and development, and fusion activities including ITER

The European Atomic Energy Community (EURATOM and EAEC) is one of the international organisations developing nuclear energy and providing financing mechanisms to fund nuclear research programs within the European Union. With an approximate funding rate of about 50 million € per year for nuclear fission and a total research budget of 2.7 billion € for projects in fission, radiation protection and notably fusion, it is a globally significant effort that Europe is making. This funding is available under the seventh Euratom Research Framework Programme and the distribution of the total budget can be seen in Fig. 7.2.

The ITER[34] fusion program has by far the largest share of the whole budget. ITER aims to build an experimental nuclear fusion reactor in Cadarache, France and was given a high priority. With the site preparation work starting in 2008, the project is now almost 10 years into construction and it is expected to achieve first Deuterium-Tritium fusion by 2027. ITER is a collaborative partnership between Europe, China, India, Japan, Russia, South Korea and the USA. With an expected project duration of over 35 years and a cost of more than 16 billion €, it is the world's biggest energy research project. ITER demonstrates the scale and importance of scientific research in the energy sector and provides a conservative yardstick for future ambitious fission reactor projects and the associated fuel cycle developments. Another notable project that received around 6 million € in direct contributions

---

[34]www.iter.org

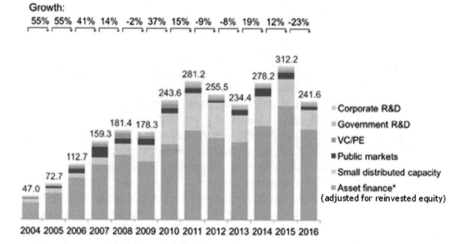

Fig. 7.3 Global new investment in renewable energy by asset classes, 2004–2016 in USD billions. Total values include estimates for undisclosed deals (UN Environment, Bloomberg New Energy Finance, Reproduced from Global Trends in Renewable Energy Investment 2017, Frankfurt School-UNEP Centre/BNEF. 2017. http://www.fs-unep-centre.org (Frankfurt am Main)). VC/PE = venture capital/private equity

through the FP7 program[35] is the MYRRHA accelerator reactor being built at Mol in Belgium (see Sect. 6.2.5 for details).

These numbers can be compared to total public and corporate investments in renewable energy R&D of USD8 billion for 2016,[36] which was 14% below its peak in 2011. Of this number, the largest contribution is government spending on renewables research that reached the record level of U.S. USD5.5 billion, with USD3.6 billion for Solar R&D, USD1.2 billion for wind and USD1.7 billion for biofuels. The investments in R&D must also to be put in perspective with respect to the investments for new power generation. The global amount of new investments in renewables (excluding large hydro) for 2016 totaled USD241.6 billion, a 23% decrease from 2015. This investment was associated with a record installation of renewable power capacity worldwide in 2016, where wind, solar, biomass and waste-to-energy, geothermal, small hydro and marine sources between them added 138.5 GW, up from 127.5 GW in the previous year. New investment in solar in 2016 totaled USD113.7 billion of investment globally. Figure 7.3 presents the evolution with time of total new investment in renewable energy per asset classes. Historical investments in developing the engineering base of civilian nuclear technology is another important point of reference, but is difficult to estimate. However, amortized

---

[35]Eur-Lex Database, http://eur-lex.europa.eu/legal-content/DA/TXT/HTML/?uri=OJ:JOC_2014_405_R_0001&rid=1#d1e8712-1-1, accessed 6 Aug. 2017.

[36]Global Trends in Renewable Energy Investment 2017, Frankfurt School-UNEP Centre/BNEF. 2017. http://www.fs-unep-centre.org (Frankfurt am Main).

over historical kWh generated (like an externality), it is unlikely to make a decisive difference in a comprehensive comparison of energy sources. Further, external benefits should also be credited.

But even though there are several joint frontiers and efforts in energy research, most of the significant energy players also have national programs aimed at the development of nuclear technology. As an example, the Russian Federation, with 35 currently operating nuclear reactors[37] and a plan to put one large reactor per year (on average) on line up to 2028, is heavily expanding and strengthening the role of nuclear power in Russia (see also the development of the BN-800 and future BN-1200[38]). Moreover in 2010, the Russian government approved a program to support fast neutron reactor development and provides more than 60 billion rubles (about USD2 billion[39]) to achieve this goal. In addition to this, ROSATOM[40] has concrete plans to further work on and test floating nuclear power plants (FNPP) and in May 2014 committed to cooperate with the China Atomic Energy Authority on the construction of floating nuclear plants for China's islands. A year later, Indonesia's National Atomic Energy Agency and ROSATOM also signed an agreement, stating that they will cooperate on the construction of further floating nuclear plants. As for 2013, ROSATOM spending for research and development reached 24 billion rubles (about USD420 million) amounting to 4.53% of its turnover.[41]

Another country with a relatively high profile nuclear program is Finland. As already mentioned in Sect. 3.2, Finland is not only generating 30% of its electricity with four nuclear power plants and completing the construction of the EPR at Olkiluoto but, also as of December 2016, started the construction of the world's first final disposal facility for used nuclear fuel at the Olkiluoto site. This is partly being financed by a State Nuclear Waste Management Fund with a size of USD2.4 billion.[42]

The Department of Atomic Energy of India has a yearly budget of USD1.4 billion and oversees six different research and development facilities, with the latest one established in 2014. These centers conduct work on a wide range of technologies ranging from light and heavy water type reactors to thorium fueled[43] and accelerator based systems, as well as high-temperature gas-cooled reactors, and also on

---

[37]WNA, http://www.world-nuclear.org/information-library/country-profiles/countries-o-s/russia-nuclear-power.aspx

[38]http://euanmearns.com/the-bn-800-fast-reactor-a-milestone-on-a-long-road/

[39]This currency comparison is probably rather deceptive since there is a significant difference in purchasing power between the USA and Russia, and these values should be adjusted for purchasing power parity.

[40]Russian State Atomic Energy Corporation. It was established in 2007 and is the regulatory body of the Russian nuclear complex

[41]ROSATOM, ROSATOM annual report 2013, http://ar2013.rosatom.ru/255.html

[42]WNN, http://www.world-nuclear-news.org/WR-Finlands-waste-fund-grows-to-over-2-billion-euro-27021501.html

[43]Motivated by the fact that India has 25% of the world's thorium reserves.

non-technical associated issues. India has some of the largest Thorium ore deposits in the world and is therefore keen to develop the Thorium fuel cycle.

In comparison, the US Department of Energy managed a budget of roughly USD980 million for nuclear energy alone in 2016. USD345.5 million of this sum is reserved for fuel cycle research and development and reactor concepts research, development and demonstration. Moreover, for 2017, the overall requested budget for nuclear energy is 0.8% higher[44] and the upwards trend seems set to continue. The accumulated share of nuclear funding in total energy funding received since 1978 until 2012 is almost 37%, compared with 25% for fossil, and 16.5% for renewables[45] and the data of funding requests confirm that funding application volume is steadily increasing. In July 2017, the US Department of Energy announced moderate investments worth over USD66 million in projects that will help to advance innovative nuclear technologies, while Westinghouse announced the formal launch of its *accident-tolerant fuel solution* after many years of development and important investments.[46]

The French equivalent of the Department of Energy is the Alternative Energies and Atomic Energy Commission (CEA[47]), which declares nuclear energy (fission and fusion) to be one of its four main areas. While public funding is only one of the income sources for the CEA, its share in investments in civil nuclear power research is significant. From 1990 to 2010, 8.8 billion € from public resources were invested by the CEA in this area. The overview for the total research expenditure on nuclear power from 1957 to 2010 in France shows that about 55 billion € were spent in total, with an average of 1 billion € per year. It is worth noting that, of this total, 43 billion € were spent on existing nuclear facilities, while 12 billion € were invested in fast neutron reactor research.[48]

A number of countries have been involved in developing new Generation IV reactors, whose key requirement is to eliminate the potential of off-site emergencies and its consequences. One promising technology is the modular high-temperature gas-cooled reactor (MHTGR), which remains one of the most innovative and challenging technologies (see Sect. 6.2.4). The solid fuel is encapsulated inside extremely small silicon carbide/carbon casings of less than 1 mm diameter, which provides highly secure containment of the fuel and fission products during operation. The high temperature leads to steam turbine inlet conditions that are similar to modern supercritical coal plants with high thermodynamic efficiency (42%), which is seen as a great advantage. Proposals exist to convert coal plants to nuclear MHTGR power stations where nuclear will replace the coal fired furnaces but will

---

[44]Department of Energy, Department of Energy FY 2017 Congressional Budget Request Volume 3 (2017).

[45]DOE Budget Authority History Table by Appropriation, May 2017.

[46]World Nuclear News, June 15/14, 2017.

[47]CEA: its initial name was "Commissariat à l'Energie Atomique" and since 2010, it has changed its name to add "et aux energies alternatives", while keeping its acronym.

[48]French Court of Audit, The costs of the nuclear power sector (2012).

then utilize the turbines and electricity production systems of the coal plant. Following advanced R&D performed in Germany and the USA with government support in the 1980s and 1990s, China and Japan have taken to build their own high-temperature gas-cooled reactor pebble-bed module (HTR-PM) demonstration power plants.[49] For China, the goal is to substitute the coal combustion units with nuclear reactors in its fleet of modern supercritical coal plants that are located close to cities, with the potential for quickly improving air quality. It is worth mentioning that South Africa invested about USD1.5 billion over a period of 10 years into the development of modular pebble-bed HTR but finally stopped this program due to lack of interest of any client.

Recent work by the Nuclear Energy Agency (NEA) suggests that global electricity generating capacity needs to increase by 2.3 times by 2050, if countries are to meet their CO_2 reduction targets. This large increase in part reflects the electrification of transport (electric cars) and heating (heat pumps). The NEA asks whether current research and development (R&D) efforts are sufficient to foster the innovation in nuclear fission technologies that are required if safe nuclear power is to play a significant role in this vast expansion of our electricity supply system.[50]

7.4   Need and Framework for an Apollo-Like Program

To summarize: History tells us that human polities are short-lived and prone to instabilities, and many people dread anything nuclear. Over the 70-year-old "nuclear age", we have created large amounts of artificially concentrated nuclear matter that needs to be responsibly managed with sustained technical expertise and resources over decades to centuries and more. Thus, we cannot walk away from nuclear. In this context, we would not bet on the fact that humans will put aside their emotional nature, become wiser at governing their polities, and form an eternally stable society. Nor would that necessarily be a good thing overall! Therefore, we should rather embrace nuclear technology to transform it into a super-safe, super-clean, reliable and engaging technology. In other words, we need nothing less than a revolution to remove the risks, the dread and the long-lived wastes from nuclear power. What follows is a bold proposition, which builds on the synthesis of this book and aims to foster a positive and constructive attitude.

---

[49]Zuoyi Zhang, Yujie Dong, Fu Li, Zhengming Zhang, Haitao Wang, Xiaojin Huang, Hong Li, Bing Liu, Xinxin Wu, Hong Wang, Xingzhong Diao, Haiquan Zhang and Jinhua Wang, The Shandong Shidao Bay 200 MWe High-Temperature Gas-Cooled Reactor Pebble-Bed Module (HTR-PM) Demonstration Power Plant: An Engineering and Technological Innovation, Engineering 2(1), 112–118 (2016).

[50]https://www.oecd-nea.org/ndd/ni2050

We must attack the key social and technical problems confronting nuclear power, which are poor economics, the danger of catastrophic accidents, radioactive waste production, and linkage to nuclear weapon proliferation.[51] By prioritizing the safety and cleanliness of civil nuclear energy, we ensure the necessary development and sustainability of the technological and human skills. In Sect. 7.3, we outlined the existing efforts to develop new generation nuclear technology and documented that the financial efforts were relatively small, of the order of hundreds of millions to billions per year for the major participants. We propose that the challenges as well as the potential gains warrant a massive increase of the fundamental research, development and industrial deployment aspects.[52] As a reference, at the Paris Climate Conference in 2015,[53] the developed countries reaffirmed the commitment to mobilize USD100 billion a year in climate change prevention and adaptation finance by 2020, and agreed to continue mobilizing finance at the level of USD100 billion a year until 2025. The commitment refers to the pre-existing plan to provide USD100 billion a year in aid to developing countries for actions on climate change adaptation and mitigation. We propose that an effort at a similar level. i.e. $100 billion per annum, be invested in attaining "totally safe" and clean nuclear energy generation.

One of us has even called for a bold 1% of global GDP per year for a decade be spent on Generation IV or V reactor R&D.[54] An additional and very powerful argument is the need to reboot innovation in modern economies. The financial crisis of 2008 and great recession that followed, leading to an under-performing economic growth since 2007, have origins that may be linked to the delusionary belief in policies based on a "perpetual money machine" using debt and equity withdrawal from assets as the engines of growth.[55] In response to the crisis, monetary and fiscal policies have been used, essentially amounting to more debt and more financialisation via massive quantitative easing policies that worked with the objective of keeping the global banking system afloat.[56]

In Chap. 1, we explained how human society developed on the back of harnessing "energy slaves" in the form of wood, human labor, wind, coal, oil, natural gas,

---

[51]Ramana, M.V. and Z. Mian, One size doesn't fit all: Social priorities and technical conflicts for small modular reactors, Energy Research & Social Science 2, 115–124 (2014).

[52]D. Sornette, A civil super-Apollo project in nuclear R&D for a safer and prosperous world, Energy Research & Social Science 8, 60–65 (2015).

[53]"Paris Agreement" (12 December 2015). United Nations Treaty Collection. Entry into force and registration 4 November 2016 (https://treaties.un.org/pages/ViewDetails.aspx?src=TREATY&mtdsg_no=XXVII-7-d&chapter=27&clang=_en, accessed 6 Aug. 2017).

[54]D. Sornette, A civil super-Apollo project in nuclear R&D for a safer and prosperous world, Energy Research & Social Science 8, 60–65 (2015).

[55]D. Sornette and P. Cauwels, 1980–2008: The Illusion of the Perpetual Money Machine and what it bodes for the future, Risks 2, 103–131 (2014) (http://ssrn.com/abstract=2191509).

[56]In an op-ed in the Financial Times on June 12, 2011, Larry Summers summarises vividly the spirit of the interventions: "[...] This is no time for fatalism or for traditional political agendas. The central irony of the financial crisis is that while it is caused by too much confidence, borrowing and lending, and spending, it is only resolved by increases in confidence, borrowing and lending, and spending."

hydroelectric and nuclear power. It is harnessing energy that makes our growing economies viable and these in turn produce society's wealth that in turn provides pensions, healthcare, welfare, security and defense. In short, money is simply a transmuted form of energy. It is, therefore, totally straightforward to argue that investing 1% of GDP in energy systems research and development is one of the most important policies that human society could implement today. Investment today may assure prosperity of future generations.

The success of the policies implemented in 2008 and thereafter has been less than convincing, which calls for new ways of rebooting the economy and ensuring long-term prosperity. This can be achieved by pursuing a bold technical innovation policy, rather than just monetary and fiscal policies. The exceptional low and even negative interest rates since 2008, with only prospects for slow and small future increases, may reflect a secular slowing down of growth[57] where high energy prices have played a role. Fundamentally, there is only one way to get out of debt and over leverage, and to generate wealth, which is growth. Countries have rarely if ever paid off their national debts. The current model is based upon growing their GDP faster than their debt, thus shrinking the debt in relative value. There are only two ways for per capita growth in the real economy as opposed to the financial virtual economy of three decades leading up to 2008,[58] which is to create and innovate and increase productivity. Low energy prices are a vital ingredient to increased productivity as witnessed by the global recessions caused by high oil prices in the 1970s. This will require special measures at the fiscal and credit levels to facilitate the drive of the energetic and scientifically educated part of the population, to raise its head once again, to be enabled to take the bull by the horns and to innovate and create technological advance and wealth for everyone.

Free markets and neo-liberal capitalism expound the values of private initiatives and investments for innovation. However, it is not sufficiently recognized and understood generally that the Government has a key role to play in catalyzing innovation. This is particularly well illustrated by the role of the Government in the development of Silicon Valley (see Annex Box 7.2). Moreover, a number of studies have clearly documented that innovation is the result of a complex "three-player game" between the state, the financial capitalists and the entrepreneurs.[59] The state can play an essential role in advancing fundamental research, as illustrated by semiconductors, the Internet and human genome research.[60] Note that these examples often have a military element but that needs not to be the case. Blue skies

---

[57]Robert J. Gordon, The Rise and Fall of American Growth: The U.S. Standard of Living since the Civil War, The Princeton Economic History of the Western World, Princeton University Press (January 12, 2016).

[58]D. Sornette and P. Cauwels, 1980–2008: The Illusion of the Perpetual Money Machine and what it bodes for the future, Risks 2, 103–131 (2014) (http://ssrn.com/abstract=2191509).

[59]William H Janeway. Doing capitalism in the innovation economy: markets, speculation and the state, Cambridge University Press (2012).

[60]Monika Gisler, Didier Sornette, and Ryan Woodard. Innovation as a Social Bubble: The example of the Human Genome Project. Research Policy, 40(10), 1412–1425 (2011).

research creates possibilities, not products. The venture capitalists play a key role in identifying attractive applications of a new technology and how to bring them to market through investment in entrepreneurs and new ventures. Innovation has an intrinsic and irreducible uncertainty that cannot be measured. Exuberant bubbles can play a positive role. In all their apparent wasteful excesses, bubbles have been necessary drivers of economic progress.[61] The strategic and political choices to invest heavily in a particular technological sector are often questioned but technology remains an extraordinary tool for spurring innovation and growth. For example, the US managed to take a man to the moon[62] and contributed within an international effort to read the human genome using this strategy.[60] In this spirit, the USA created the Defense Advanced Research Projects Agency (DARPA) in 1957, with the mission of ensuring that the USA would avoid any technological surprises after the Soviets launched Sputnik 1. DARPA, with an annual budget of USD3 billion, plays a key role in the defense sector, but has also made technological inventions whose benefits to the overall population are immense,[63] such as: liquid-crystal displays (LCD), global positioning systems (GPS), multi-touch screen or artificial intelligence with a voice-user interface.

It is with this understanding that we propose the massive effort in making civil nuclear energy safe and clean, and a major contributor of energy for future generations. Chapter 6 documents many interesting ideas and ongoing developments which, in our view, lack sufficient funding and focus to achieve revolutionary breakthroughs. Compared with the paltry 1.35 billion of 2013 USD per year in total spending of the US DOE, a courageous commitment would consist in a 100-fold increase in each of the major nuclear energy constituencies (USA, Europe, Japan, India, and China). This may seem huge, but is a tiny fraction of the bailout program launched in 2008 that involved trillions of dollars to save financial institutions during and after the financial crisis. It is also less than 1% of US 2014 GDP.[64] It is also a fraction, less than 20%, of military spending. In the case of Japan, 18 trillion Yen may seem astronomical but is dwarfed by the expansion of the Bank of Japan bond buying program at a rate of 80 trillion Yen of bonds a year, as of 31 October 2014.[65] Rather than supporting the "virtual economy" of finance, our proposition targets the real world directly, epitomized by the physics, chemistry, engineering of nuclear energy, with well-defined targets. Energy is so-to-speak transmuted to

---

[61]Monika Gisler and Didier Sornette. Bubbles Everywhere in Human Affairs. In L. Kajfez-Bogataj, K.H. Müller, I. Svetlik, and N. Tos, editors, Modern RISC-Societies. Towards a New Paradigm for Societal Evolution, pages 117–134. Echoraum (2010).

[62]Monika Gisler and Didier Sornette. Exuberant Innovations: the Apollo Program, Society, 46 (1):55–68 (2009).

[63]Mariana Mazzucato. The Entrepreneurial State: Debunking Public vs. Private Sector Myths. Anthem Press (2013).

[64]100 and 50 billion euros is more than half the present GDP of Greece, but it amounts to just 2 months and a half of the European Central Bank Quantitative Easing program launched in March 2015 on the tune of 60 billion euros per month to buy European sovereign bonds.

[65]https://www.boj.or.jp/en/announcements/release_2014/k141031a.pdf, accessed 8 Aug. 2017.

money, and money must pay its way through supporting the energy system that creates it. By investing vigorously to obtain scientific and technological break-throughs, we create the spring of a world economic rebound based on new ways of exploiting nuclear energy and other possible energy sources, both more safely and more durably.[66] In the same spirit, we also welcome and support the "Global Apollo program" to make renewables less costly than coal within 10 years.[67]

We suggest that only WWII-level investments[68] in technological innovations will deliver new sustained and sustainable growth, and address the pressing issues of CO_2 emissions reduction targets, as well as the pressing needs and aspirations of the growing human population. Some have even gone as far to suggest that "*the lack of major wars may be hurting economic growth*",[69] and that the desire to prepare for and the drive to prevail in wars spurred technological invention and also brought a higher degree of internal social order[70,71,72] Thinking of wars as engines of growth in the future is, however, unpalatable. We propose rather to focus on global energy scarcity and high energy cost problem and the enormous challenge of decarbonizing energy supplies. These perils manifest a common enemy which should help mankind transcend its internal differences and conflicts. Mankind has shared enemies in the forms of potential scarcity of affordable energy, environmental degradation, water shortages, the availability of affordable food, plastics pollution of the oceans and over-fishing. Focusing on these issues may catalyse the human spirit to address them via technological innovation. International concern over rising CO_2 emissions and the possible impacts on Earth's climate combined with international resolve to address this issue creates an imperative for an abundant supply of reliable and affordable low CO_2, clean and safe energy. New extremely innovative nuclear fission technologies are in many ways the most promising and viable solutions that we know of. This is why we now call for a super Apollo-scale increase in funding of RD&D into these new technologies.

---

[66]Kovalenko, T. and D. Sornette, Dynamical Diagnosis and Solutions for Resilient Natural and Social Systems, Planet@ Risk 1(1), 7–33 (2013) Davos, Global Risk Forum (GRF) Davos (http://arxiv.org/abs/1211.1949).

[67]David King, John Browne, Richard Layard, Gus O'Donnell, Martin Rees, Nicholas Stern, Adair Turner, a global Apollo programme to combat climate change (June 2015) (http://cep.lse.ac.uk/pubs/download/special/Global_Apollo_Programme_Report.pdf).

[68]It should be remembered that the end of the great depression was obtained only by WWII. The extraordinary "trente glorieuses" years of economic growth and wealth creation after WWII was the result of the spillover of the immense investments in innovations, both in technology as well as governance and management developed during WWII.

[69]T. Cowen, The Lack of Major Wars May Be Hurting Economic Growth, The New York Times, June 13, 2014 (http://www.nytimes.com/2014/06/14/upshot/the-lack-of-major-wars-may-be-hurting-economic-growth.html?_r=1).

[70]A. Gat, War in Human Civilization, Oxford University Press; 1 edition (April 15, 2008).

[71]I. Morris, War! What Is it Good For? Conflict and the Progress of Civilization From Primates to Robots, Farrar, Straus and Giroux (15 avril 2014).

[72]K. Kwarteng, War and Gold: A 500-Year History of Empires, Adventures and Debt, Bloomsbury Publishing PLC (2012).

The proposed program will not only support reaching the stringent requirements for safer and cleaner nuclear technologies defined in Sects. 5.6 and 5.8 but will enable revolutionary designs and solutions, whose feasibility is made credible by the flurry of creative technical solutions described in Chap. 6.

Annex

> **Box 7.2   The Case of Public Spending (In This Case Military) and Its Role in Fostering Innovation in Silicon Valley**
>
> It is useful to reflect on the history and raison-d'être of Silicon Valley, arguably the most innovative place on Earth. Most people associate Silicon Valley with the electronic revolution and the rise of the personal computer. However, its true origin is very different, going back to WWII. The following summarises the synthesis of Blank.[73]
>
> The story starts with the efforts of the Allies to counteract the German Air defense system, the most sophisticated in the world at the time, aimed at destroying the Allies' planes flying from England towards Germany, through a network of early detection radars. Using this system, the Germans managed to destroy about 4–20% of the Allies' planes on each mission. In response, the Harvard Radio Research Lab (HRRL), with 800 workers, was secretly founded in 1942 to understand and shut down the German's radar system. In this way, a drastic change was initiated in the relationship between militaries and universities. Militaries began to directly fund some universities to complement their R&D programs. Frederik Terman, director of the HRRL, set up his own lab in the Stanford School of Physical Sciences in 1945 to research microwaves with 11 key members of the HRRL, which closed after the war. In 1949, the US military approached Terman and doubled the size of the electronic lab in Stanford at the beginning of the Cold War. The goal was the same as during WWII, to understand the enemy's electronic system and beat it.
>
> The first satellite was launched in 1957 by the Russians. As the race for space played a key role in the Cold War, Silicon Valley was developing the systems for these satellites and acquired valuable expertise in this domain too. Silicon Valley begun to change in 1955 as Terman encouraged his students to leave the university and his lab to create their own companies. The valley became a "Microwave Valley" with the government, including CIA and NSA being the biggest clients of many private companies. The military was funding

(continued)

---

[73]Steve Blank, Hidden in Plain Sight: The Secret History of Silicon Valley. https://steveblank.com/secret-history, 2016, accessed 8 Aug. 2017.

Box 7.2 (continued)

entrepreneurs to develop and produce technological systems and products for the Cold War. At that time, the main motivation was not profit but winning the technological battle against the USSR. William Shockley, known as the other founder of Silicon Valley, was the inventor of the transistor and had a military background in weapon R&D. He founded Intel and 65 other chip companies, all involved in semiconductors. He really helped to implement new technologies and an entrepreneurial spirit in the Valley. Today, the Valley is full of private capital and the companies financed directly by the military are few.

In 1955, the first companies went public, which attracted the first venture capitalists and some angel capitalists to Silicon Valley. Regulations were adapted such that private capital could take over the role of public funding. And by the end of the Cold War 1991, civilian technology had become the main activity of Silicon Valley. This brief history clearly shows the massive influence of the (American) government investment and legal apparatus on the formation of this highly innovative and productive region.

Printed in the United States
By Bookmasters